Henry Drury Noyes

A Treatise

Diseases of the eye

Henry Drury Noyes

A Treatise
Diseases of the eye

ISBN/EAN: 9783337036126

Printed in Europe, USA, Canada, Australia, Japan

Cover: Foto ©berggeist007 / pixelio.de

More available books at **www.hansebooks.com**

A TREATISE

DISEASES OF THE EYE

BY

HENRY D. NOYES, A.M., M.D.

Professor of Ophthalmology and Otology in Bellevue Hospital Medical College; Surgeon to the New York
Eye and Ear Infirmary; President of the American Ophthalmological Society; Member of the
New York Ophthalmological Society; Permanent Member of the Medical Society of the
State of New York; Member of the New York Academy of Medicine, etc.

NEW YORK

WILLIAM WOOD & COMPANY

27 GREAT JONES STREET

1881

TROW'S
PRINTING AND BOOKBINDING COMPANY
201-213 *East* *Twelfth Street*
NEW YORK

PREFACE.

In this treatise the writer has attempted to condense into the limits assigned to him the substance of modern ophthalmic knowledge. The standpoint is clinical and the purpose is practical; but, whatever is necessary to a correct understanding of a disease is presented, because sound treatment can only be attained by a proper knowledge of causes, connections, and processes. Brief statements of anatomy, including microscopic structure, have been introduced. In microscopic pathology, the writer has been obliged to depend upon the labors of others, and has endeavored to present views which are most recent, in so far as he has offered any.

Knowing how largely disturbances of refraction and of muscular function enter into troubles of sight, he has recognized their importance by discussing them before beginning the consideration of inflammations and other structural changes. They are considered in Part First. In Part Second the diseases of the eye, as they are usually understood, are presented. It is perhaps worth remarking that in each part we are conducted to the brain: in the first, by the motor nerves; in the second, by way of the optic nerve. It is, therefore, unavoidable that there should be some reference to the physiology and pathology of the brain and nervous system. Nor can diseases of the eye be discussed without taking into account diseases of remote organs and of the general system. To them, therefore, brief reference is often made. Ophthalmology, as a field of special cultivation, has in late years attained a wonderful development, but its relations to and dependence upon general pathology are more manifest now than ever. It has shed light of no feeble kind upon the pathology of other parts of the body, but it none the less seeks to be illuminated by knowledge procured from the study of general medicine.

The writer has not only contributed what his own experience may have fitted him to present, but has largely made use of the labors of others. Besides the standard works and journals enumerated on page v, and to which reference is made in the text, he has consulted many monographs and smaller treatises which are not mentioned. Many of the illustrations have already been published, while a few have been especially prepared for this work. In the list which is given, proper acknowledgment is made.

The author's warmest thanks are due to his friends, Dr. James L. Minor and Dr. A. H. Buck, for assistance in preparing this work for the press.

NEW YORK, 253 MADISON AVENUE.

BIBLIOGRAPHY.

Books.

Graefe and Saemisch (abbreviated to G. and S.): Handbuch der gesammten Augen-
   heilkunde, in 7 vols.   Leipsic, 1880.   (The names of the contributors are Profs.
   Arlt, Arnold, Jr., Aubert, Becker, Berlin, Förster, Alfred Graefe, Hirsch, Iwanoff,
   Landolt, Leber, Leuckart, Manz, Merkel, Michel, Nagel, Saemisch, Sattler, Schir-
   mer, Schmidt, Snellen, Schwalbe, Waldeyer, v. Wecker.)
Klein : Lehrbuch der Augenheilkunde.   Vienna, 1879.
Schweigger : Handbuch der Augenheilkunde.   Fourth Edition.   Berlin, 1880.
Mauthner : Vorträge der Augenheilkunde.   Wiesbaden, 1879–1881.
Zehender : Lehrbuch der Augenheilkunde.   Stuttgart, 1879.
Stellwag : On the Eye.   Fourth Am. Ed.   New York, 1873.
Soelberg Wells : On the Eye.   Edited by Dr. Bull.   Philadelphia, 1880.
Carter : On the Eye.   Edited by Dr. Green.   Philadelphia, 1876.
Nettleship : Diseases of the Eye.   Am. Ed.   Philadelphia, 1880.
Donders : Accommodation and Refraction of the Eye.   London, 1864.
Alt : Lectures on the Human Eye (Histology and Pathology).   New York, 1880
Jaeger : Beiträge zur Pathologie des Auges.   Vienna, 1856.
Liebreich : Atlas d'Ophthalmoscopie.   Paris, 1863.
Magnus : Ophthalmoscopischer Atlas.   Leipsic, 1872.
Bouchut : Ophthalmoscopie Médicale.   Paris, 1876.
Galezowski : Traité d'Ophthalmoscopie.   Paris, 1876.
Abadie : Traité des Maladies des Yeux.   Paris, 1876.
Sichel fils : Traité d'Ophthalmologie.   Paris, 1879.
Galezowski : Traité des Maladies des Yeux.   Paris, 1875.
Meyer : Maladies des Yeux.   Paris, 1880.
Landolt : Examination of the Eyes.   Translated by Dr. Burnett.   Philadelphia, 1879.
Wecker and Landolt : Traité complet d'Ophthalmologie.   Tome premier, première
   partie.   Paris, 1879.
A. Robin : Des Troubles oculaires dans les Maladies de l'Encéphale.   Paris, 1880.
Bernhardt : Hirngeschwulste.   Berlin, 1881.
Wernicke : Lehrbuch der Gehirnkrankheiten.   Band I.   Cassel, 1881.

Gowers : Medical Ophthalmoscopy. London, 1879.

Albutt : The Ophthalmoscope in Diseases of the Nervous System **and of the Kidneys.**
   London, 1871.

JOURNALS.

A. v. Graefe's Archiv für Ophthalmologie. Berlin.

Annales d'Oculistique. Brussels, Belgium.

Monatsblätter für Augenheilkunde. Zehender. Stuttgart.

Centralblatt für Augenheilkunde. **Hirschberg.** Leipzig.

Archives of Ophthalmology. Knapp and Hirschberg. **New York and Wiesbaden.**

Archives d'Ophthalmologie. **Panas, Landolt, Poncet. Paris.**

Transactions of the American Ophthalmological Society. **New York.**

Ophthalmic Hospital Reports. London.

Brain : A Journal of Neurology. London.

American Journal of Medical Sciences. Philadelphia.

Medical Record, New York.

TABLE OF CONTENTS.

PART I.

PART II.

CHAPTER I.

Anatomy—Diseases of the Lachrymal Gland—Diseases of the Excretory Apparatus—Stricture—Phlegmonous Inflammation of the Lachrymal Sac—Lachrymal Fistula.

viii   CONTENTS.

CHAPTER II.

CHAPTER III.

CHAPTER IV.

CHAPTER V.

CHAPTER VI.

CHAPTER VII

CHAPTER VIII.

CHAPTER IX.

CHAPTER XVII.

CHAPTER XVIII.

LIST OF ILLUSTRATIONS.

ERRATA.

Page 279, **second line** from top, read, **instead** of ethmoidal, *sphenoidal.*

Page 287, tenth line from bottom, read, **instead of occur,** *concur.*

Page 293, seventeenth **line from bottom, footnote omitted:** A *negative scotoma* is one of which the **patient is unaware in bright or ordinary light,** but which can be **brought out under a feeble** illumination, and which is also a color scotoma, and chiefly **for red.**

DISEASES OF THE EYE.

PART I.

GENERAL ANATOMY OF THE GLOBE.

THE *eyeball* is a spheroid contained within the **orbit**, whose protection is completed by the eyelids. **It rests** upon a cushion of fat and fibrous tissue, and is rotated by six muscles. It is lubricated behind by fluid upon the layer of fibrous membrane with which it is in contact, and which is called the *oculo-orbital fascia* or *capsule of Tenon*. In front it is moistened by the secretion from its covering membrane, *the conjunctiva*, and from the lachrymal gland. It consists externally of the *cornea* and *sclerotica* or *sclera*. A line drawn perpendicularly through the centre of the cornea is its *antero-posterior diameter* or *axis ;* a line perpendicular to this, in a plane parallel to the median plane of the body and through its geometrical centre, is its *vertical axis ;* and another in a horizontal plane perpendicular to **both these,** and passing through the same centre, is the *horizontal* or *transverse* **axis.** A plane which shall pass through both vertical and antero-posterior diameters will touch the surface of the eye on its *vertical meridian.* A similar plane, passing through the transverse and antero-posterior axes, will form at the surface of the globe the *horizontal meridian.* The plane passing transversely through the vertical meridian forms at its surface the *equator* of the globe, and the anterior and posterior **extremities of the diameter** perpendicular to this plane, which is its axis, are the *poles* **of the eye.** All planes going through the geometrical centre will form *principal meridians* or *great circles.* All planes not passing through this centre will form *lesser circles,* or those of latitude. These terms and all others common to spherical geometry are made use of in the topography of the eye.

In the measurements of the eye it has become the custom to employ the metric or decimal system ; but other systems are to some degree retained, and it will be proper to indicate their relations to each other.

We have to know the proportions of the English, Paris, Prussian, and Austrian inches to the metric system. The inch of the United States is identical with the English inch except in the fourth decimal, which difference may be disregarded. The metre is divided into 10 decimetres, 100 centimetres, and 1,000 millimetres.

1

The metre contains 39.37 English **inches.**

"	"	"	37.	Paris	"
"	"	"	38.23	Prussian	"
"	"	"	38.	Austrian	"

One English inch contains 25.4 mm.

"	Paris	"	"	27.07	"
"	**Prussian**	"		26.16	"
"	**Austrian**	"	"	26.28	"

All inches are divided into 12 lines ; therefore—

	One English	line	= 2.1116 mm.
	"	Paris	"	= 2.256	"
	"	**Prussian**	"	= 2.18	"
	"	**Austrian**	"	= 2.19	"

Very many of the measurements in use, until within late years, were based on the Paris foot, inch, and line. **A few examples of their English** and metric value will be convenient.

20 Paris	feet (20′)	= 21	English feet 1 inch (253″)	= 6.496 metres.
1 "	foot (1′)	= 12.65 "	inches	= .3248 "
1 "	inch	= 1.056 "	"	= .02707 "
1 English foot		= 11.3408 Paris "		= .3048 "
1 "	inch	= .9467 "	"	= .0254 "
6 Metres		= 18′ 6″ "	"	= 19′ 8″ English.

If a horizontal section **of the eye be made, we** find, going from before backward, the following parts, **viz.:** *The cornea ;* the space called *aqueous chamber* and filled by aqueous humor, and which contains also the *iris,* which divides the aqueous chamber into the *anterior and posterior chambers,* and is itself perforated by an opening, *the pupil ; the crystalline* **lens,** enclosed in a capsule **which** by certain fibres is attached by its edge to **the** tips of the ciliary processes ; behind the lens the *corpus vitreum* or *vitreous humor ;* in contact **with the vitreous** is the *retina,* into which passes the *optic* **nerve ; external to the retina** is the *choroid,* which, at a place near the corneal **edge, takes the name of** *ciliary body,* and is raised into folds called *ciliary processes,* **and is also** continuous with the iris ; outside of the choroid and in front, joined to the cornea, is the *sclera,* which behind is continuous with the sheath of the optic nerve. The optic nerve passes through **the sclera and** choroid and joins the retina.

The eye has been spoken of as a spheroid, its antero-posterior diameter being the larger. This and other facts in regard to its dimensions appear in the following table compiled from various authors, but chiefly from Ed. Jaeger.

Eye-ball	Antero-posterior axis	= 24.3 mm.
"	Transverse diameter	= 23.6 "
"	Vertical diameter	= 23.4 "
Cornea	Thickness at apex	= 0.9 "
"	" at margin	= 1.2 "
"	Radius of front surface	= 7.5 "
"	Diameter **at** base	= 12. "
Sclera	Thickness at ciliary region	= 0.7 "
"	" behind	= 0.9 "
Pupil	**Average** diameter	= 4. "

Lens .... Radius of anterior surface........... = 8.2 mm.
" .... Equatorial diameter................. = 8.7 to 10.3 mm.
" .... Axis (thickness)...................... = 3.7 to 4.7 "
Vitreous..... " .... = 15.1 mm.
Optic disc .... Diameter.......................... = 1.4 "
Ciliary body .. Breadth.......................... = 7. "
Retina....... Thickness at fovea................ = 0.1 "
" .... " " macula lutea........... = 0.49 "
Dist. from posterior surface of cornea to front of lens = 2.6 "
Dist. from centre of macula lutea to centre of optic disc = 4. "

For practical purposes it is important to understand very correctly the relations of the parts composing the anterior half of the eye. The subjoined diagram, from Merckel, presents them, in most respects, satisfactorily.

The edge or limbus corneæ is indicated by the shading and by the attachment to it of the conjunctiva. A point to be noted is that the extreme limit of the transparent cornea does not reach back to the place from which the iris springs; hence, a puncture can be made into the anterior chamber through the anterior edge of the sclera. The periphery of the iris is more retracted than the figure shows, while its pupillary margin comes forward in advance of the peripheral attachment, because it rests on the convex surface of the lens. A clear perception of these facts is indispensable in operating at this region. The existence of that congeries of vessels called the *canal of Schlemm*, or the *circular venous sinus*, is also to be noted. It has important relations to the physiology of the anterior chamber. It is the outlet by which the aqueous humor finds its way into the circulation, and is supposed by Schwalbe to have in its wall minute clefts for this

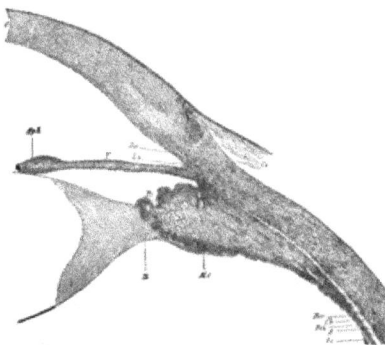

Fig. 1.—Section of the Anterior Part of the Globe, magnified six times. The iris has been shrunken and displaced by the preserving fluid. It should lie more in contact with the lens, and would thus reduce the size of the posterior chamber. *C*, cornea; *Co*, conjunctiva; *Sv*, sinus venosus, or canal of Schlemm; *Lp*, lig. pectinatum; *S*, sclera; *Mc*, musculus ciliaris; *Pc*, processus ciliaris; *Ch*, choroidea; *I*, iris; *Sph*, musc. sphincter iridis; *PcA*, perichoroidal space; *Os*, ora serrata retinæ; *Z*, zonnla or suspensory ligament of lens; *L*, lens; *Per*, ciliary part of retina.

purpose; but this is disputed by Leber. Immediately in front of the sinus are to be found delicate fibres passing from the cornea across the angle of the anterior chamber to the iris. They are insignificant in man, and are called the pectiniform ligament. In lower animals, viz.: in the ox and in swine, etc., they are more developed, and constitute the *canal of Fontana*. Upon theories of intra-ocular pressure, the parts now alluded to have an important value. The aqueous humor is derived from the vessels of the iris and of the ciliary processes. The posterior chamber is entirely shut off from the anterior chamber by the contact of the iris with the lens, and even when the pupil is well dilated the contact continues. A remark about the anterior chamber may be made: that, while its greatest depth is from 2.6 mm. to 3.5 mm., it does not seem to be so deep, because the refraction by the cornea and aqueous humor makes the iris seem nearer to us than it really is—just as, in wading a brook, the water looks more shallow than we may have found on trial, to our regret, it actually proved to be. The sphincter of the iris makes the pupillary part of the membrane thicker than the rest of it. Another point to be observed is that the ciliary processes do not touch the rim of the crystalline. There is always a separation between them. The zonula of Zinn, or suspensory ligament of the lens, comes from the posterior surface of the ciliary processes, and is attached to the lens-capsule. It splits into fine fibres, which go more to the anterior surface of the lens than to the posterior.

Passing to the deep part of the eye, we have the *retina*, beginning at the optic nerve and lining the concavity of the globe to the posterior edge of the ciliary body. Because its edge is irregular it is called the *ora serrata*. The retina is transparent and thicker near the optic nerve than at any other part. At a point 4 mm. from the centre of the nerve, on its temporal side, and about 1 mm. below it, is a depression called the *fovea centralis*. Around it the retina has a faint yellowish or tawny color over an ill-defined elliptical space, and this region is called the *macula lutea*, or *yellow spot*. Its greatest diameter, which is horizontal, is about 0.8 mm. The thickness of the retina near the nerve is 0.3 mm. The *fovea centralis* is 0.2 mm. in diameter. Outside of the retina is the *choroid*, which is perforated by the optic nerve, and consists chiefly of blood-vessels and pigment and connective tissue. The pigment is of a dark brown color, and varies in amount in different persons. We find a layer of hexagonal epithelium,

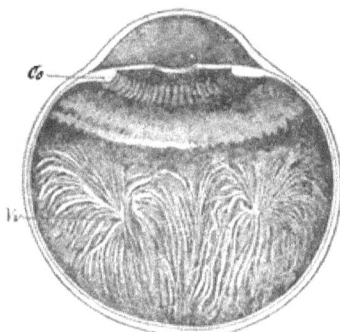

filled with pigment-granules and containing each a nucleus, which was formerly assigned to the choroid, but is now regarded as the most exterior layer of the retina. In the choroidal stroma are irregular cells with stellate processes and nuclei filled with pigment-granules. There is also free pigment scattered among the vessels. The choroidal vessels will be mentioned hereafter. At the point where the retina terminates, or no longer possesses nerve-elements, we have the beginning of that part of the choroid called the *ciliary body* (Fig. 2). It is divided into the *pars non plicata* behind, and the *pars plicata* in front. The plicæ or folds are some seventy in number and of unequal length. They consist of a congeries of vessels, which in front lift themselves up into projecting masses, and are called the *ciliary processes*. The great abundance of blood-vessels aggregated together in the choroid and the ciliary processes is required to secrete the pigment and to furnish nutritive material for the vitreous body and lens, which have no blood-vessels.

Outside of the ciliary body, inserted between it and the sclera, is a mass of muscular fibres, known as the *ciliary muscle*. Its most exterior fibres run in the meridians of the eye; those which lie next run in oblique directions which slant more and more as we go deeper, until we come to the innermost set, which take a circular direction. The whole mass in meridional section has a triangular form whose apex and point of attachment is at l.a.,

Fig. 3 (from Gerlach: "Beiträge zur Normalen Anatomie des Menschl. Auges," 1881).

The place of attachment is called by Gerlach the *ligamentum annulare*. The anatomy of this region of the eye was long misunderstood, and there is likely to be confusion from the variety of terms which have been employed at different periods. The insertion of the ciliary muscle is upon the choroid, and its effect is to relax the fibres which pass from the tips of the ciliary processes to the margin of the lens, and which fibres are known as the zonula of Zinn, or, suspensory ligament of the lens. This name is also extended to a transparent membrane which lies between the ciliary body and the vitreous. The purpose and effect of the ciliary muscle is to permit the crystalline lens to become more convex. The space between the ciliary processes and the margin of the lens has lately attracted special attention in reference to the way in which effete matter

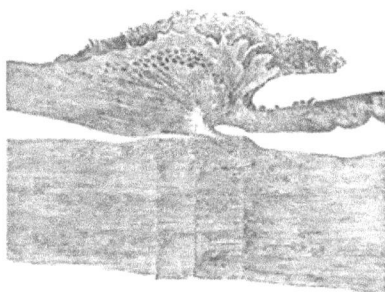

Fig. 3.

from the vitreous can escape from the eye. The iris has pigment, blood-vessels, epithelium, and also two sets of muscular fibres which regulate the size of the pupil; iris, ciliary body, and choroid are together known as the *uvea*.

Experiments upon living animals, made first in 1876, by Prof. Boll, of Bologna, and subsequently pursued by Prof. Kühne, of Heidelberg, have demonstrated the existence of a pigmentary substance in the retina, which is called the *visual purple* or *visual rose*. It is a secretion from the hexagonal pigment-epithelium of the retina. Its properties are summed up by Dr. Ayers (in the *New York Medical Journal*, May, 1881, p. 582), who says that it is an albuminoid compound belonging to the rods in their outer segments, not to the cones. Its extraction requires a ten per cent. solution of sodium chloride, or a two per cent. solution of gall, and other steps which a foot-note describes. It is a photo-chemical substance, sensitive to light,

Fig. 4.

and in man becomes bleached to a yellow hue. In some fishes, chiefly the deep-sea varieties, it is not changed in color by light, but remains purple. Its secretion in animals is increased by pilocarpine and muscarine. We know of no drugs or nerves whose action can diminish its quantity. When a person is for a long time kept in darkness, it becomes abundant, and chemical rays of light have the greatest effect; hence, if then bright light be let in on the eye, it is greatly dazzled. On the other hand, being bleached by light to a yellow hue, this tint is the greatest obstacle to the action of chemical rays, and, being in this condition, the eye does not see well on passing into a dark room. The purple seems needful to the appreciation of dim light, and its conversion into yellow is a defence of the retina against the injurious influence of bright light. It is seen that the retina, in its chemical properties, bears out the analogy of the eye to a photographic camera in the most surprising and complete manner. Indeed, by confining rabbits in darkness for a length of time and then exposing them to a bright window crossed by bars, decapitating them in a room lighted only by a sodium flame, and treating the retina by a solution of alum, and in a manner similar to the usual processes of photography, a picture or optogram can be developed and fixed in the retina and preserved for future study. Such a picture is given in the diagram (Fig. 4) copied from the *New York Medical Journal*, March, 1881, and taken by Dr. Ayers, who worked with Prof. Kuhne in his laboratory.

GENERAL PHYSIOLOGY OF THE EYE.

Having these general facts of the anatomy of the eye, we are prepared to understand its function, while the details of structure of the separate parts will be postponed to the several chapters in which their diseases will be considered.

We find the eye constructed of lenses, between which is interposed a diaphragm perforated by an opening—the pupil—which becomes larger or smaller as circumstances require, and thus cuts off the peripheral rays. These lenses gather the rays of light into an image which falls upon the retina, and reflections are in great measure prevented by the absorbent action of the pigment-layer. The retina is made up of nerve-elements of peculiar structure, and of the fibres coming to it from the optic nerve, and of connective tissue. The only elements we now need to consider are the *bacilli* or *rods and cones*. They are upon the outer surface of the retina, next the epithelium, and may be likened to the pile of velvet, because they stand perpendicularly to its surface. At the fovea centralis they are most numerous and elongated, the cones alone existing here. The minute structure of the retina at the fovea centralis is shown in Fig. 5, which is taken from Schultze's schematic section, given in *Stricker*. The fibres of the optic nerve are the innermost of the nerve-elements of the retina, and at the fovea cen-

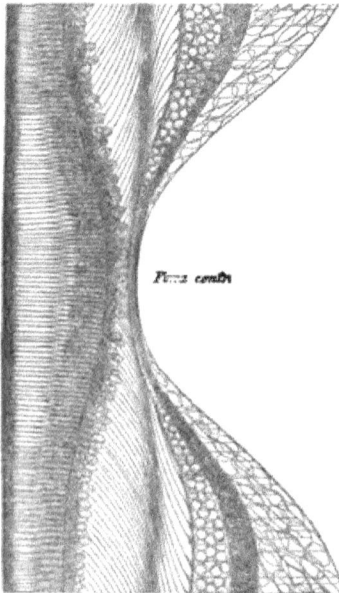

Fovea centralis

Fig. 5.

tralis are not to be found. They convey to the brain the impressions excited in other elements of the retina, and are themselves not capable of being stimulated by light. On this account the optic disc is insensitive to light and constitutes the blind spot in the visual field. If, with the right eye, one look at the cross in Fig. 6 (from Helmholtz) placed at twelve inches' distance, the circular white spot will correspond to the size of the vacancy in the field of most persons. The cross is above the level of the centre of the circle, because the fovea centralis is lower than the middle of the optic disc.

All parts of the retina up to the ora serrata are capable of perceiving light. The impressible surface is not quite a hemisphere, and if it stood out beyond the obstruction of the nose and the other surroundings of the eye, each would be able to include within its scope a corresponding ex-

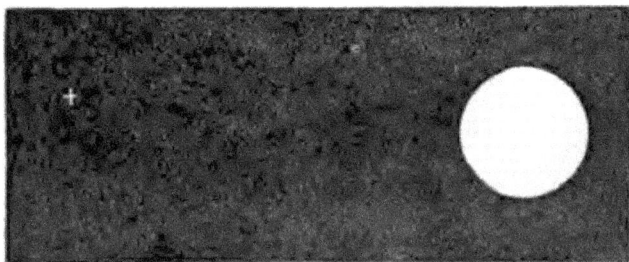

FIG. 6.

terior hemisphere or field of vision. The space which the eye at rest can cover in vision varies according to the facial peculiarities of each person and the prominence of the eye from the orbit ; we shall return to the subject of the visual field farther on. The middle of the retina is its most sensitive part, and when we give attention to an object, i.e., look at it, we turn the eye so that the object shall be imaged on the fovea centralis. When from a luminous point rays pass through a biconvex lens, they are again gathered together on the opposite side of the lens to a point, and if we place before the lens a luminous object like a candle, from each of its points rays diverge which converge on the other side of the lens to corresponding points ; and as point answers to point, a picture or image of the object is formed which can be presented on a screen. This image, which we are all familiar with in the photographic camera, is small and inverted as respects the object. If we construct it in a diagram on paper, we find that the size of the image is given by lines drawn from the extremities of the object and passing through the centre of the lens to the extremities of the image. The nearer the screen is to the lens, the smaller will be the image. For a single biconvex lens the optical centre is the same with its geometrical centre. But, if we have two lenses acting together, the optical centre of the combination will be away from the centre of each. Now, in the eye we have a combination of lenses, viz., the cornea and aqueous humor composing one lens, the crystalline being the other lens. The vitreous humor has no refractive effect. It is important to know where is the optical centre of the eye. We have also another centre which is to be considered, viz., the centre of rotation. And thirdly, the geometrical centre differs from both the preceding. The optical centre lies very near to and just within the posterior surface of the crystalline lens. Therefore, to get the size of the image of a given object, lines only need to be drawn from its extremities through this point to the retina, and the space which they include upon it encloses the image. The measure of this image is the angle which the limiting lines form at the optical centre. This we know as the *angle of vision*. All measurements of vision are necessarily taken in terms of angles (see Fig. 7). If now we wish to find out the extreme degree of sensibility possible to the retina, we may determine it experimentally by using a variety of minute objects as tests, viz., the pollen of plants, or the capacity for resolving stars which are near together. But the understanding of the acuteness of sight requires a measurement of the angle which the object subtends at the visual centre. Different eyes are found to vary considerably, and much depends on the brightness of the illumination. For a correct understanding of the subject we must know the exact situation of the optical centre or nodal point. Its place, according to Donders, is 14.858 mm. from the bacillary surface of the retina and 0.3602 mm. in front of the posterior

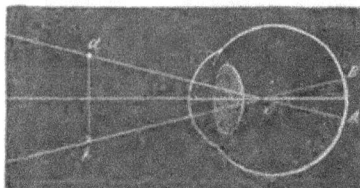

surface of the crystalline lens. It is taken generally at 15 mm. from the retina. The centre of rotation in normal eyes varies with the position of the eye, but for the look straight forward it is, in emmetropic eyes, according to Donders (p. 181), 13.54 mm. behind the cornea in an axis of 23.53 mm. If we subtract the thickness of the membranes, we have it at about 8

FIG. 7.

mm. from the retina, and 7 mm. behind the common nodal point. The geometrical centre will be at half the axis, viz., 12 mm. It thus appears that the optical value of the eye, if it were reduced to a single lens, would be equal to one whose focus would be three-fifths of an English inch in focal length.

For investigations upon sight, we need to know what is the average acuity of vision in good light of a large number of persons. This problem has been worked out by Ed. Jaeger, and more perfectly by Snellen. The latter found that a visual angle of one minute was easily within the ability of most persons, and for a practical test he constructed a series of test-types, in which the letters should have for each stroke a visual angle of 1' and the whole letter should form an angle of 5', or be included within a square whose side should have an angle at the optical centre of 5'. These letters are made of various sizes, and are each designated by a number which indicates at what distance they must be placed from the eye to subtend the angle of 5'. They number from CC. (200) to X. (10).

The most useful size is that known as No. XX., and this is to be placed in a fairly lighted room at 20 feet from the eye. The expression for the acuity of sight is $V=\frac{d}{D}$—a formula in which V stands for visual acuteness, d stands for the distance at which the card is hung, and D for the number attached to the type read. The fraction indicates the required expression. A person who reads print No. XX. at 20 feet has $V=\frac{20}{XX}$, which is 1 ; one who reads XL. at 20 feet, has $V=\frac{20}{XL}$, which is $\frac{1}{2}$; one who reads CC. at 20 feet, has $V=\frac{20}{CC}$, which is $\frac{1}{10}$; one who reads XII at 20 feet, has $V=\frac{20}{XII}$, which is $\frac{5}{3}$, and is better than 1. Critical study of Snellen's series of letters has shown that there is some error in going up the card to the larger letters, and that the requisite proportion is not perfectly observed. To remedy this small error, Dr. Green, of St. Louis, devised a series of types in which the cube root of 0.5 = 0.795 is taken as the factor, and he fills certain gaps in Snellen's scale. See the accompanying table :

Green.	Snellen.	Green.	Snellen.
200	200	40	40
160		32	30
126		25	
100	100	20	20
80		16	15
64	70	12.5	12
50	50	10	10

There are many other series of test-types, viz.: by Schweigger, by Giraud Teulon, and others, and some in which symbols are employed, which may be described by those who do not know how to read. A set of test-types, copied from Snellen, has been reproduced by Messrs. Wm. Wood & Co., under the supervision of Dr. Cutter. Since the metric system has been taken as the standard of measurement, a little alteration has been made in the types. What was formerly taken as 20 Paris feet, or 6.496 m., is now reduced to 6 m., which is 18½ Paris feet, and consequently a reduction of $\frac{3}{16}$ of the former distance. This has compelled a slight reduction in the size of the test-types, and, as now numbered in the metric fashion, they run as follows:

XII. XVI. XX XXX. XL. L. LXX. C. CC.
IV.   V.   VI. IX.   XII. XVIII. XXIV. XXXVI. LX.

The upper line gives the old numbers in **feet**; the lower line the **new** numbers in metres. We may, therefore, if we have the old set in **feet**, read $V = \frac{20}{XX}$, and in the **new** set will have for the same degree, $V = \frac{6}{VI}$. It must be noted that 20 **Paris feet** amount to 21 feet and 1 inch English measure. Having gained a standard for visual acuity, we also need to have a standard for vision at short distances, such as we are accustomed to, in reading and writing. This was first provided by Jaeger, and for this purpose his types, numbered from 1 to 20, are quite as much used as are Snellen's.

It will be seen that Snellen II. and Jaeger 2, that Sn. III. and J. 7, that Sn. V. and J. 13, that Sn. VII. $\frac{1}{7}$ and J. 14, and Sn. XVI. and J. 18 almost exactly correspond.

If we compare the types of Snellen and Jaeger we have the following equivalents:

Snellen. } I. $\frac{1}{7}$ II. III. IV. V. VI. VII.$\frac{1}{7}$ X. XIII. XVI. Paris feet.
{ 0.5 0.6 1.0 1.3 1.6 2. 2.5 3.2 4.2 5.2 Metres.

Jaeger. } 1 2 7 11 13 14 18 19 20 Feet.
{ 0.325 0.65 .975 1.3 1.6 2.3 5.5 8.8 12 Metres.

It has already been stated that vision, as thus measured, is merely an average. It by no means expresses what acuity is possible for many persons, and with strong light; $V = \frac{20}{16}$ or $\frac{20}{12}$ is not at all uncommon. I have lately tested a lady who had $V = \frac{20}{10}$, measured by Green's types.

Helmholtz found the visual angle in one case equal to 50″ in the distinction of parallel wires. Instead of assuming the bodies of the cones as the sensitive elements of the fovea, Hansen locates the sensation in their tips or stems, and they measure, according to Schulze, 0.0006 mm., which corresponds to a visual angle of 10″ (see "G. & S.," II.,

2, p. 584). Some further details respecting the fovea centralis are here in place. H. Müller, quoted by Becker, gives its diameter as 0.2×0.4 mm., and Becker calculates that, if it be taken as a rectangle, 0.4 mm. $\times 0.2$ mm., we shall have in it 13,000 percipient elements (cones). This gives in the schematic eye a horizontal angle of 1° 31' 1'', and in the vertical, 1° 8' 1''. An object at 15 ctm. distance (6'' English), which should be 3 mm. high and 4 mm. across, would just fill this space; and at 6 m. an object 12 ctm. high and 16 ctm. wide, would also fill it. It follows that for 12 inches we can include at one view print contained within 8 mm. space, which will be such a word as, pencil, of the type of this page. In reading we therefore run the eye forward over the line, and while the impression of one word is being transmitted to the brain, the following word is making its impression on the retina. The time for a retinal impression to be made depends upon the intensity of the light, and is quite appreciable with weak illumination. Because time is required for the retina to respond to light, this is called its adaptation. The retina attains a certain maximum of reaction to light, and then this declines, so that it becomes less sensitive. The lessened sensibility to light is called *fatigue of the retina.* On the other hand, impressions on the retina remain for a certain time when the light is withdrawn. Such impressions of objects are called *after-images.* The brighter the image, the longer will the after-image remain, and the brighter will it be. The duration of after-images varies much in different persons (see Aubert in "G. & S.," III., 2, p. 508).

The fovea has been declared to be destitute of blood-vessels, but Wadsworth figures one capillary within it, and thinks he saw transverse sections of others. Mention has been made of the optic nerve blind spot which occupies a position at an angle of about 12° to the nasal side of the centre of the retina and 1° above it. But there are also other insensitive parts of the retina, and these are such as are covered by the retinal vessels. By looking into a dark space and letting the light of a lamp fall into the eye while a screen perforated with a pin-hole is moved rapidly to and fro before it, the vessels cast shadows on the underlying parts of the retina in rapid succession, and as the image persists for a short period, the effect is to raise before the eye in the darkness a phantom representation of the vascular distribution of the retina. This is called the *vascular image of Purkinje.* It has been used to prove the sensitiveness of the rods and cones, and that they are the primary organs of light-perception. At the middle of the macula lutea, no considerable vessels are found. H. Müller's injection of an old man's eye proved this non-vascular space to be 0.41 mm. $\times$ 0.31 mm. Becker also asserts that he can perceive the blood-corpuscles streaming through the capillaries of the vessels, provided he excite retinal hyperæmia by holding the head down and looking through a blue glass at the sky. The capacity for exciting this susceptibility is very variable in different persons.

Pressure on the retina, or the galvanic current, excites subjective luminous impressions, which may be golden or crimson, etc. The membrane, when irritated, responds only by luminous sensations, never by those of pain. The sensations caused by other excitants than light are called *phosphenes,* or subjective sensations. They may be caused by disease.

Having spoken of visual acuity at the fovea and how it is measured, and also having alluded to other functions of the retina, we must attend to a function of the eye which is purely optical, *viz.,* the *accommodation.* If we stand before a window, and, using only one eye, hold at 8'' before it the point of a pencil in the line of some object at which we look—say it be a distant tree or a window-shutter—we will notice that when the tree is distinct the pencil-point is blurred, and if the pencil is well defined the tree is hazy. If we look through a minute hole in a piece of dark paper at the tree, and hold a pin at the same distance as before in the line of sight, both the pin and the tree can be seen distinctly up to within a few inches, simultaneously. But if we look through two pin-holes, say at 2 mm. apart, and get the pin in the centre where the holes seem to overlap, we see two pins in looking at the tree, and only one in looking at the pin when it is brought to a distance suitable to the age of the person.

The first and last of these experiments demonstrates that the eye must possess a faculty of accommodating for different distances, or, in other words, of focussing for near objects. The experiment with two pin-holes is known as that of Scheiner, and may be illustrated by a diagram as follows :

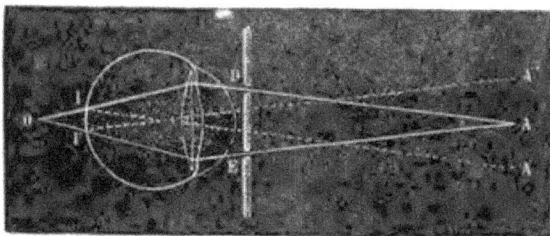

Fig. 8.

It has some value in determinations of the refraction of the eye as well as in accommodation, as will be noticed later. In Fig. 8, the object at A is nearer the eye than the distance for which it is focussed ; hence the bundles of light going through the holes D and E impinge on the retina before coming to their focus, which would be at O. Two images are seen, each somewhat indistinct, and they appear to lie in the directions I A' and I' A''.

In Fig. 9 the object is more distant than the place for which the eye is focussed, and the beams, after passing the holes, form a focus at O, but

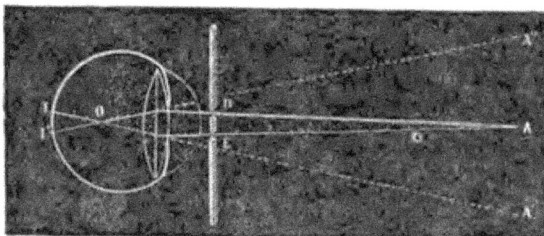

Fig. 9.

cross and strike the retina in two spots at I and I' ; they are seen to lie in the direction I' A'' and I A'. It will be noted that the places where the images are seen are reversed in the two figures. If over the hole D a slip of red glass were placed in each figure, for **Fig. 8 the red image** would be on the opposite side of the optic axis as **respects the red hole,** while in Fig. 9 it would fall on **the** same **side of the optic axis with the red** hole.

Evidently, **therefore,** the eye must adjust itself **to** the **distance of the** object it wishes to view. In this the eye is subject to the laws which prevail **in all similar** optical instruments. For **a** biconvex lens, when a luminous object approaches, **the image** retires, and as the object retires **the** image approaches the lens. Now, **in** the eye the provision made to secure the invariable location of the image on the retina **is** called the *faculty of accommodation.* This faculty has its limits in that there is a point so close to the eye that a distinct image cannot be formed, while for a normal eye the remote limit is infinity, **or a distance** at which rays become parallel.

The experiment with **a** single pin-hole **proves that** a minute aperture renders the accommodation to a considerable degree superfluous; but the experiment with two pin-holes proves that the faculty of accommodation is absolutely indispensable to sight. Because, while one pin-hole will give a tolerably clear image of a close object, two or more pin-holes multiply the object, thereby proving that the rays from it do not all coincide upon the same spot of the retina.

How is the accommodation produced, *i.e.,* what is its mechanism ? If **a** lens will **bring parallel rays to a focus** at **a given distance, rays which** are divergent **will not decussate at this** point because the bending power over light **which the lens possesses will not be sufficient.** If a lens focusses at 2 inches from itself, an image of a gaslight 20 feet away, when **that** light approaches the lens within 6 inches, the image is not at 2 inches, but farther off, *viz.,* **at 3 inches.** But, suppose the **image of an** object at 6 inches distance **must be** located at 2 inches from the lens, what **must** be done ? **Evidently the lens** must be made stronger, that is, its power of bending light must be increased, which is only another way of saying that its focus must be shortened. Now, the refractive parts of the eye may be compared **to** a lens of ⅗ **of an** inch or 15 mm. focus, and this is adequate to bring parallel rays to **a focus on** the retina. If an object approach to within 6 inches, **or 150 mm., the image,** if there be no suitable adjustment, will fall **beyond the retina to a distance of 1** mm., and be indistinct. Now, there are **evidently two ways of meeting the** difficulty in the case of a lens whose **focus is so short as 15 mm. If it were** advanced toward the object only 1 mm. the error in focus would be substantially corrected, and this mode was long held to be probable. But the other solution is to **increase** the curvature of the lens and thereby shorten its focus. By measurement of the images reflected from the cornea and from the front **and** back surfaces of the lens, when the eye looks at a distant and again at **a near** object, it was found that the curve of the cornea undergoes no change. The front of the lens becomes more convex, and therefore pushes forward, while the back of the lens undergoes a scarcely appreciable change. It was Helmholtz who first made these measurements by an instrument which he invented, called the ophthalmometer. His calculated **results** have been repeatedly confirmed. We may sum up changes in the **eye during** accommodation **as** follows: the pupil becomes **smaller, the front of the lens** becomes **more convex, and,** by advancing a little, carries **with it** the iris and reduces the **distance between it** and the cornea ; the posterior surface **of** the lens becomes inappreciably more convex. The lens **is** thus increased in thickness at its axis, and its equatorial diameter is shortened. Its edge becomes more rounded. **The** ciliary processes swell and project a very little nearer to the axis of the eye. These changes of the ciliary processes have been proved by examining **the** eyes of albinoes, and eyes in which iridectomy has been performed. A considerable magnifying power is needed to see the swelling, **and some observers have** denied that it occurs ; but Coccius, Becker, and Hjort have established it. The **active agent in the process is the** ciliary muscle, and **the way in which it is** brought **about is** regarded **to be as** follows : **the** ciliary muscle, whose fibres are meridional, oblique, and circular, having its origin at a point exterior and in front, when **it contracts becomes thicker, and presses** the **ciliary processes nearer the optic axis and enlarges their volume.** At the same time **the fibres draw upon the zonula (suspensory ligament of the lens) and release the crystalline from the tension under which it is kept, and it,** by its **elasticity, increases its** anterior convexity while **its border grows more** rounded. The distance between lens border and ciliary process is not altered, neither does the lens increase in volume. The aqueous humor becomes slightly displaced toward the periphery of the chamber, and **the pupil dimin-**ishes. The actual increase of the **axis of the lens,** in accommodating from infinity **to five inches, is** 0.4 mm. The radius of the anterior **surface of the** lens is shortened from 10 mm. to 6 mm., the radius of the posterior surface from 6 mm. to $5\frac{1}{2}$ mm. The

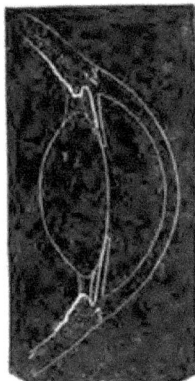

FIG. 10.

lens increases from a central thickness of 3.6 mm. to 4 mm. (see Mauthner : "Vorlesungen," 1872, p. 20). The changes are figured above as copied from Landolt.

Let us now study more carefully the functions of the eccentric parts of the retina. The first fact which we notice is, that outside of the fovea the acuity declines very rapidly. For instance, at 1° outside of it, acuity of vision is reduced to $\frac{1}{3}$; at 2° to 3° V $= \frac{1}{7}$ (Königshofer). If the fingers be spread widely, they can be counted at almost the outer limit of the field of vision. But, for the peripheral parts of the retina, we confine our examination to the recognition of form, without attempting to ascertain discriminating power. This investigation is called taking the field of vision. It is done for each eye alone, the other being covered. To do this properly, an arc of a circle must be placed in front of the patient, which shall be not less than 90° nor more than 180° in extent. Its radius should be about 12 inches or 30 ctm. The eye to be examined must be at the centre of the circle, and fixed steadfastly upon the point directly in front. An object, the size of which will be chosen according to the accuracy demanded —generally a white object, $\frac{1}{2}$ inch square, is suitable—will then be moved along the arc from its centre to its extremity, or *vice versa*. When the perception has been determined with the arc in one meridian, it must be turned to another, until the whole field has been explored. Beginning with the arc in the horizontal position, it will be carried around to the vertical position, and a determination made for each meridian at intervals of 15° or 30°. The examination is easily made by an instrument called a perimeter, of which the first and most generally used is that made by Förster. Simpler ones have been contrived, and one which is to be commended has been made by Dr. Carmalt, of New Haven, of which a figure is introduced. It is to be screwed to a table by a clamp E. The arc, which extends to 90° on one side of the upright stem and to 50° on the other, is laid out in spaces of 10° each, rotates on a pivot at zero, where is a bright, round spot, on which the patient is to look. His chin rests on the crutch C, and this can be raised or lowered as desired. The patient's eye is at 35.5 ctm. (14 inches) from the arc, and he is asked to note the appearance of a bit of card on the end of a rod, which is passed along from the extremity of the arc to its centre, from either side. On the reverse of the arc a series of radii are drawn, beginning at zero and going around to 360° from left to right, as the patient faces the instrument. An index shows at what meridian the arc is placed. The findings of the perimeter are recorded upon a chart, which is laid out in radii and circles.

Before the perimeter was invented, the custom was to take the field of

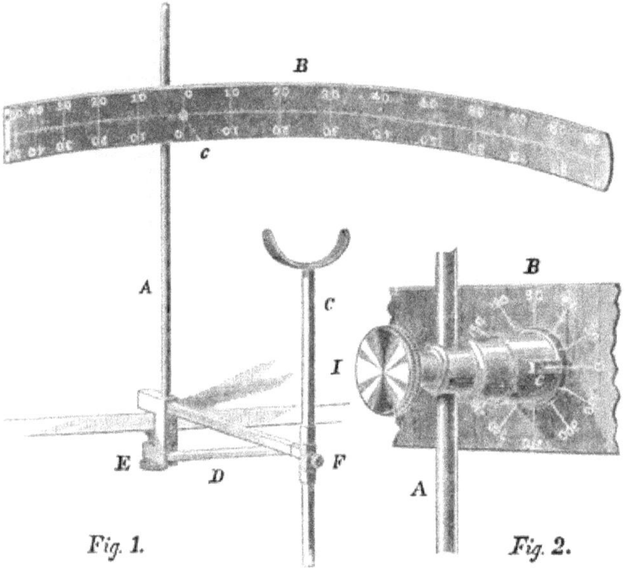

Fig. 1.   Fig. 2.

Fig. 11.

vision on a blackboard or any convenient plane surface. This is manifestly very inadequate, because it is impossible to denote the extreme limit of the field, and when an angle greater than 45° is passed, the size of the test-object also becomes considerably reduced by being further removed. A plane surface will serve to map out defects in the field located near the centre, but will not, like the perimeter, answer for all requirements. Snellen has a perimeter, whose centre is a plane surface, extending to 45°, and beyond this is a movable arc, which completes the whole field.

The outlines of the visual field are far from symmetrical. Its greatest extent is on the horizon and to the temporal side. On the opposite side the limit is determined by the height of the nose, while above, the eyebrow, and below, the cheek, fix the extent of its boundaries. The position of the eye in the orbit, the configuration of the face, the size of the pupil, and the length of the optical axis, are factors which enter into the form of the field. Usually the extent on the temporal side is 90°; on the nasal side, 50°. Above it is 50°, and below it is 65°. These figures are liable to great variations in different persons. To be sure that the full limit belonging to each case is recognized, the observer may sight across the arc from extreme positions on the outer, inner, or upper sides, and note upon it the place across which he is able to catch a view of the patient's pupil. This marks the limit to which the field ought to extend, and should be noted on the chart as the proper boundary within which the peculiarities of the actual field are laid out (Fig. 12). The figure gives a chart of

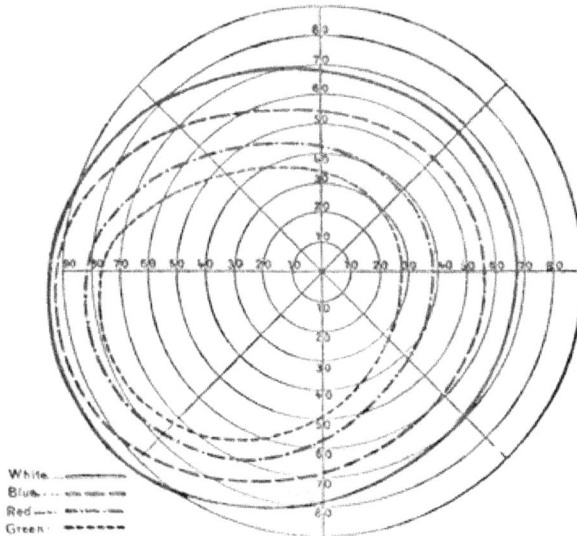

White ⸻
Blue ⸺ ⸺ ⸺
Red ⸺ ⸻ ⸻
Green ⸻ ⸻ ⸻

Fig. 12.

the field as found in a normal left eye. The conditions which we examine by the perimeter are defects of perception (scotomata) in any part of the field, or restrictions of its extent. In examining the visual field, one sometimes finds that its outlines are of normal extent, that is, not unduly "limited," but that over a certain area there is diminished or possibly no perception of the test-object, and there may be no perception of light. Such spaces of dim vision or blindness are called *scotomata*, and spoken of as partial or total, according to the degree of loss of perception. Förster has made a distinction between positive scotomata and negative scotomata. By *positive scotoma* he means a defect of light-perception within a certain area, to such a degree as to occasion a dark cloud or spot which the patient can recognize; by *negative scotoma* is meant an area in the field of vision, where color-perception is wanting, and where light-perception is not noticeably dulled. This defect the patient is not conscious of.

By the same means we are enabled to examine the faculty of color-perception. This has importance in two classes of cases: 1st, those in whom the color-sense has been impaired by disease ; and 2d, those in whom it is naturally defective. It must be premised that there is a natural limit in the extent of field over which color can be recognized, and that this extent varies according to the color. Landolt has given us the latest studies in this matter, and it appears that the limits over which the four fundamental colors of green, red, yellow, and blue can be perceived are as delineated in the chart, page 15. While within the boundaries of green all other tints can be known, and for the red, also yellow and blue ; and for the space within yellow, also the blue ; outside of yellow, only blue can be seen (Fig. 12); and outside of blue, no color is recognizable. These color limits cannot be held to be uniform even among normal eyes, but they have a value as a general statement.

In making this test, a bit of card-board, 2 ctm. square, of the proper tint, is put on the end of the rod and brought across the field, always beginning from the end of the perimetric arc. It is well to have different colors on its opposite sides, that, by turning the card around occasionally, we may guard against mistake or deception. One may have several rods, each having cards of different color, and thus be more sure of correct answers.

It happens that patients may have a dimness of color-sense for a certain hue all over the field. This will be discovered by finding out that the fainter shades of a special color fail to be correctly noted. But for this kind of error it usually happens that more than one color is dimly discerned. It also happens that there may be a scotoma in the field of a particular color, as, for example, red. The best way of finding this out is to have two rods, each with the same color-card. One is to be held at the centre of fixation, the other at a peripheral place. The one at the middle ought to be most brilliant, while it happens in certain cases that the eccentric card seems to have the brighest hue. The extent and boundary of the scotoma can, by this means, be made out, provided the test be not continued for a period so protracted as to fatigue the color-sense. This examination belongs to cases of tobacco or alcoholic amblyopia, etc.

A different kind of examination of the color-sense is that undertaken to ascertain whether it exists to the normal degree as a congenital endowment ; that is, whether there be *original color-blindness*. The importance of this investigation is found in its application among seamen, railroad officials, members of a signal corps, and other persons who are required to look out for and be guided by colored lights or signals. Both in Europe and in this country the importance of testing such persons in this respect is being widely appreciated. There are three kinds of original color-blindness : the red-green blindness, the blue-yellow blindness, and that which includes all colors. The first is by far the most common, and other combinations than these do not occur. The approved method of detecting this visual defect is by means of skeins of worsted of assorted colors, there being duplicates and various shades of each color. The test consists in laying out a color and asking the person to put beside it similar shades. If there be any hesitation or mistake, the error is at once perceived. This method is Holmgren's. Another mode consists in requiring the person to read letters on a card, which has numerous small squares of such colors as are interchangeable and in which a letter or figure is traced. The person who is color-blind would not see any letters, because there would be no contrast. This method is highly effectual in detecting persons who wish to conceal their error. It has been worked out by Stilling. Tests for this purpose are contained in the set of test-types issued by Wm. Wood & Co.

The proportion of color-blind is about five per cent. or less among men, and two per cent. or less among women. This includes all varieties and degrees of the defect. A method employed on the Pennsylvania Railroad, and devised by Dr. Thompson, has proved highly practical, and is described by him in the "Transactions of the American Ophthalmological Society for 1880," p. 142.

Another kind of examination, which belongs to subjects who cannot see enough to discern test-letters, is to determine their acuity of light-perception. If they can see forms, *i.e.*, test-letters or other objects, they are said to have *qualitative perception*; if they can only discern light, as happens in mature cataract, or with occluded pupil, etc., they have *quantitative perception*. The problem is to measure their degree of quantitative perception of light. We do this roughly by turning down the gas and throwing the light into the eye by the ophthalmoscopic mirror; or we may withdraw the mirror very far away from the full light, and state at what distance the patient recognizes it. Förster has contrived an instrument for making such examinations with precision. He has a box containing a standard candle, which illuminates a square, whose size can be altered at pleasure, and whose area is read off on a scale. The same instrument is used for somewhat different cases where qualitative perception remains, but the quantitative is likewise reduced. Test-objects, viz., strokes of certain sizes, are within the box and illuminated by the candle, whose light is varied by the contrivance before mentioned. If these objects are visible to a normal eye through a square of 2 mm. area, a diseased eye may require a square of 20 mm. area to see the same. By this apparatus a distinction is made between diseases of the percipient apparatus, viz., the retinal bacilli, and those of the conducting apparatus, viz., the optic nerve layer in the retina and the optic nerve. As between optic neuritis, retinitis, hemorrhage, turbid vitreous, alcoholic amblyopia; and, on the other hand, syphilitic choroiditis, pigmentary retinitis, and some cases of detachment of retina, it is claimed that in this way a diagnosis can be made. The distinction is theoretically well founded, but practical experience has not yet decided upon the value of the method, because it has been little used.

Another matter demands our attention, viz., the degree of resistance of the eye to pressure, or what is called its *tension*. If we press upon it with the tip of one or two fingers, we find it firm, yet to a slight degree impressible, and this because, while almost full, it does not contain all the substance it can hold. The cornea and sclera are both incapable of being stretched, and the contents of the eye being essentially like water, are incompressible. Hence, variations of tension depend simply on variations in quantity of the ocular fluids. These variations occur in the aqueous and vitreous humors. When in the former, it is seen by the depth of the anterior chamber, and the iris will be pushed farther from or nearer to the cornea. When in the vitreous, the same variations in the position of the iris will ensue, but to a less degree and with less rapidity. Endeavors have been made to measure these variations in resistance with accurate instruments, but thus far with only moderate success, and none have come into general practical use. We still employ the method by the fingers, and designate by degrees and symbols, suggested by Mr. Bowman, viz., normal tension T_n, subnormal or reduced tension, T_{-1}? T_{-}, T_{-2}, T_{-3}; supranormal or increased tension, T_{+1}? T_{+1}, T_{+2}, T_{+3}. Only by practice can one

2

learn to designate small differences with any assurance of correctness. The tip of the forefinger is gently laid on the ball until the full degree of resistance is felt. If one finger do not convey an adequate impression, apply the index of each hand gently and firmly on the eye.

BLOOD-VESSELS OF THE GLOBE AND THEIR RELATIONS TO THE DIAGNOSIS OF DISEASES OF THE EYE.

As preliminary to an understanding of how to diagnosticate its inflammations, it lies next in order to speak of the mode in which the eye and its appendages are supplied with blood. Our present knowledge is largely derived from the skilful injections of Professor Leber, whose diagram in Stricker's "Handbuch der Gewebelehre" is introduced below.

The vessels of the palpebral conjunctiva, and of the portion which passes from the lids to the globe, are derived from the vessels of the lids, viz., the median and lateral palpebral arteries and their accompanying veins. As we approach the cornea the vessels of the conjunctiva unite with those from another source, viz., the terminal branches of the anterior ciliary arteries. The relations of the vessels of the eye will be made plain by examining the following diagram, taken from Leber's paper in Stricker's hand-book.

There are four systems of vessels, which may be distinguished from each other : 1st, the arteria centralis retinæ, e, which enters the eye through the optic nerve, is destined exclusively for the retina and optic nerve, and forms almost no anastomoses with other vessels. This system is remarkably separate, and by Cohn is classed as "a terminal system ;" 2d, the posterior, or short ciliary arteries, a, which perforate the posterior part of the sclera and supply the choroid, and, with the long ciliary arteries, b, are the chief source of the elaborate vascular system of the choroid, of the ciliary body, and of the iris ; 3d, the anterior ciliary arteries, c, are derived from vessels which come from the recti muscles and perforate the sclera about 4 to 6 mm. behind the cornea. They are visible to the naked eye, are more or less conspicuous, and supply the ciliary body, the iris, and the anterior part of the sclera, and furnish the plasma which nourishes the cornea. These vessels join with the branches of the posterior ciliary arteries, and at the border of the cornea send off loops, which constitute the peculiar vascularity of this region. These vessels here anastomose with, 4th, the vessels which have come from the ocular conjunctiva. It thus happens that for a zone about the cornea there is a system of vessels which have communication with the face and with the deep and the superficial tissues of the eye. The vessels proper to the conjunctiva are of darker hue than those more deep, and they can be moved about as the membrane is slipped over the sclera by traction of the lids. This statement of the anatomy of the vessels shows how untrustworthy is any attempt to make a diagnosis of the locality of an inflammatory process, by fixing attention chiefly on the kind of hyperæmia. The vascular phenomena are important as auxiliary evidence, but do not take the first rank in deciding a diagnosis.

For this purpose we must look at the condition of the several tissues, their structure and their function : 1st, alteration of tissue, and 2d, perversion, disturbance or loss of function, are the trustworthy signs. Let us take up the external parts in succession, with a view to emphasize this point. The conjunctiva is a thin membrane covered with epithelium, whose purpose is to

supply lubricating fluid for lids and globe to move upon each other without
friction. The function is the same which the pleura performs between the
lung and the walls of the thorax. If the membrane be inflamed, it be-

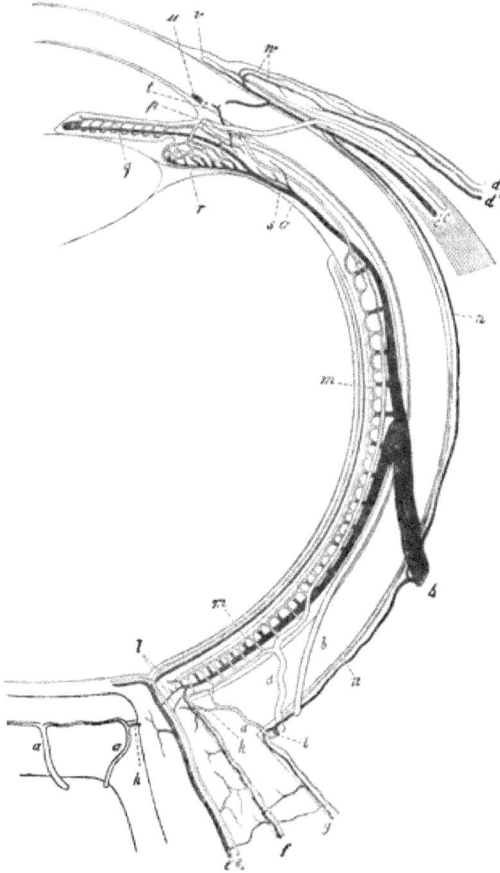

Fig. 13.—From Leber, in Stricker's "Handbuch der Gewebelehre," p. 1030. Diagram of the vessels of the
eye, horizontal section—veins dark, arteries light. *a*, Art. ciliares post. breves; *b*, art. ciliares post. longæ;
cc, art. and ven. cil. anter.; *dd*, art. and ven. conjunctiv. post.; *ee*, art. and vena centralis retinæ; *f*, vessels of the inner sheath of the optic nerve; *g*, vessels of the outer sheath of the optic nerve; *h*, venæ vorticosæ; *i*, ven. ciliar. post. brev.; *k*, twig of the art. post. ciliar. brev. to the optic nerve; *l*, anastomosis of
the choroidal vessels with those of the optic nerve; *m*, the chorio-capillaris; *n*, episcleral branches; *o*, art.
recurrens choroid.; *p*, circulus arter. iridis major (transverse section); *q*, vessels of iris; *r*, ciliary process;
s, twig of the venæ vorticosæ from the ciliary muscle; *t*, twig of the anterior ciliary veins from the ciliary
muscle; *u*, circulus venosus; *v*, vascular loops at the border of the cornea; *w*, arter. and ven. conjunctiv.
anterior.

comes less transparent, also thickened, because of infiltration, and may be
elevated above the subjacent parts. The moderate quantity of clear fluid

which it furnishes when in health is notably increased and changed in quality by disease; in other words, abnormal secretion appears, which makes the lashes adhere to each other in bundles, or may flow in quantity from the eye. The typical features of conjunctivitis, then, are œdema in or beneath the membrane, and unnatural secretions.

For the *cornea* the essential quality is that it be transparent and maintain a polished surface and correct curve. When inflamed, we find that opacity, of every possible degree, appears in the structure, and its surface may become eroded or ulcerated. Other morbid conditions, such as vascularity of its structure, infiltrations of pus, and many other changes, can occur, but all are included under the head of departures from transparency or from correctness of form.

A patient having inflammation of the cornea will have a free discharge of tears, but this fluid will not have the glutinous, mucoidal, or purulent quality which belongs to the secretion from an inflamed conjunctiva. The two kinds of inflammation are often combined, in which case we are to look for the characteristics of each.

The *iris* is an opaque diaphragm with a central hole, whose purpose is to exclude all light except what shall pass through the pupil, and to regulate its quantity by rendering the pupil larger or smaller, according to circumstances. To do this the iris is loaded with pigment, and has two sets of muscular fibres. The pigment gives to its surface an extremely diversified look, and the muscular fibres stand out in distinct curves and lines. The surface has epithelium, and therefore is polished, and there are numerous blood-vessels. The first effect of inflammation is to restrain the action of the muscular fibres, to cause swelling of the membrane both by distention of its vessels and by infiltration of its tissue. The result is seen in alteration of color of the iris and in inactivity of the pupil. The change of color is in part due to the effusions which are poured into the aqueous humor, and from this the iris acquires a dull and washed appearance, as if it had been smeared. The pupil becomes inactive and it also becomes small, because the swelling crowds the iris into the space which is left free; and this reduced size and fixity is made permanent by the formation of adhesions between the pupillary border and the lens. The color of the iris and the behavior of the pupil are the conditions which signalize iritis.

Now, for all these kinds of inflammation there are certain types of hyperæmia which are in a measure distinctive; but, to regard hyperæmia as indicative of inflammation, is to cause confusion and mistake. The vessels of the iris are taken from the anterior ciliary as well as from the posterior and long ciliary arteries, and when it is inflamed one sees turgescence of the region of the sclera forming a zone around the cornea about 4 to 6 mm. wide. But certain cases of keratitis exhibit the same kind of injection, and in addition to the turgid vessels of the sclera there is engorgement of the conjunctiva both in iritis and keratitis. Therefore we must leave the vessels out of view and scrutinize the tissues themselves, and their function, in order to learn where an inflammation may be seated.

It remains to say a word about scleritis. In this case we have almost nothing to guide us except close observation of the character and locality of the redness. One must first exclude all the signs of implication of the conjunctiva, cornea, or iris, and these being set aside, a deep-seated and usually circumscribed hyperæmia may warrant the diagnosis of scleritis. In addition it will sometimes be true that thickening and infiltration of episcleral tissue will be present. Scleritis is of much less frequent occurrence than the other maladies cited, and therefore embarrassment will not often arise.

Inasmuch as such importance attaches to critical inspection of the tissues, it is needful to have every advantage in attempting to do it. A good light and a fair exposure of the eye are to be secured, but great assistance is gained by resorting to oblique or focal illumination by means of a convex lens whose focus is about $2\frac{1}{2}$ inches. This may be used in ordinary daylight, the patient being at a little distance from the window; or, still better, the examination may be made by gaslight in a dark room.

The lens is held about two inches from the eye, condensing the light on one side of it while the observer looks from the other side.

The focus of the lens is made to play over the eye in all directions, deeper and more superficially as the various parts are to be examined. The contrast between the intense light of the focus and the shadow which surrounds it constitutes the chief advantage of this proceeding. Caution must be used not to subject cases to this method which are likely to suffer harm by the strong glare, but experience will soon indicate what patients are not to be thus investigated.

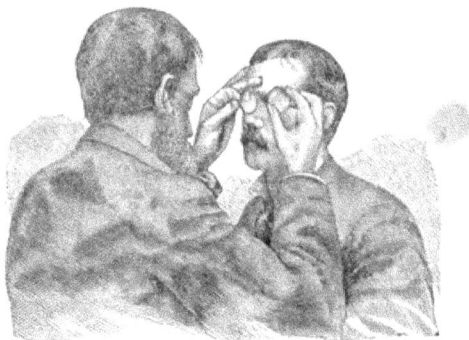

FIG. 14.

For slight lesions of the cornea, for examining for foreign bodies upon the cornea, or for studying the iris and pupil, and for exploring the crystalline lens and anterior portion of the vitreous humor, oblique illumination is indispensable.

One may also use a magnifying-lens, both without and with the help of the illuminating lens, holding one in one hand and the other in the other hand. There is no great difficulty in managing two lenses if the patient be tractable.

Another device of value is illumination by a weak ophthalmoscopic mirror, or by reflecting a weak light. If we have no other than the mirror usually made for the ophthalmoscope, which is concave and of about seven inches focus, it must be held at twelve or fifteen inches from the patient's eye to properly reduce the light. The examination must be made in a dark room, the lamp placed about six inches behind the head, so as to leave the face in shadow. If we view the eye through the hole in the mirror, at the distance of a foot or more, whatever opacities may exist in the cornea or in the lens are easily revealed. By little tilting movements of the mirror the light plays over the eye, and if opacities exist they flit like shadows across the illuminated pupil. A shadow appears where before the surface looked transparent, and again a clear surface comes out as the shadow glides to another spot. The same phenomena occur when the cornea has lost its natural curvature, and has become conical or bulges in any manner. The shadows caused by irregular reflection sometimes are very striking. More will be said on this point when we deal with diseases of the cornea and of the crystalline lens.

Before passing from the consideration of the **external diseases of** the **eye, let** us revert to the consideration of how the vessels of the eye appear **when** inflammation occurs. The conjunctival vessels are more tortuous, **of darker** hue, and the hyperæmia grows more intense as we approach the **equator of** the globe—and upon the palpebral surfaces we see, as we evert the lids, that this **membrane** has a much more intense redness than natural. **In the normal state the everted** lids show many blood-vessels, especially at **the outer and inner angles.** In cases of keratitis the conjunctival vessels **partake of the congestion, but a deeper layer also** becomes conspicuous. **In many cases only a portion of the** ocular surface exhibits hyperæmia, as **the process is or is not of a** limited extent. A case of chronic scleritis will **show a patch of redness in** which the **vessels** bordering the cornea will be **so fine and** so numerous as to give a uniform red hue, and this shades off **into a** number of larger vessels, increasing **into** a few considerable and tortuous trunks which run back to the equator.

In iritis, if of a mild type, or as an **attack** is fading away, we have sometimes a beautiful exhibition of the **zone** of anterior ciliary vessels. They cluster around the cornea in fine and nearly straight lines, radiating **from its** circumference, and compose a crown singularly well defined. **The** breadth is about five millimetres, except in the vicinity of the recti muscles, where **the breadth increases,** because the affluent vessels come in at these places.

This distinctive vascularity is, however, overlaid, and to some degree hidden in more severe conditions of iritis by the presence of many other injected vessels, and which may be as numerous as in any case of severe conjunctivitis. In fact, as inflammations of the eye grow in severity, it becomes impossible **to draw** distinctions between them according to any peculiarities in **hyperæmia.**

Serous, **or rather** œdematous effusions, occur in many inflammations, and find lodgement **in the** loose connective tissue which unites the conjunctiva to the sclera. **This occurs more** readily in conjunctivitis than in other ailments, but it **will appear** whenever any inflammatory trouble has reached a sufficient activity. **This** symptom, therefore, must be held in subordination, like the hyperæmia, and the true groundwork for diagnosis be sought, 1st, in tissue-changes, and 2d, in disturbances of function.

METHODICAL **EXAMINATION OF THE** EYE.

We begin with the external parts. A. When we examine the eye and its accessories, we note first, the lachrymal sac, and press it with the finger; the lids—their edges, the cilia, the Meibomian follicles, the lachrymal puncta, their cutaneous and their mucous surfaces; the width to which the lids separate, their mobility—whether insufficient or spasmodic, the length of the palpebral slit; the cornea—whether transparent or affected by opacity, its shape or curvature; the ocular conjunctiva—its color, the appearance of **its** vessels; the depth of the anterior chamber; the pupil—its size and **mobility, its** clearness; the iris—its hue and brilliancy—is it adherent to **the cornea or** to the lens? is its periphery retracted? is its tissue healthy or atrophied? do both irides look alike? the crystalline—is it clear or smoky, or more positively opaque? Next attend to the mobility of the eye, that its range of motion is sufficient in all directions, without tremor or spasm, or lagging; that the two eyes move in harmony, both for near and for distant objects, and in all directions. Second, examine the tension of the globe by pressure with the finger—is it elastic, yet firm, like the normal eye, or too resisting, or softer than normal? does pressure cause pain, especially if made upon any spot of the ciliary region? Third, ascertain if the cornea and conjunctiva, if touched by a fleck of cotton, or by a hair, exhibit their proper acuteness of sensibility.

B. Examine next the function of the eye, viz., first, the acuity of vision, both for remote and for near distances. At this point it may become needful to employ glasses, either for correction of distant or near vision, or for both. Second, the field of vision, whether of proper extent or in any way limited, or whether certain regions present scotomata or spots of blindness, leaving out of view the blind spot of Mariotte. Third, the perception of color, if deficient either for certain colors, or over a certain part of the field, and not over the whole.

C. Ascertain the condition of the interior of the eye behind the crystalline, viz., the vitreous humor, the retina and optic nerve, and the choroid. To do this demands the use of the ophthalmoscope and of oblique illumination. Also determine objectively by the ophthalmoscope whether the eye is normal in its refractive construction, thus correcting or corroborating the choice already made of glasses for vision at a distance.

D. Another topic of inquiry is the condition of the orbit, which includes the presence of tumors, the occurrence of protrusion of the globe, and the function of the muscles, as exhibited by the existence of diplopia. Exploration of the border of the orbit with the finger, and especially of the foramina of exit of the supra-orbital and intra-orbital nerves, belongs to this proceeding.

By following the line of examination thus indicated, a thorough knowledge of the condition of the eye can hardly fail to be obtained, and in this, as in every other case of diagnosis, a regular system rigidly adhered to is the only safeguard against oversights or mistakes. There will be two conditions which render such a kind of exhaustive inquiry unnecessary; the first, when a malady presents itself which is so obvious and compromises so slightly the function of the eye as not to justify the needful expense of time; and the second, in which pain or distress from exposure to light forbids a rigid examination, or in which the testing of sight or the use of bright light for inspection is liable to cause harm. When dealing with cases which come under the second head, it is not always admissible to forego the examination because of either pain or photophobia. Examples of this sort are cases of acute keratitis or iritis, and most strikingly cases of hysterical photophobia. It is sometimes imperatively necessary to know how much mischief has been done, and what evil threatens; and the pain which a patient is made to suffer cannot counterbalance the advantages which a complete inspection will give. This likewise appears in many cases of acute inflammation of the conjunctiva, where the swollen lids make it difficult to get a view of the globe; but it is absolutely necessary to know what is the state of the cornea. A burn of the eye by lime, or a wound, may cause much suffering under examination, but without doing it properly the physician might with propriety decline to attend the case. For these cases a combination of firmness and gentleness, with dexterity in manipulation, will often gain the object, while, if it be necessary, one may unhesitatingly resort to anæsthetics. In children the use of anæsthetics is especially to be commended, and I have no hesitation in advocating the employment of chloroform for those under ten years of age. In fact, I would use it rather than ether in many older subjects, when a brief inspection or quick manipulation is the only requirement.

The value of anæsthetics is greatly to be insisted upon in dealing with young children who have acute conjunctivitis and acute keratitis. Not only are they spared the infliction of pain, but the eye is less likely to sustain injury. It is not necessary to give anæsthetics in every examination, after the condition has been once determined, but ofttimes efficient treatment can be practised in no other way than by their help. I once treated a child, aged about five years, with granular conjunctivitis, during nine months, and gave her chloroform about eighty times, to enable me to evert and touch the lids. She not only got well of the disease, but evinced no ill effects from the use of chloroform, and has grown up to be healthy and have good eyes.

In certain cases patients complain that the use of their eyes in reading test-types inflicts great pain, and they may allege that light is so obnoxious that they cannot look at any tests. The only answer to their objections is, that if they cannot be examined they cannot be relieved, and for all tests at close distances they must be willing to bear some pain. For tests at more remote distances, such as the standard of 6 m., or of 20 feet, the use of sulphate of atropia will often give great relief, by setting aside spasmodic or excessive efforts of accommodation. The sulphate of atropia is to be used in many cases where the patient objects to submit to the inconvenience which suspension of accommodation entails, because without it a correct conclusion as to the state of refraction may be impossible. This subject will be referred to in more detail at a future time. On the other hand, some caution is to be observed in using solutions of sulphate of atropia in a few special cases. For example, atropia is liable to aggravate the trouble in subacute, or, still more, in acute inflammatory glaucoma. The instillation of such a solution has been known to originate an acute inflammatory outbreak, where the glaucomatous process had hitherto been of the simple variety. Again, atropia has been known to excite irritation, and sometimes severe irritation, when its solution is dropped into an eye which, having had chronic iritis, has a complete posterior synechia, the whole pupillary border being glued to the crystalline. This is not likely to occur in old cases in which the fibres of the iris have undergone atrophy, but in such as yet preserve a good degree of contractility in the muscular fibres. In the class of cases now under consideration, caution is not so needful as in glaucoma, but the suggestion may become apposite.

We now must penetrate the interior of the eye, and this brings us to the use of the ophthalmoscope.

EXAMINATION OF THE INTERIOR OF THE EYE BY THE OPHTHAL-MOSCOPE.

The ophthalmoscope, invented by Helmholtz in 1851, has revolutionized the doctrine of the internal pathology of the eye. Its principle is simple, being a device for illuminating the interior of the eye, and at the same time for enabling the observer to catch in his own eye whatever light may be reflected from the eye examined. That light returns from an illuminated eye is well known in the case of certain animals, viewed under peculiar circumstances, and also in human eyes which are the seat of tumors. It was noted that the eyes are to be viewed in a dark room, and that the reflected light must be nearly in a line with the observer's eye and with the axis of the examined eye. Helmholtz perceived that, to generalize the experiment and make it effective for all eyes, the problem consisted in putting the eye examined, the eye observing, and the source of light, all upon the same line. The solution was found in using a transparent reflector between the observer and the eye inspected, and this was afterward changed for a perforated metallic mirror, the light being placed beside the patient's face. The eye is thus illuminated; the rays coming back from its interior constitute a beam the size of the pupil, which remains for some distance but little larger. In normal eyes adjusted for distance they are parallel, and the observer receives them, through the hole in his reflector, into his own pupil. The emergent beam is determined in size by the patient's pupil, and that received by the observer is determined in size by the hole in the mirror. Of necessity the surface covered by the eye of the observer is most extensive if we bring the mirror as close as possible to the patient, just as we would peep into a room through the key-hole. To get a view of the details of the bottom of the examined eye—provided its refraction is normal and its accommodation is suspended—the observer must put his own eye into a state for reception of parallel rays, i.e., look as if the object were far away, notwithstanding he knows it is only about an inch distant. He can use but one eye; what the other sees must be disregarded, or it must be closed. The practical details are as follows:

We darken the room and use a single light—an Argand gas-burner or a student's lamp. The object to be sought for is the optic disc, and the

FIG. 15.

patient is bidden to look straight forward, while the observer looks in from the temporal side at an angle of about 15°. For the examination of the left eye the observer's left is used, and for the right eye the observer's right; the place of the lamp being shifted and the instrument put into the corresponding hand, the observer comes as close to the eye as possible, and this may be within one inch or even within fifteen millimetres.

If now the eyes of both be normal in refraction, and in both the accommodation be entirely at rest, the details of the eye-ground will be easily seen.

The examination thus described is known as the *direct method* or *by the upright image.* Another method, called the *indirect,* or *by the inverted image,* is as follows : the observer holds the mirror twelve or fourteen inches from the patient, and brings before the latter's eye and within two inches of it a biconvex lens of two and one-half inches focus.   This lens

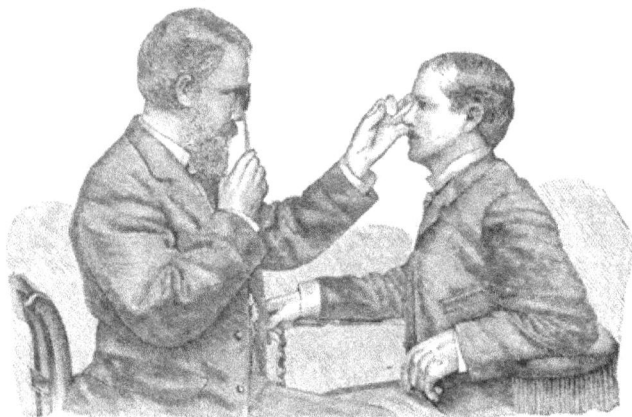

Fig. 16.

condenses the light from the mirror, and also collects the emergent light into an inverted image which lies at about two and one-half inches from the lens, between it and the mirror. The observer examines this aërial image, and not the eye. It is bright, small, and covers a larger surface than is to be seen with the indirect method, and shows better the relation of the parts.

To know where to direct the light, the observer should keep both his eyes open, and rest the upper edge of the mirror on the inner end of the brow.   When he has thrown it on the eye, he will be attracted and embarrassed by the reflection from the cornea. This annoyance is greatest when the region of the macula is under inspection. One learns, after a time, to look beside this reflection and ignore it. When using the indirect method, the biconvex lens furnishes in addition two reflections of the mirror as small, round spots, and these are gotten out of the way by giving it a slight inclination.   In this kind of examination the corneal reflex sometimes seems to cover the whole field.   A little change in the position of the lens or mirror will remove it.

The direct method of examination presents fewer difficulties of instrumentation than does the indirect method, but it offers a more complicated problem than the other, because the refractive condition of the eye must be determined, and, if erroneous, must be corrected by proper glasses before the inspection of the fundus can take place. What in the beginning is a difficulty, becomes, after a time, a most valuable quality of the direct method.

It will soon be noticed that the eye is best illuminated when a dark spot lies right over the place to be inspected. This spot is made by the hole of the mirror, from which no light is reflected. The valuable part of the mirror is that portion immediately next the hole. The hole, too, must not form an appreciable canal, but have a sharp edge. Its diameter must not be too small, because it will cut off too much light from the observer and act upon the rays coming from the patient's eye as a stenopaic opening, to limit their tendency to deviate from parallelism. If it be too large, the quantity reflected into the eye is seriously reduced. The size found by Dr. Loring to be most useful is 4 mm. in diameter, and the edge made very thin and blackened. Dr. Wadsworth, of Boston, first indicated the inutility of a large mirror for the upright image, and that one only 18 mm. in diameter suffices for the inverted. Dr. Loring has chosen a mirror which is 18 mm. wide by 32 mm. long. It is hung on trunnions, as Helmholtz's mirror was placed, obliquely, to avoid the need of turning the whole instrument to a decided angle. This provision has utility, when strong corrective lenses have to be put behind the mirror, viz., of less than 8" focus.

A. *For examining the normal eye when the observer's eye is normal.*—The only requirement is that both eyes lay aside all efforts of accommodation. The patient usually does this, because he has no object to inspect, and his eye is dazzled by the glare; yet too much dependence cannot safely be placed on this assumption, as will be dwelt upon hereafter. The inexperienced observer never does this, but looks as he always would at a near object, and not as he would at a distant one—in other words, he calls in play his accommodation. To prove this and to enable him to see, he may put behind the mirror a concave glass of 10 inches focus. Then he will see the bottom of the eye, just as he would read a book ten inches away. But let him weaken this glass to 20 inches, and again he will see; and then to 40 inches, and perhaps he still will see. His problem is to see clearly, without any glass and with no effort. He must cultivate this habit. Let him practise looking with a convex glass of 8" focus before one eye at a page 7½ inches away, or as much farther as he can read, keeping the other eye open. He will finally find whether he can, at pleasure, utterly abandon accommodation, or what fraction of it he is obliged to use. Whatever that may be, he is to allow for it as his personal equation of error. If, however, the observer do not have normal eyes, he must put behind the mirror the glass which corrects his sight for distance plus or minus the glass which his habit of accommodation compels him to employ. Then he is in position to examine abnormal eyes. In doing this he will have to add to his correcting-glass, or subtract from it, the glass which corrects the error of the patient's eye. On a later page a slight modification of this statement will be made.

B. To be prepared to meet the contingencies of all eyes which can possibly come for examination, saying nothing of the errors of the observer's eyes, it is needful to have a large series of glasses behind the mirror. In the early days of ophthalmoscopy, when Liebreich's instrument was in general use, not more than six were provided. Now, the cheaper instruments of Dr. Loring have seventeen, and these serve only for a limited range of cases, or for limited needs in practice. This will suffice for the latter, because, while not provided for every case by the direct method, the indirect method and the mirror alone can always be used. By the latter methods a considerable number of cases will be diagnosticated. The simple instrument of Dr. Loring is figured in the text. Another one, containing a large number of lenses which, in its optical parts, is copied from Dr. Loring's, is figured on page 29. It was brought into use by the writer, for the purpose of putting at command, in rapid succession, the full series of glasses which can be required. The mechanism for doing it is by cog-wheels, and there are two lens bearing discs. There is no interruption in

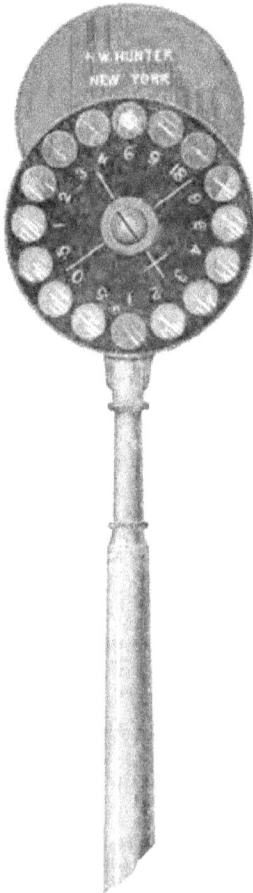

the succession of the glasses as they may be needed. There are 38 convex and 38 con-cave lenses, beginning with O .5 D, and going to 37 D, or from 80 inches to about 1 inch focus.

It remains to be stated what amount of surface can be inspected at one view by the upright and by the inverted image respec-tively ; also what magnifying power does each method give us. The surface covered by the upright image is chiefly determined by the size of the patient's pupil and by the proximity of the observer's eye. If the ob-server place his eye so that his pupil shall occupy the place of the anterior focal plane of the observed eye, *i.e.*, be at 15 mm. dis-tance from its cornea, his ophthalmoscopic field will exactly equal the patient's pupil. But if the observer must go farther away than this, it follows that rarely so much as two nerve-diameters are included in the field. The nerve-diameter may be taken at 1.5 mm., and the pupil may be as large as 5 to 8 mm. with atropine. This applies to normal eyes. For myopic eyes, whose error is so great that they may be viewed in an inverted image without using a convex lens, the field is really limited by the pupil, and grows larger with the in-crease of the myopia. For hyperopic eyes the opposite is true in every sense. If now we take the indirect method, the field of vi-sion is made larger the farther the lens is held from the eye, up to the limit at which the pupil grows larger. When the lens is at a distance from the pupil equal to its focal length, the field is the largest possible, and is limited only by the opening (diameter) of the objective lens. If the diameter of the lens is equal to half its focal length, the field will be 7.5 mm. (Helmholtz : "Phys. Optics," p. 179).

As to the enlargement conferred by the ophthalmoscope : 1st, by the upright image. There is no absolute standard by which to measure the enlargement, because the image is virtual, and seems larger or smaller as each observer may project it. It can be measured only in terms of angles, and will vary as the observing and observed eyes are respectively normal or abnormal in refraction. Suppose both to be normal, *i.e.*, em-metropic. For simplicity of calculation the reduced eye of Listing is em-

FIG. 17.

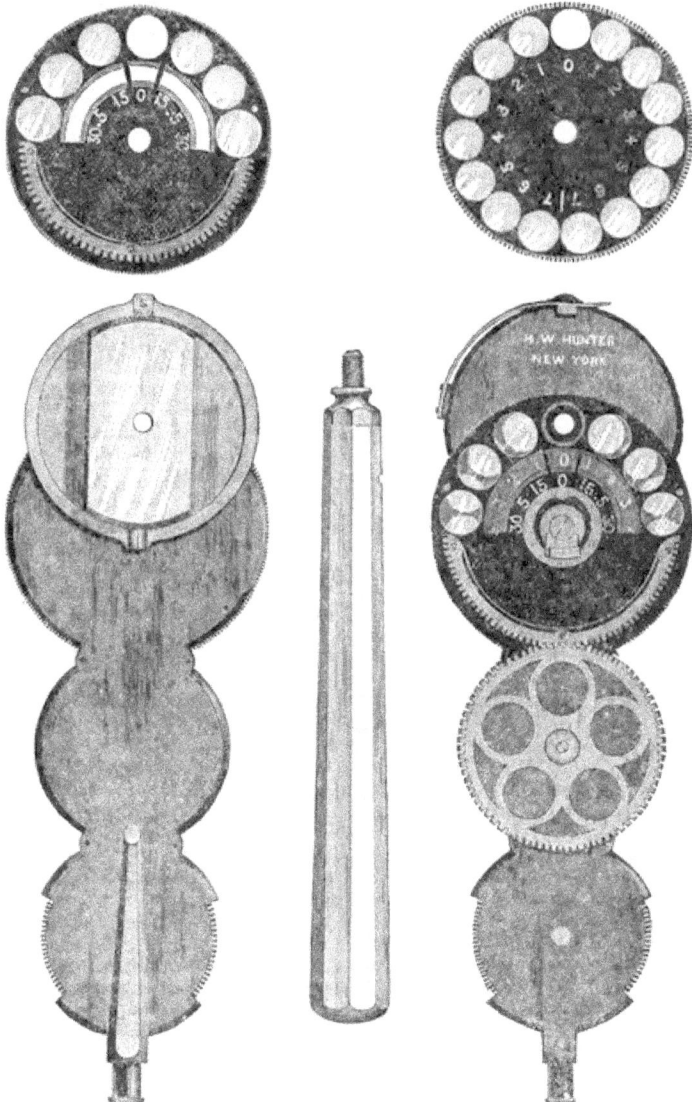

FIG. 18.—The mirror, besides swinging in the trunnions, may be rotated in a circular direction and thus assume any angle. The front disc is moved by the lowest wheel and the back disc by the upper and exposed wheel. There is a spring clip on the back of the instrument which will carry a cylindric glass. In other particulars the instrument is copied after Dr. Loring's latest model. It gives command of a complete set of spherical glasses, both positive and negative, amounting to seventy-six in number, and cylindric glasses may be inserted at pleasure from the spectacle-box.

ployed, and this is a mathematical fiction in which the refracting media are reduced to a single curved surface, whose index is $\frac{4}{3}$, the length of the eye is 20 mm., and the optical centre is 15 mm. from the retina. Without introducing the formulæ (see Schweigger: "Handbuch," p. 103, 4th edition, 1880), it may be said that the optic nerve forms in the observer's eye an image which is exactly of the same size as itself, and forms an angle in the observer's retina of 5.73°. To get this in linear measure we must adopt a conventional standard of projection. If this be taken at 300 mm. or 12″ English, the linear enlargement will be 20 diameters.

If a normal eye examine a myopic eye whose axis is too long, a concave lens corresponding to the myopia must be used, and the problem becomes complex. The size of the angle is dependent on the distance of the correcting-glass from the eye examined, and on the focal length of the glass. If the myopia equal $-\frac{1}{5\frac{1}{2}}$, or about -7 D, and the correcting-glass be $\frac{1}{7}$ at 15 mm. from the eye, the angle of the image of the optic nerve is 5.4°, which is less than in a normal eye. If with the same myopia the correcting-glass be $-\frac{1}{9}$ or -11 D, held at 50 mm. from the eye, the angle will be 8.59°. It thus appears that in myopia the angular enlargement increases with the myopia, and increases too as a stronger correcting-glass at a greater distance from the eye is employed; at the same time the field diminishes.

If a normal eye examines a hyperopic one whose axis is too short, the case is the reverse of myopia. Suppose an eye with hyperopia $\frac{1}{8}$ or 8 D, it needs a convex glass for examination; and let this be $+\frac{1}{8}$ at 15 mm. from the eye, the angle becomes 5.7°. If the glass be $+\frac{1}{8}$ at 50 mm. from the eye, the angle is 4.76°. Then it seems that in hyperopia the enlargement is less, and is still further reduced when a weaker correcting-glass at a longer distance from the eye is used. Schweigger further remarks that, inasmuch as the optic disc varies in size by Henle's measurement at the choroidal level between 1.2 mm. and 1.6 mm., this constitutes too large an error to make it a fit standard of measurement as to refractive qualities of different eyes.

2d. For the inverted image. The magnifying power, as already said, depends entirely on the focal length of the objective lens and on the refraction of the eye examined. If this lens be 2″ and the eye normal, the enlargement is 3.6; if it be 3″, it will be 5.3. But if the examined eye be myopic, say $-\frac{1}{8}$, and the lens 3″ focus, the enlargement will be 4.6, that is, less than for a normal eye. If the examined eye be hyperopic, say $+\frac{1}{8}$, and the lens 3″, the enlargement becomes 6.1, that is, it grows larger than in either the myopic or the normal eye. It remains also to be said that if the examined eye is made myopic $\frac{1}{8}$, not by lengthening of its axis, but by effort of accommodation (spasm), the magnifying power by a lens of 3″ is not 4.6, but increases to 5.2. So a hyperopic eye having H. $\frac{1}{8}$ by shortening of axis, has enlargement to the inverted image with objective of 3″ of 6.10; while if such an eye overcome its error through effort of accommodation, the magnifying power is reduced to 5.9 (see Schweigger).

As to the extent of surface illuminated, it will depend on the form of the mirror, whether plane, concave or convex, on its focal length, its distance from the eye, and the size and distance of the source of illumination. If we use a concave mirror of seven inches focus and an Argand burner of one and one-half inch in diameter, about eight inches from the mirror, and the latter close to the eye, the surface illuminated will be about equal to the size of the patient's pupil. The farther the mirror up to eight or twelve inches, the more intense the illumination. The weaker the illumination the better will one be able to note faint opacities in the media, and the less will the retina be irritated. By the upright image the light is usually feeble enough to be borne by any eye. Gas or lamp-light is commonly used, but daylight may be employed. The shutters are to be closed and a small hole four inches square left to admit light from the sky, and not too far from the patient. If daylight be used, the fundus will be found to have a more yellowish look, and it is said that opacity of the retina and nerve can be more easily discerned. To be sure of our judgment, such illumination must first be used with sufficient frequency to make one familiar with the normal look of the parts. As a rule, artificial light is used. Direct sunlight is proper only to totally blind eyes, and the heat is even then objectionable. For patients in bed, a kerosene lamp may be held behind them as they are propped up by pillows, and the revolving as well as tilted mirror of the instrument figured on page 29 is very useful. A long handle is convenient, but to be able to shorten it by unscrewing the ivory part is sometimes desirable.

The following method of conducting an examination with the ophthalmoscope is suggested as being sure to cover all the points of a case.

First, illuminate the eye with the mirror from a distance of sixteen or eighteen inches, and let the light play from side to side over the cornea. This will show opacities in the cornea or lens and the degree of luminosity of the fundus. If the eye be of decidedly abnormal refraction or ametropic, retinal blood-vessels will be visible. They may indicate that the eye is either near-sighted or far-sighted. If the former, the vessels will move in a direction opposite to the motion of the observer as he moves his head from side to side, while for far-sightedness the vessels will move in the same direction with the motions of the observer.

Having this preliminary idea of the state of the eye, the biconvex or objective lens may then be put up and the inverted image examined, beginning with the optic nerve. The lens is held by the thumb and forefinger, while the little finger takes a support on the edge of the temple. The lens is to be moved a little from side to side, which of course carries the image with it; and it will be noticed that parts upon deeper planes, as in the case of excavation of the optic nerve, have a greater range of movement than do the more superficial parts. For instance, the edge of the nerve will move less extensively than its deeper-lying bottom if there be excavation. The little finger may be allowed to press on the eye, at the same time lifting the lid, and thereby determine whether a little increase of tension will cause pulsation of the retinal vessels. After inspecting the nerve, the patient should be directed to look in every direction, to bring all parts of the eye-ground into view. The region of the macula will also be noted, although this will often not be well seen unless the pupil has been dilated by atropia.

Next, the eye should be inspected by the upright image, the observer coming so close to the face as even to touch it, and bringing the light to the requisite position to permit close approach. Now, it will be needful to put behind the mirror such glasses as neutralize any refractive errors, and the details of the fundus will be more fully appreciated, besides learning what is the state of refraction. I do not mean to be understood as intimating that the diagnosis of the state of refraction will easily be made by the beginner—on the contrary, he will meet not a few difficulties; but at a later portion of this treatise the subject will be considered.

After having studied the bottom of the eye, a strong convex lens, say of three inches focus, may be put behind the mirror to enable one to inspect the crystalline, the anterior part of the vitreous, and the cornea, the patient being told to look in different directions, to throw into view the periphery of the lens or vitreous opacities not in the field. Finally, turn the patient to face the light and use the focal illumination already described (page 21). Of course, regard must be had to the sensitiveness of the patient's eye and its liability to injury by intense light. In most cases no harm results, and this is specially true of lesions of the optic nerve, retina, and choroid. In very many cases only the direct method need be used, and to the fundus as thus seen we will now call attention.

The fundus oculi as seen by the ophthalmoscope, and especially by the upright image.—The object first sought is the optic nerve, which appears as a circular disc on which the retinal vessels are seen. Its color varies from a pinkish white to deep red; often the whole surface is not of the same hue, a part being red and the rest pale, and this may be respectively the nasal side contrasted with the temporal side, or the circumference contrasted with the centre. The whiter parts reflect light more brilliantly because they are sunken and concave, and the paucity of fibres in the depressed part favors the penetration of light to, and its reflection from, the lamina cribrosa. The depression or so-called excavation often found in the nerve may be central and small, or in extent it may exceed half its diameter, or it may be a slope on the temporal side, or more rarely downward; or the outer half may be almost flat and below the level of the inner, like a step. The nerve is sometimes a true papilla, and the highest part may be central or on the nasal side. In all cases the tissue is

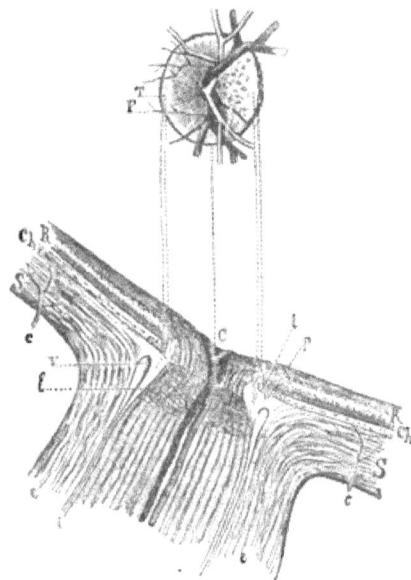

FIG. 19.—*i*, internal sheath of optic nerve; *e*, *e*, external sheath of optic nerve; *v*, the intervaginal space; *l*, lamina cribrosa; *c*, *c*, posterior ciliary arteries; *S*, *S*, sclera; *Ch*, choroid; *R*, retina; *t*, *T*, tendinous or scleral ring; *p*, *P*, choroidal ring; *C*, optic papilla.

translucent, so that one looks through a depth of substance, and the limit of inspection is the lamina cribrosa. The latter when seen is densely white, and is often mottled with dark spots. As the nerve-fibres come through its meshes, they lose their neurilemma and become transparent axis-cylinders. The nerve is sometimes oval, with its long axis vertical, and, even when truly circular, may by reason of astigmatism seem to be oval in any direction. It sometimes has an irregular outline. The border is well defined, being sharply cut by the edge of the choroidal aperture, and often a black pigment deposit extends more or less about it. Sometimes the choroidal opening is appreciably larger than that in the sclera, and a narrow ring of the latter is to be observed. If the optic fibres are

heaped together in a certain space, they will be easily recognized as they
cross the edge of the disc and extend into the retina, sometimes to a considerable distance.   In eyes deeply pigmented the optic nerve is always
by contrast more red, and the nerve-fibres are more distinct.   Sometimes
they make a complete fringe or aureole of hair-like radiating lines.

The conspicuous feature of the nerve is the network of vessels which
appear upon it.   They emerge and enter near its centre, and present many
varieties of arrangement and subdivision.   A single arterial trunk usually
comes up from the bottom **of the** disc and sends branches above and
below, the veins taking a course nearly parallel with the arteries.   **It**
would **be** useless to attempt to describe all the varieties which the **vessels
present.**   The diagram from Leber gives the **vessels and their nomenclature.**—G. and S.   Besides these main branches, there **are many finer twigs**

which pass from the nerve in the horizontal meridian, and they are most
numerous on the temporal side.   It may happen as a rarity that a single
vessel of considerable size emerges from the very edge of the disc.   The
number of the vessels on the disc is exceedingly various, and sometimes
they spring forward in large curves and take a sinuous course, or may even
curve around each other in complete or partial spirals.   Such peculiarities
will have relation **to** the vascularity **of** the general system, and due allowance must be **made.**   Sometimes the walls of the arteries are of **unusual**
thickness for a **certain** distance beyond the disc, and then **they have a**
whitish border.

At the region of the yellow spot there are never any large vessels, **but**
it will be seen **that from** the transverse branches above and below numerous small twigs are sent down which run almost to the fovea.   So fine are
these that for a long time it was declared that the region of the macula

3

was the most poorly supplied with vessels of all the fundus. This, however, is erroneous, as has been shown by Nettleship, Becker, Loring, and others. The diagram from Nettleship Ophthalmic Hospital Reports, vol. viii., p. 261, indicates what vessels may here be seen in an injected preparation. In some diseases, as in embolism of the retinal artery, they become conspicuous.[*]

Again, certain anomalies appear at the optic nerve, in that a vessel may come out at its margin, or at a point beyond the margin, and go back to the retina. These have been called cilio-retinal vessels (see Nettleship : *Royal London Oph. Hosp. Reports*, vol. ix., part 2, p. 161, December, 1877). Mr.

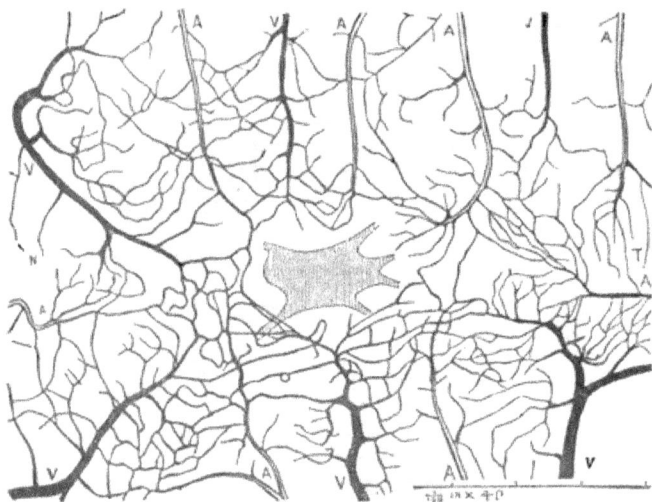

FIG. 21.—A, A, A, arteries, $\frac{1}{100}$ inch $\times$ 40. V, V, V, veins, the macula has an unusually irregular form.

N. found one such vessel in a microscopic examination of the optic nerve, and proved that it passed from the sclera, at the level of the lamina cribrosa, into the nerve and to the retina, and such vessels seem in all cases destined to the supply of the region of the macula lutea. The opportunity of seeing them is most often given in the choroidal crescent of myopic eyes.

The retina is to a slight degree discernible as a tissue, notwithstanding its transparency, and near the disc its optic nerve-layer usually appears, with greater or less conspicuousness, as fine hair-like lines radiating from the margin. Above and below they are most marked, and they cluster around the principal vessels. The visibility of the retina, as well as the tone of the fundus, depends chiefly on the quantity of pigment in the epi-

---

[*] Nettleship **says** : " On comparing different **parts of** the retina, I find that while in an area of $\frac{1}{2500}$ square **inch in** the yellow spot **region**, forty complete capillary meshes can be counted, not more than from six to nine are included in the same area at a spot $\frac{1}{5}$ inch behind the ora serrata, the injection being equally complete in both places. The area of the *fovea centralis*, which is destitute of vessels in the specimen here figured, is equal to about $\frac{1}{500}$ square inch, and is irregularly oblong. It is scarcely larger than the single capillary meshes at the *ora serrata*."

thelium and in the choroid. In blue-eyed persons the retina seems very transparent, and the fundus of a brilliant red. In dark-eyed, and especially in dark-skinned persons, negroes, indians, etc., the retina seems opalescent, and the hue of the fundus is dull, and of a dun or tan color. The pigmentation is always deepest about the central region, because the epithelium is more saturated, while the remoter parts permit the choroidal vessels to be seen as light red stripes with irregular islets of pigment. The surface of the retina sometimes shows a flashy, silvery reflection, which glances along the vessels and plays about the macula lutea. It alters in place and form, on the slightest movement of the eye or of the mirror, and the spot which it has left has a perfectly normal look. This is seen in dark eyes and in young children most frequently. It is not pathological. Another phenomenon is a circle which sometimes appears at the middle of the fundus, around the fovea as a centre, and has a diameter varying from one to two discs, as seen by the upright image (see colored plate, Fig. II.). This is also visible by the inverted method, and is evidently an annular reflection. Probably in these cases the source of reflection is the membrana limitans. The reason why the macula should be the special seat of such appearances is its convexo-concave surface. The fovea centralis often shows as a small glistening dot, more or less completely circular as the light plays over it. Its concavity favors its action as a reflector. In a myopic eye, where this was seen, I have observed it to be most brilliant before the perfectly correcting glass was employed, and that when this was used it disappeared almost entirely.

It has been said that the arteries of the retina are smaller and brighter than the veins. It must be added that they exhibit a well-defined line of light along their centre, which, in the veins, is much less conspicuous. This is an optical effect whose cause has been disputed, and a most valuable paper upon it was published by Dr. Loring in " Trans. Amer. Oph. Soc." That it is due to the refractive action of the column of blood in the vessel condensing the light which passes through it and is again reflected from the underlying surface, has been proved to myself by two cases. In one of them there was an effusion of blood beneath the choroid, which made a dark patch. This was crossed by a vein on which no light streak was present while it traversed this dark surface, but where situated upon the normal choroid the usual streak was distinct. As the blood-patch became absorbed and a white scleral surface came to view, which was caused by rupture of the choroid, not only did the vessel recover its usual light streak, but this became much more decided than upon the adjacent portions of the vessel. A second case bearing on this point was one of extreme colloid deposit upon the choroid, having all the brilliancy of the most marked patches of fatty degeneration, as found in albuminuric retinitis. This glittering surface was about two discs long and one disc wide, and was behind one of the transverse retinal arteries. As the artery crossed this spot, the whole vessel was a bright ribbon of light—the central streak being intensified and widened so as to equal the diameter of the vessel. On either side of this spot the artery had the usual appearance. It is therefore evident that the light streak depends chiefly on the reflecting properties of the surface over which the vessels pass, and on the nature of the blood-column. That some reflection comes from the surface of the vessel is true, but it is excessively slight, as proved by my first case while the blood-patch was fresh and dark. The " light streak " is, therefore, a phenomenon of refraction and reflection, and the light must pass through the vessel from in front and penetrate to the sclera, to be then reflected from the latter and again acted upon by the blood-vessel, which condenses it into the bright, luminous streak. This is essentially the view first announced by Loring (see "Trans. Amer. Oph. Soc.," 1881).

Pulsation of the veins upon the optic disc is quite common. It is explained by Donders as the effect of the arterial tension communicated to the veins through the vitreous, and causing pulsatory movement on the optic disc, because here the column of venous blood is just escaping from the intra-ocular pressure. It is most apt to be seen when the veins are large. Schoen was able to study this in a patient whose pulse was only 16 to 23 per minute. He concluded that the venous pulse is merely the effect of the pulse of the artery upon the vein as the two vessels lie in juxtaposition in the optic nerve.—*Klin. Monatsblätter* (Zehender), Sept., 1881. Pulsation of the arteries occurs when the intra-ocular pressure rises to an abnormal degree, or in cases of disease of the heart (aortic valves) or large vessels, and under some other morbid circumstances. Pulsation of both arteries and veins can always be caused by pressure with the finger, and, if it be made very strong, the circulation can be entirely suspended.

In observing the fundus closely, if the tissues are normal and the retraction perfectly corrected, the retinal epithelium is seen as a granular surface, like the finest emery-paper, and its molecular look is perfectly distinct. A few glistening dots are sometimes seen near the macula, which appear to have no special importance.

The fovea centralis is always the most difficult spot to examine, especially with undilated pupil. It has a dull, red look, or may return, as above stated, a gray reflection, which may be a partial or complete ring, which flickers at the slightest movement, and is about one-fourth or one-sixth of a disc in diameter. In young persons it is widest and most distinct. The very centre is so deep in color as to be almost brown.

The degree to which the choroid can be seen varies with the pigmentation of the eye. In albinoes the vessels are visible, even about the macula. In greater degrees of pigmentation, some vessels may appear between the nerve and macula, and in all persons they are distinct at the eccentric parts of the fundus. They are of a light pink hue, appearing like flat stripes, and have a curvilinear arrangement and interlacement in distinct meshes. No distinction can be made in them between arteries and veins. Sometimes the place of beginning of the venæ vorticosæ is recognizable. Between the meshes of the choroidal vessels the pigment-stroma is seen in more or less dark patches of irregular shape. The visible choroidal vessels are always broader than the retinal trunks. Immediately around the optic nerve the choroidal pigment is often quite abundant over a considerable breadth of surface, and, as above said, the central part of the fundus is overspread with a uniform layer, which usually completely hides the choroidal vessels. For verification of above description, see colored plate at the back of the book, Figs. I., II., III.

GENERAL NATURE OF DISEASES OF THE EYE AND THEIR ETIOLOGY.

We have to distinguish between optical defects and impairment of perception. *Optical defects* include errors in the form of the globe and of its refractive surfaces, such as conical cornea, abnormities in the lens, myopia, hypermetropia, and astigmatism ; also, opacities in the cornea, aqueous, lens, and vitreous. *Defects of perception* include diseases of the retina, whether direct or indirect—the latter by lesions of the choroid; also, diseases of the optic nerve and of the brain. We have also fatigue and decline in power of the muscular apparatus, viz., of the accommodation and of the extrinsic muscles, and disharmony of the latter. Besides these, we have diseased conditions of purely local origin, which are mostly external and caused by foreign bodies and injuries, by cold or heat, by smoke or dust, by extreme light, natural or artificial, or by use of eyes with insufficient light. We also have contagion, and the influence of contiguous parts, such as the nasal cavity in diseases of the tear-passages and of the conjunctiva. We have diseases dependent on remote organs, such as the brain and spinal cord and sympathetic nerves, on the heart, the kidneys, and the uterine system, and the arteries and veins, and the skin, etc. We have diseases of the eye which are incidental to general disorders or dyscrasiæ, such as syphilis, rheumatism, gout, scrofula, scorbutus, anæmia, pyæmia, septicæmia, leucocythæmia, diabetes, malaria, and general poisons, such as lead, tobacco, alcohol, quinine, osmic acid. We have also the morbid products of cancer, tubercle, amyloid and fatty degeneration. We also have diseases which are recognized as the expression of debility, exhaustion, and malnutrition.

In short, the eye in its pathology embraces disorders peculiar to its own structure and function, and, in addition, another quota, which are outcroppings of ailments of the rest of the body, and subject to the same laws.

It follows that the relief of disorders of the eye must comprise the correction of errors special to its own function and structure, and sound knowledge of general diseases and their influence upon it, coupled with skill in understanding both the general laws of therapeutics and how they may need modification to be brought to bear on the eye. The skilful eye-surgeon adds to the qualifications of the general practitioner the peculiar fitness gained by acquaintance with the facts of ophthalmology. Necessarily he must prepare himself to do well the work of the former before he ventures to assume the delicate responsibilities of the latter service.

TREATMENT OF DISEASES OF THE EYE.

Treatment of eye disease resolves itself into the local and the general. Under *local treatment* is included correction of optical errors and opacities, for which glasses, and medical and operative treatment may be needed. We must speak of protection of the eye from hurtful influences, viz.: from contagion, by antiseptics, by occlusion, by removal to a pure atmosphere ; from dust, smoke, glaring light, and extreme heat, by colored or transparent glasses, by shades, by seeking another locality, by a bandage, by seclusion in a dark room or in bed. Protective glasses are known usually as coquilles, are shaped like a watch-glass, and tinted either London smoke or blue, in various shades, known by letters A, B, C, D, etc. Very dark shades are objectionable to most cases, because they so diminish the light that the eyes are strained in groping about. The neutral tint is generally better than the blue. Workmen exposed to injury by chips of metal may wear large glasses of mica, if they will, but they are seldom inclined to accept them. Eye-shades may be single or double ; they should be shaped according to their purpose ; if to cut off light from above, as in reading, they should flare like a cap-front ; to cut off light in all directions, they should lie flat and come around well on the temple. To lie flat, they should have a notch for the nose, be three inches wide, come to the temples, and will be kept flat by having the strings fastened three-fourths of an inch below the corners; they must go twice around the head. A bandage should be made of thin flannel (*i.e.*, merino, which is a texture of both wool and cotton), be three and one-half yards long, and two and one-half inches wide, for an adult. The width will be less in some cases, and always less for children. It goes about the head like a figure 8, and presses the eyeballs through a packing of absorbent cotton laid upon patches of muslin. To adjust a single or double bandage smoothly and firmly requires a little practice. It is usually employed where some pressure is to be exerted on the eye. When patients are kept in dark rooms it is important not to have streaks of bright light at the edges of the shades or in the shutters. It hardly need be said that a patient wearing a bandage need not be imprisoned in a dark room; the moral influence on him is bad, and the physical effect on his attendants equally bad. I have known delirium produced by no other cause, in old people, after cataract extraction. With dark rooms, unusual care must be given to ventilation and cleanliness. Many serious eye diseases require a patient to be kept in bed, and often it is difficult to make him submit to the hardship. The object is quietude of the whole body and absolute rest of the eyes, which a patient sitting in a chair or walking about under a bandage will not and cannot so perfectly maintain. I advocate this only during the active period of acute disease —never in case the general health suffers or is unfavorably influencing the eye trouble. Even photophobia, which is usually the symptom necessitating seclusion in darkness, is sometimes aggravated by such confinement, especially in hysterical persons, in weakly or scrofulous children, and when the fear of light has outlasted the cause which originally excited it. To this point Dr. Agnew has called especial attention. Such persons must be provided with smoked glasses, and sent outdoors to navigate for themselves. A proper understanding of hygiene and of the conditions of healthy nutrition in food, clothing, exercise, and air and occupation, is of the utmost importance in ophthalmic treatment. I shall have to emphasize this repeatedly.

Of medicines which have a special applicability to the eye, are those which act on the pupil and the ciliary muscle, viz., mydriatics and myotics. Of the former we have atropiæ sulphatis, duboisia, homatropiæ hydrobromatis, daturiæ sulphatis. All of them are poisonous, and can exert toxic effect when used in sufficient strength as collyria, because they go through the cornea by endosmosis and enter the circulation by solution in the aqueous humor. They also pass down the tear-passages to the throat, and are there absorbed. Sulphate of atropia is the most common of these remedies. It affects the dilator iridis before it affects the ciliary muscle. It likewise is an anodyne to the sensitive nerves of the cornea and iris. It is used in solutions from one-fourth grain to sixteen grains to the ounce. In rare cases, by prolonged use, it causes a peculiar irritation of the conjunctiva. It was thought by Graefe to relieve intra-ocular tension, and when it fully paralyzes the ciliary muscle and iris, the eye often feels much relief. On the other hand, in certain cases of glaucoma, atropia intensifies the intra-ocular pressure. If toxic effects appear, they will come as dryness of the fauces, which need not be heeded; but more important are, quickening and weakness of the pulse, flushing of the face, palpitation of the heart, headache, nausea, prostration, garrulous delirium, desire to urinate, and sometimes muscular violence. The antidote will be brandy and morphia. Duboisia has about the same effect on the eye as atropia, except that it does not irritate the conjunctiva. Its toxic effects come more quickly and are more alarming, the prostration being extreme. Homatropia acts more feebly than the preceding. It dilates the pupil, if used in the strength of gr. iv. ad. ℥ j., in about half an hour, and has moderate effect on the accommodation ; but in twenty-four hours its influence is gone. It is serviceable for purely ophthalmoscopic work, leaving out refractive determinations. The full effect of atropia, whenever obtained, will last for from seven to twelve days. Of daturia nothing need be said. Of myotics, we have pilocarpine hydrobromate, the alkaloid of jaborandi, and eserine sulphate. The former is quite mild and painless, toning up the ciliary muscle and contracting the sphincter iridis. Its effects in doses of gr iv. ad. ℥ j. pass in 36 to 48 hours. Sulphate of eserine, the alkaloid of Calabar bean, is a powerful myotic. As a solution of gr. iv. ad. ℥ j. it causes spasm of accommodation with considerable pain, and the pupil is reduced to very small size. It is used to draw the iris away from a peripheral wound of the cornea. It is used by some prior to cataract extraction. It is also used to counteract paralysis of the ciliary muscle of the sphincter iridis. It has the property, in some cases of glaucoma, of relieving the intra-ocular tension by drawing the periphery of the iris away from the angle of the anterior chamber and opening the canal of Schlemm (circular venous sinus). It is used to counteract the undesired effect of mydriatics. Both are sometimes used in alternation to tear adhesions of the pupil to the lens.

The only other remedy to be mentioned in this connection is strychnia, by internal administration, which is thought to have a specific stimulating influence on the fibres of the optic nerves.

Of external applications none is so common as water of various temperatures, and its effect is modified in the most remarkable manner by the mode of its use. For violent inflammatory attacks, as after wounds or in severe purulent conjunctivitis, a block of ice is kept beside the patient, and bits of muslin transferred from the ice to the eye every minute so long as the symptoms demand such extreme cold. We may use the water of higher temperature until it has no effect upon the surface, but serves merely to soften the secretions. From this we make it warm, i.e., above 98.4° F., until we get to 104° or 106°. To keep the water cold the compresses must be constantly renewed ; so, too, in attempting to keep it warm. To avoid such frequent change various contrivances have been adopted. I sometimes let a patient hold a small piece of ice, wrapped in muslin, upon the eye as long as it feels good, and I have used a small rubber bag as large as a hen's egg, filled with ice, and stopped by a cork ; but neither of these is very satisfactory. For most cases we need moist cold or moist heat, and this we get best by compresses wrung out of water. Eye-douches are useful for certain chronic cases and are easily contrived, and may be of warm or cool water. They are used for only a few minutes at a time. For continuous moist heat, a good appliance is a poultice of ground slippery-elm bark. Spongio-piline dipped in hot water, covered by oiled silk, is cleanly and serviceable. It is an old rule, which measurably holds good to-day, that applications to the eye should be of such temperature as shall be grateful to the patient. This cannot be accepted absolutely. For example, while to the early stage of many external inflammations hot water is a relief, if kept up for several hours, or if, as too often is done, a hot poultice be bound on the eye, an œdematous effusion is promoted which ensues in possible ulceration of the cornea and in such relaxation of tissues as to protract the attack. Some cases reject all moist applications ; these are apt to be such as have little or no secretion, except of tears by reflex irritation, viz., scleritis and iritis. Dry heat, like a folded and warm napkin, is often most satisfactory. On the other hand, when secretion is abundant, moist applications wash it away, and by their temperature control the exudation to some degree as they influence the contractility of the vessels. It is for the great majority of cases proper to use local application for only a portion of the time—say for ten minutes, or for thirty minutes, three, four, six, ten times a day. Intermittent use is the rule in moderate cases. Continuous use applies only to severe cases. Details in this matter will come up in special diseases.

Antiseptic applications, such as chlorinated water (five per cent.), carbolic acid water (one to two per cent.), or best of all, boracic acid (four per cent.), are, in purulent and secretive cases, of the highest value. They are to be poured into the eye from a rag or sponge, or injected by a dropping-tube or small syringe, and as often as is needful to wash away secretion. Necessarily the fluid must get beneath the lids, although it is not proper to disturb them greatly in severe cases; but we must trust to the copiousness of the fluid to irrigate the eye, aided by gently drawing the lids apart.

We next come to the so-called collyria, whose name is legion, and whose utility is regarded by the public as of the highest moment. They are to be given almost exclusively to cases of conjunctival disease. They are soothing, stimulating, astringent, and caustic. The indication for them will be found in the presence of secretion which comes ordinarily from the conjunctiva, although the primary lesion may be in another tissue. This secretion is serum, epithelium, fibrine, pus- and blood-cells. The remedies are chosen according to their power of causing contraction of the vessels and coagulation of the secretion, or as they soothe the irritated nerve-fibres. We do not know enough of the modus operandi of medicines to reason exactly on this subject, and we act according to the results of experience. It is simply my purpose in this place to speak a warning against the misapplication of such remedies. To apply to iritis, cyclitis and pure scleritis, such remedies as tannin or alum, or nitrate of silver, or sulphate of zinc, is utterly mischievous. So too they do harm in many, if not in almost all cases of acute keratitis. Before any "drops" are ordered, a diagnosis of the disease must be made, and if this be not made, no drops capable of mischief are to be thought of; better temporize by ordinary lukewarm water, or a weak solution of borax, or, best of all, frankly state the difficulties of diagnosis, and seek futher light. Such conduct will save many an eye which rashness or false pride would ruin.

Again, we are called upon to apply leeches, as, for example, for severe inflammations, and for inflammations of the deep textures. In reality they are not frequently employed. They should be placed on the temple, and not too near the lids—never on the lids or in their near vicinity. The artificial leech of Herteloup is a cupping instrument which draws blood rapidly, and is useful for deep-seated congestions. It has quite superseded the ordinary cupping apparatus. As a matter of fact, the abstraction of blood is resorted to in visible ocular inflammations to a much less degree than formerly, and only in those which are attended by great pain and hyperæmia. For post-lental diseases it is used in a way advised by Graefe. From one to two ounces of blood are withdrawn rapidly from the temple, and the patient remains in a dark room for twenty-four hours afterward. This proceeding is repeated once in three, seven, or fourteen days, according to the character of the case.

Blisters and external stimulants are not as much used as they formerly were. Their value as antiphlogistics is almost nil, and they were formerly in favor because the cases were too often incorrectly diagnosticated. As remedies for neuralgia they sometimes are useful, and in a few other special conditions. The hypodermic injection of the muriate of pilocarpine (gr. $\frac{1}{8}$ to gr. $\frac{1}{4}$) has within a few years found favor in eye practice. It has seemed to do good under certain peculiar conditions ; for instance, in the late stages of chronic keratitis or scleritis, especially in gouty subjects, and also in the late period of gouty or rheumatic iritis : Virtue is claimed for it in subretinal effusion. On the whole, the remedy has seemed to me to be overrated, although its powerful action on the salivary glands and on the skin gives it influence over local disease which doubtless can be sometimes successfully applied. As yet the indications for its use are not precisely formulated.

The Turkish bath is a similar measure, and is to be employed in similar conditions. It has decided value, but it is also capable of mischief if not properly regulated.

An indication of the highest importance in diseases of the eye is the regulation of its tension, especially to reduce it when excessive. The

Fig. 22.

cases where it is below par are usually of a chronic character, and are less amenable to improvement. To reduce increased tension we have, first, the doubtful claim in behalf of atropia. We also have eserine to effect reduction under special conditions of glaucoma. But the chief means are mechanical, viz., puncture of the cornea, and often not more than two drops of aqueous fluid will be removed. Again, free division of the cornea to let off all the aqueous fluid, and with it morbid products like pus or lymph in the anterior chamber. Thirdly, we have sclerotomy, which is done at the margin of the anterior chamber by a peculiar method. Fourthly, we have iridectomy, which is done at the sclero-corneal junction, and includes excision of a piece of iris. Fifthly, under special conditions after chronic iritis and loss of the lens, division of a mass of agglutinated tissue (iridotomy) relieves extreme tension. Sixthly, I have seen two cases where removal of the whole iris through a small wound reduced the size and tension of a staphylomatous globe. In ordinary practice, paracentesis of the cornea, and section of the cornea are proceedings which claim adoption by physicians who do not regard themselves as skilled operators, provided they cannot refer their patients to more experienced hands. The other proceedings need surgical training before they should be attempted.

This introduces us to the operative treatment of the eye. It has always commanded great attention, and its scope has been largely extended. A better knowledge of pathology, the invention of new methods of operating, and higher skill in the manufacture and adaptation of instruments have conspired to make the operative surgery of the eye one of the most brilliant chapters in medicine. Some general remarks are here in place. Shall anæsthetics be used? For operations which are painful, and which do not invade the cavity of the eye, the same rules apply as in ordinary surgery. A decision is to a great extent determined by the courage of the patient. For prolonged and painful operations anæsthetics are often as useful to the surgeon as they are comforting to the patient, because they remove the tense muscular resistance which the most courageous patient cannot avoid. Therefore plastic operations, many lid operations, enucleation, neurotomy, and ordinarily strabismus, will need anæsthesia. As to operations which enter the globe, paracentesis seldom needs it : iridectomy may often be done without it, but it is more satisfactory to have the patient passive. As to the extraction of cataract, operators differ greatly according to their personal preference, and the matter will be discussed under its proper heading. Dr. Norris, of Philadelphia, has lately called attention to cases of fatal results of ether in patients having Bright's disease. I do not think great importance attaches to this caution, but it should not be passed in silence. For children under ten I use chloroform, for older persons ether by preference, but not seldom chloroform. When a long operation is expected, and the person is feeble, ether is to be chosen. For a quick operation I often administer chloroform. The primary stage of anæsthesia, during which no muscular relaxation has come and consciousness is not fully destroyed, and which lasts only part of a minute, requires a very small quantity of either ether or chloroform, and can often be seized as the happy instant for making an incision which will perhaps be all of the operation that the patient would not be well able to bear. Mr. Priestly Smith has suggested a full dose of bromide of potassium an hour before an operation, as a means of allaying excitement, and rendering a patient more submissive to the anæsthetic. I have employed it for a year with much satisfaction. I often give sodii bromid. ℨ ss., chloral hydrate gr. xv., the previous night, and repeat the dose an hour before the operation. The anæsthetic is more willingly accepted, and vomiting is less liable to ensue. In eye operations the ill-effects of vomiting are more serious than in general surgery, by promoting prolapse of iris, loss of vitreous, and intra-ocular hemorrhage. For several years I have used an ether-inhaler which is valuable because it takes up the least space and offers the least obstruction in operating about the eye. It has a rubber face-piece, and has a dried bladder at the distal end of the box, which affords space for vapor (see Fig. 22). Eye operations should be done with the patient upon a table or a narrow bed. Operating-chairs are not well fitted to the administration of anæsthetics, and are more circumstantial without being as convenient. A head-rest or clamp is useful, but an assistant can do this service. An operator who is ambidextrous will always sit behind the patient's head, on whichever eye he may operate. But he will usually have to take his right hand for scissors, and sometimes, therefore, come to the front. One who is not equally apt with both hands will change his place as the position of the eye or the place of operation requires. It is of great convenience to enjoy perfect use of both hands, but to many the accomplishment is never sufficiently realized to warrant risking the patient's sight by a clumsy hand. As to brilliant display before spectators, no conscientious man would harbor the thought to the peril of his patient. How to gain needful skill? There must be an original endowment of facility of hand and a mechanical bent of mind. Practice upon fresh cadavers will teach something, but in them the eyes are too soft to be suitable. Pigs' eyes mounted in an operating mask, or, in lack of this, fastened into the mouth of a bottle by a section of rubber tubing of proper size, or by strings, will teach one how to manipulate in the anterior chamber, and the resistance in cutting the cornea. A light touch and steady hand, and sensitive appreciation of weight and resistance, are essential qualities.

It is desirable to have the least number of assistants. In most cases but one is needed; sometimes a second, to give the anæsthetic and keep the head steady, is desirable. In manipulating the eye the operator should steady it by fixation forceps, and not let the assistant do it when avoidable. He thus keeps the command, and can co-ordinate his hands with accuracy. A fit speculum to keep the lids apart is an important instrument. It must open them *ad maximum;* it must not press on the globe; it must be out of the way of the operator. I have experimented extensively with these contrivances, and find none perfectly adapted to all cases. I have made one which opens from the nasal side and leaves the temporal side free, but finally have given it up, although it has gained the approval of Mr. Critchett, and of the late Mr. Wells, of Moorfields Hospital, London. It is shown on Fig. 23, page 44. I have for many years used another which, when well made, answers my purposes better than any other. But other and less expensive instruments will serve in many cases. This is less in the way than others, is easy to remove quickly, can be regulated in its expansion, and gives a large field for work. (Fig. 24.) But for deep eyes all such contrivances are imperfect, and one should have a smaller one for children and a larger one for adults. In case a speculum is impracticable, the operator may lift the upper lid himself by the point of his index finger. He draws up the lid by the skin as far as may be, then places the tip of his index beneath the edge of the lid, and pushes it back into the orbit. He does not drag or lift, but presses it under the orbital roof as he would push a sliding cover into the grooves of a box. If he does not choose to do this because the finger takes up room, he may use Desmarres' elevator, which is often necessary in examinations of the eye in children. It is made in sets of two and three sizes. Another elevator which I use in operating when a speculum has been taken out, or when I wish to expose the upper part of the globe to the fullest extent, is made of fine steel wire and presses the lid farther under the orbital roof than anything else can, and it need not make pressure upon or even touch the globe.

Fixation-forceps are made with and without clamping springs. They should be used so as to *turn* the eye, not to *drag* it; the line of push must be at a tangent to the globe. If, for instance, the eye is to be turned down, the forceps will be attached just below the corneal margin, and be held perpendicularly to the globe while the latter is rotated down and the forceps take a direction approaching a tangent with the distal end uppermost. Another way, often useful, and which is very convenient for the operator, is to apply the forceps at the same place, to turn the eye by the same manœuvre, and then bring the top of the forceps up to the root of the nose, where it rests truly in a tangent, and a light push keeps the eye down

Fig. 23.

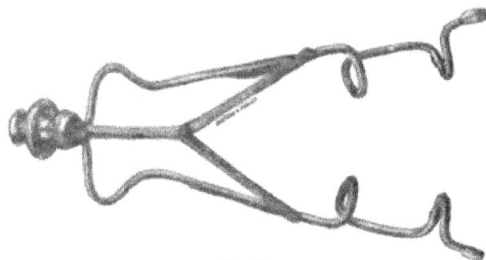

Fig. 24.

and exerts the least pressure. (See figure showing Cataract extraction.) With anæsthetics the forceps may even be left to fall obliquely over the supra-orbital notch and keep the eye down while the operator uses his hand for another purpose. He may give the forceps to an assistant, and its position will not need to be altered. Such forceps give pain and must be used with gentleness, and avoided as much as possible. The conjunctiva is liable to tear, and the patient must be exhorted rather than forced to turn the eye as desired, while the forceps shall simply maintain the position desired.

As regards other instruments, they will be considered when their special uses are to be described.

THE USE OF GLASSES.

A great number of maladies which were formerly attacked unavailingly by medicines are now corrected by glasses. A large chapter in ophthalmology was formerly described by the word asthenopia (weak sight). This term is still in use, but a vast proportion of such cases are now assigned to refractive errors; others are classed as muscular defects; still others belong to faults of accommodation. The term asthenopia has therefore a vague meaning, and has come to be used for any kind of fatigue of the eye; or, if confined to a strict class we are driven to some mixed cases and to those called hyperæsthesia retinæ, and do not feel sure that we are fully justified in holding the term as fitted to them.

We distinguish between faults of accommodation and those of refraction. The latter are inherent in the form of the eye; the former belong to the performance of its function. Both frequently are mingled in the same case, but each can be distinguished from the other. Errors in the form of the eye are as follows: its axis is too short (hypermetropia, hyperopia), H; its axis is too long (myopia), M; its curves are not symmetrical, or the composition of its refractive media is not uniform (astigmatism), As. The last-mentioned error may be combined with either of the two preceding.

The eye whose axis is normal and whose curves are correct is called emmetropic, E. Its optical qualities have been discussed in the most exhaustive manner, and, without using formulæ, a few facts may be stated. The eye constitutes a compound lens whose properties depend on the curve and number of its refracting surfaces and on their refractive quality or index. The refractive index of the cornea, of the aqueous humor, and of the vitreous, is the same, viz., 1.3379 or $\frac{199}{...}$; and that of the crystalline is 1.4546 or $\frac{49}{...}$ (Helmholtz), air being taken as 1. Knowing the curve of the cornea and of the two surfaces of the lens as they have been ascertained by measurement with the ophthalmometer, and the distances of these surfaces from each other, all the elements are given to determine the course of a pencil of light through the eye, and the theory has been fully worked out by mathematicians. Gauss established the law of the cardinal points in a compound system of lenses and thereby furnished the means of solving all problems in physiological optics. Listing simplified the calculations by giving the elements of an ideal eye, which, having but one refractive surface, viz., a cornea, should be the optical equivalent of the real eye, and Listing's diagrammatic eye is generally the basis of calculations upon the size of images and other questions. For our purpose it is only needful to speak of the nodal point, that is, of the optical centre of the eye and of the principal foci. There are, indeed, two nodal points; but, for objects at a distance, the posterior is the only one to be regarded, and for our purposes this is understood. It is the place where the rays cross as they pass to the retina, and its situation determines the refraction of the eye and the size of retinal images. In the normal eye it is at about 7 mm. behind the front of the cornea and at 14.858 mm. from the retina (Donders); the optical axis is thus taken at about 22 mm. Mauthner takes the real optic axis at 24, and the outside length of the eye at 25.3. The nodal point lies just behind the crystalline lens. The anterior focus is 13 mm. in front of the cornea, the posterior focus is in the retina. We have also a centre of motion in the eye, around which its rotation occurs. This manifestly has nothing to do with optics, but simply with anatomy and mechanics. It is situated behind the optical centre. It necessarily varies with the length of the globe, and therefore is different in E., H., and M. respectively. For E. (emmetropia) it is taken at 13.7 mm. from the centre of the cornea. In H. it is farther forward; in M. it is farther behind. The curves of the crystalline are those of a sphere, but the cornea is a section of the large end of an ellipsoid. Its radius in the vertical meridian is shorter than in the horizontal. It happens, moreover, that the anatomical axis of the cornea does not coincide with the visual line and the difference is called the angle a (alpha). In optical treatises the following lines or axes are distinguished: 1st, the visual line, which is drawn from the object viewed to the fovea centralis; 2d, the line of fixation, which is drawn from the object viewed through the centre of rotation; 3d, the line of sight, which is drawn from the object through the centre of the pupil. The place of the nodal point is not only unlike in eyes of different lengths of axis, but is changed by the accommodation, and also changed by glasses. If a convex glass be put before the eye, the nodal point advances and the retinal image is larger.

If a concave glass is so placed, the nodal point retires and the retinal image becomes smaller.

As we shall have to consider the proper use of glasses in errors of accommodation and refraction, we may say a few words in general upon their varieties and properties.

We have to deal with glasses of spherical curvature which shall be convex or concave, and we have glasses of cylindrical curvature, also convex or concave. We also have glasses whose surfaces are plane, but not parallel to each other, viz., prisms.

In spherical glasses we have the following forms :

The convex are called positive or collective or magnifying glasses, and are denoted by the sign +. The concave are called negative or dispersive or minifying glasses, and are denoted by the sign —. The focus of a glass is the place where the rays from a given object cross each other on the axis of the glass. For parallel rays the place of crossing is called the principal focus, and this is understood when no adjective is used. If an object be near enough to the lens to emit diverging rays, these, if they cross, do so at points called conjugate foci. For convex lenses the foci are real and positive, and on the side of the lens opposite to the object. For concave lenses the foci are negative, imaginary, or virtual, and on the same side with the object. But for convex glasses, if the object be situate at the principal focus, the rays after passing the lens will not converge, but be parallel; hence there will be no focus. If the object come still nearer, the rays will be divergent, and the focus virtual. For concave glasses the rays become more divergent as the object approaches the principal focus, and at this point rays cannot pass through because the divergence becomes too great. In Figure 25 we have the principal forms of lenses, viz.: the

FIG. 25.

plano-convex A, the biconvex B, the convex meniscus C, also the plano-concave D, the biconcave E and the concave meniscus F. The first three are all positive, and the last three are all negative lenses. The biconcave and biconvex are supposed to have curvatures the same on each side, but this may not be, and frequently is not, the case.

Images in plus (i.e., convex) glasses are inverted and smaller, if the object be beyond the principal focus. If the object be at the principal focus, no image is formed. If it be nearer than the principal focus, the image is not real, but is virtual and erect, and larger than the object; the lens then becomes a magnifier (loupe). Images in minus (i.e., concave) glasses are always small, erect, and virtual, provided the object be farther than the principal focus. If an object lie at or nearer than the principal focus, no image can be formed.

Cylindric glasses are ground by a cylindric tool, and have a curve whose maximum is at right angles to the axis of the cylinder, while in the direction parallel to the axis there is no curve. Such glasses cannot form images, although they may be said to have foci according to the laws of spherical lenses. Cylinder glasses are shown in Figure 26 and represented as with square outlines. In practice they are cut in oval outlines like other glasses. The axis is shown to be along the middle of the curve and parallel to the edge.

Prisms used in ophthalmic practice are of only moderate angle. They deflect rays, and, leaving out of view a slight dispersion and decomposition of rays, they do not cause them either to converge or diverge. They therefore have no foci, and form no images. When an object is viewed through them, it is apparently displaced. The direction of deflection of rays by a prism is always toward its base; the apparent displacement of an object is always toward the angle. The degree to which prisms deflect light is in proportion to their angle and their index of refraction. For prisms of small angle, *i.e.*, not more than 6°, the amount of deflection is one-half the angle.

The arrangement and nomenclature of glasses. — Formerly, when one needed glasses, one could get no better aid in determining what one should use, than the advice of the optician from whom they were purchased. At present, ophthalmic surgeons find their function to consist largely in advice upon this subject. They require a trial-case more or less complete, which must contain spherical and cylindric glasses, both convex and concave; also prisms from 2° up to 20°, and a suitable frame. A slip of red glass and an opaque screen are usually added. In giving numbers to glasses it was formerly the practice to do so upon the assumption that the index of refraction of the material was 1.5 ($\frac{3}{2}$), and, for a double convex or double concave glass both whose curves were alike, the focus was found by the rule that the focus was equal to the radius of curvature. *

Another embarrassment is the want of uniformity in the inch measure among different nations. The following are samples: as before stated, on page 2 the English inch is 25.3 mm.; the Austrian is 26.34 mm.; the Prussian is 26.15 mm.; the Paris inch is 27.07 mm. As between the English and the Paris inch the difference is $\frac{1}{12}$. In the numbering of glasses, therefore, two things need to be readjusted: first, the error arising from regarding the radius of a bi-spherical lens as the equivalent of its focus, and vice versa; secondly, the discrepancy as to the standard measure. The latter objection is overcome by abandoning the use of inches and employing the metric system of expression. The first difficulty

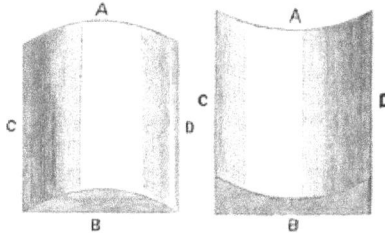

FIG. 26.

---

* This results from the formula $F = \frac{r}{2(n-1)}$; in which F stands for focus, r for radius and $n-1$ for index of refraction. If now $r = 12$ and $n = 1.5$, the formula becomes $F = \frac{12}{2(1.5-1)} = \frac{12}{1} = 12$, that is, the focus is equal to the radius. It turns out that the glass now in use does not have the index 1.5, but a higher degree varying between 1.52 and 1.55 (Nagel). Javal assumes it to be 1.54. If we substitute this in the formula, we have $F = \frac{12}{2(1.54-1)} = \frac{12}{1.08} = 11.1$. That is, the focus is less than 12 inches, and very nearly 11 inches.

is obviated by numbering glasses according to their refractive power and not according to their focus. Refractive power is the reciprocal or inverse of the focus. Thus, a lens of 30 inches focus has a refractive power of $\frac{1}{30}$. This fraction may be expressed in decimal form and it becomes .033. A lens of 20 inches focus has a refractive power of $\frac{1}{20}$, or .05. A lens of 4 inches focus has a refractive power of $\frac{1}{4}$, or .25.

The glasses in actual use begin at the numbers with long foci, and come down to those of short foci—that is, from the weak to the strong; but there is no regularity in the progression; no common interval is observed (see column 1 of the table on page 49). Now, for purposes of scientific study and convenience in examinations, regularity of interval is highly convenient. Attempts have been made to secure this desideratum, and various intervals have been suggested, viz., the fractions $\frac{1}{10}$, $\frac{1}{12}$, $\frac{1}{15}$, $\frac{1}{20}$, $\frac{1}{40}$. When, however, the metric measure was substituted for the inch, it was also resolved to establish a metric interval which should become the unit of measure and the standard of gradation between numbers. Facility in calculations and uniformity, both in gradations and in nomenclature, were the objects sought. The unit is a glass of one metre focal length, which in English measure equals 39.37 inches, and is called a dioptrie (D). But this interval is too great, and therefore the metre is again divided into fractions. By the old method a lens was known by a number which was its radius of curve, and this was assumed to be the same as its focal length. It is now known by its refractive power, and this is expressed by the number of dioptries contained in it. It is seen that a metric measurement of glasses may be quite distinct from the system of dioptries. But, where the metric system has been adopted, the dioptric interval has also been accepted. It is noteworthy that in France glasses were measured in the obsolete Paris inches, and not in a metric way, until within six years. In Europe the dioptric nomenclature is widely adopted; in this country it has not extensively made its way. Its most strenuous advocate has been Nagel, who gives in Graefe and Saemisch (B. VI., p. 310) the mode of converting the old into the new system of measure. He assumes the index of refraction at 1.528, which is German glass, and with this he finds the equivalent of a dioptry to be 41.5 English inches. If the index be that of French glass, at 1.54, the dioptry becomes 42.5 English inches; with index of 1.53 it becomes 41.7 English inches. Nagel proposes that, in transmuting the old to the new system, 40 inches be taken as the equivalent of the dioptry, and with this Javal concurs, the error not being very large. Therefore, an 80 inch=5 D : 40″=1 D : 20″=2 D : 16″=2.5 D : 10″ = 4 D. The table on the next page, modified from Mauthner, gives a sufficiently complete series, and according to both systems, accepting 40′′ as the dioptry.

It is seen that at the upper end of the scale the interval between glasses is small, viz., 0.25 D, but that beyond 3.5 D (11 inches) the interval is .5 D, and again becomes still greater. The reason for a large interval among the strongest glasses is that a slight alteration in their distance from the eye greatly modifies their refractive value, and any little change can be thus effected. The special claim of advantage on behalf of the dioptric system is the ease with which calculations can be made in adding and subtracting lenses. For instance, put two positive lenses, + 2 D and + 3 D, together, and their result is + 5 D. If + 3 D and − 1 D are united, + 2 D results. If − 4 D and − 2 D unite, − 6 D results. One need only deal with simple numbers, and remember the effect of the precedent signs of + or −. If lenses are to be united which are designated only by their foci, the calculation must be made in fractions, viz., + 3 D and + 1 D, becomes $\frac{1}{3}$

Focus in Inches.	Refractive power in Decimals.	Number in Dioptries.	Focus in Inches.	Refractive power in Decimals.	Number in Dioptries.
160		0.25	9	.111	4.5
80	.013	0.5	8	.125	5.
60	.017	(0.67)	7	.143	5.5
50	.020	0.75	6½	.154	6.
40	.025	1.00	6	.167	6.5
36	.028	(1.11)	5½	.182	7.5
30	.033	1.25	5	.200	8.
24	.042	1.5	4½	.222	9.
(22)	.045	1.75	4	.250	10.
20	.050	2.	3¾	.267	10.5
18	.056	2.25	3½	.286	11.
16	.063	2.5	3¼	.308	12.
14	.071	2.75	3	.333	13.
13	.077	3.	2¾	.364	14.
12	.083	3.25	2½	.400	16.
11	.091	3.5	2¼	.444	18.
10	.100	4.	2	.500	20.

$+\frac{1}{16}=\frac{1}{16}=\frac{1}{16}$. A lens of six inches focus added to one of twelve inches focus $=\frac{1}{6}+\frac{1}{12}=\frac{1}{4}$. If, however, we use the second column of the table above, which gives the refractive power in decimals of lenses of every focus, we have only simple addition and subtraction to perform, and, when the result is given, to compare it with the figure in the column of focal lengths which is the nearest approximation. Thus, for $\frac{1}{10}$ and $\frac{1}{8}$ add .100 and .125, giving .225, and the nearest figure is a lens of $4\frac{1}{2}$ inches focus, or 9 D. All such problems are in this manner easily solved.

The use of the system of dioptries makes calculations simple, but it is a mistake to imagine that it makes any practical difference in selecting glasses. We are compelled to take such glasses as the patient actually needs, and whether we express ourselves in terms of dioptries, or of focal length, or of refractive power, is not of strenuous importance. Certainly one dioptry is too large an interval, a half dioptry is often too small, and we may be obliged to choose glasses which do not come under this rubric at all. The essential thing is to know the real refractive worth of our glasses, and whether we reckon it in dioptries or by other means is indifferent. If we should measure foci in centimetres instead of inches, the valuable part of the metric system would be gained and the utility of dioptries might be left to the test of time.

Attempts have been made to make available a fewer number of glasses than the list above given, by using a spectacle-frame which may carry simultaneously a combination of three for each eye. Dr. E. G. Loring, jr., and Dr. John Green, have gotten up such a series, and where cheapness is more to be regarded than convenience it will answer the purpose. It must, however, be admitted that three glasses put in the place of one single glass will not in practice be the same, however correct the calculation, because by six reflecting surfaces the loss of light is three times greater than by two surfaces, and for strong glasses allowance must be made for their respective distances from the nodal point. If, however, the choice lies between an abridged series and no spectacle-box, the former alternative is much to be preferred.

The power of glasses depends not only on their focus, but on the distance at which they stand from the nodal point. This varies according to the depth of the globe, and the height of the nose, and the kind of frame.

4

When, however, a glass is worn **at** the anterior focal point of the eye, which is about 13 mm. from the cornea, it has no influence on the size of the **retinal** image—an important fact first pointed out by **Dr.** Knapp. But at any other place nearer or farther **its** influence **is** potential. Usually the distance is about $\frac{3}{4}$ inch, or 2 **ctm.** This is to be added **to** the power of a convex lens and subtracted **from** the **power of** a concave lens. The practical importance of this is **not** felt except among **the** stronger glasses. One who wears a glass **as** high as $+\frac{1}{6}$ finds that by slipping **it** down upon the nose it becomes stronger, and with persons who **use** cataract glasses this manœuvre is often of advantage. On the other hand, **if** a "deep" concave glass is worn, say $-\frac{1}{4}$, its power is diminished by holding it away from the eye, and increased if pushed nearer to it.

ACCOMMODATION.

The retina upon which the image falls is not a mathematical plane, but **is a tissue** of measurable thickness, and so is its bacillary layer. The length **of** a cone offers an appreciable variation in the localization of an image, and necessarily permits a variation in the distance of the object not inconsistent with good sight. Between infinity (∞) and one hundred feet (thirty metres) this fact renders accommodation unnecessary, and at fifteen inches a variation of two-fifths of an inch is allowable consistently with correct accommodation. Moreover, **if the rays** which converge upon **the retina** do not come to an absolute **and exact** point, but form a minute **circle (the** so-called diffusion or dispersion circle), correct vision is still **possible,** provided **the** dispersion circles **do not** too greatly exceed the **diameter** of a cone.

But, making allowance for these comparatively slight departures from accuracy, it is essential **that** when an object approaches near enough for the rays **to** become appreciably divergent, provision be made to increase the refracting power of the eye. The greater the degree of visual **acuity,** or, in other words, the more minute the diameter of the cones in **the fovea,** the more urgent is the need of accurate focussing. The increase of **refraction** may be most simply represented by supposing a supplementary lens of requisite focus to be added to the crystalline. It may be supposed to be a positive meniscus. The nearer an object approaches, the stronger must be the refracting power of this meniscus, or, in other words, the greater the **degree** of accommodation. As an actual fact, we find that in early life the **degree** of accommodation is highest, and that it afterward steadily di- **minishes.** Donders, to whom we owe most of our knowledge on this subject, showed that if at ten years of age the nearest point of distinct vision is at 2.8 inches, at twenty it has receded to 3.9 inches, at thirty to 5.7 inches, at fifty to 16 inches. This means that the supplementary meniscus which the eye is, at these respective ages, capable of constructing, becomes more and more weak.

At first thought the diminution which **occurs at** the age of thirty does not seem important. We have already seen (page 48) that lenses are to each other inversely as their focal length. A lens 4 inches is to one of 12 inches as $\frac{1}{4}$ is to $\frac{1}{12}$. The former is three times as strong as the latter, and the difference between them is $\frac{1}{4} - \frac{1}{12} = \frac{1}{6}$, that is, it equals a lens of 6 inches focus. Now, in comparing the accommodation at ten years of age with that present at thirty, we are to use the formula $\frac{1}{2.8} - \frac{1}{5.7} = \frac{1}{5.5}$. The first has a refractive power of .357, the second of .177. Subtraction gives refractive power of .180, which in the column on page 49 gives a focus of $5\frac{1}{2}''$. The same calculation can be made in dioptries. In other words, by thirty years of age the eye has lost one-half its power of accommodation, at fifty years we have $\frac{1}{2\frac{1}{8}} - \frac{1}{1\frac{1}{6}} = \frac{1}{3\frac{1}{4}}$, which is a loss of almost $\frac{3}{5}$, its original accommodative power.

The nearest point to which the eye can adjust itself is called the near point of accommodation, denoted by the symbol P (*punctum proximum*). The farthest point of accommodation is denoted by the symbol R (*punctum remotum*), or far point. The breadth or range of accommodation is expressed by the formula, $\frac{1}{p} - \frac{1}{r}$, and may be taken as the difference in refractive power of lenses whose foci shall be respectively P and R. The range of accommodation becomes, therefore, a lens of definite focus, whose refractive power is expressed by $\frac{1}{A}$. Now, in normal eyes, up to about fifty-five years of age, R is at an infinite distance, and the refraction is denoted by $\frac{1}{p} - \frac{1}{\infty}$, that is, it equals the near point. But, beyond this age, the far point goes still farther away than infinity, an expression not absurd in mathematical language, and which means that the eye can now bring to a focus rays which are slightly convergent, and, as light from natural objects never travels in converging lines, a convex lens is needful to enable the eye perfectly to see distant objects. The course of the accommodation is given in the subjoined table, constructed by Donders, and taken from Nagel (G. und S., Bd. VI, p. 466,) and is given both in metres and in English inches :

Age in Years.	Distance of P in Metres.	Distance of R in Metres.	Distance of P in English Inches.	Distance of R in English Inches.	Breadth of A.	
					Metres D.	Inches.
10	0.071	∞	2.8	∞	14 D	1 : 2.8
15	0.083	∞	3.32	∞	15.	1 : 2.3
20	0.100	∞	4.	∞	10.	1 : 4.
25	0.128	∞	5.1	∞	8.5	1 : 5.1
30	0.143	∞	5.7	∞	7.	1 : 5.7
35	0.182	∞	7.2	∞	5.5	1 : 7.2
40	0.222	∞	8.88	∞	4.5	1 : 8.8
45	0.286	∞	11.44	∞	3.5	1 : 11.44
50	0.400	∞	16.	∞	2.5	1 : 16.
55	0.666	−4. (H 0.25)	26.64	−160.	1.75	1 : 41.
60	2.	−2. (H 0.5)	80.	−80.	1.	1 : 49.
65	−4.	−1.33 (H 0.75)	−160.	−57.	0.5	1 : 80.
70	−1.	−0.8 (H 1.25)	−40.	−32.	0.25	1 : 160.
75	−0.571	−0.571 (H 1.75)	−25.	−23.	0.	1 : 0.
80	−0.4	−0.4 (H 2.5)	−16.	−16.	0.	1 : 0.

The above measurements relate to the accommodation of one eye by itself ; they are not strictly true when both eyes, working simultaneously, are considered. The binocular accommodation is rather less than the monocular A. In binocular sight the visual lines converge upon the object, and a suitable amount of A is exerted, according to the distance of the object. There is, therefore, a relation between convergence of visual lines and A. This relationship is of great importance in dealing with objects near to the eye, and we speak of it as the relative accommodation. For a given angle of convergence it is possible for the eyes to put forth a greater and also a less degree of A than the distance of the object requires. To illustrate by a diagram, in which, upon the line A B, the visual lines converge at a point O, which is at the same time the place at which the eyes are accommodated : while the visual lines remain at the same angle of inclination, it is possible for the eyes to see O correctly when it is viewed either through a convex glass, which will by so much diminish the effort of accommodation and place it virtually at A, or through a concave glass, which will compel greater effort of A, and make the object seem to be at C. If, with a person fifteen years old, O be taken

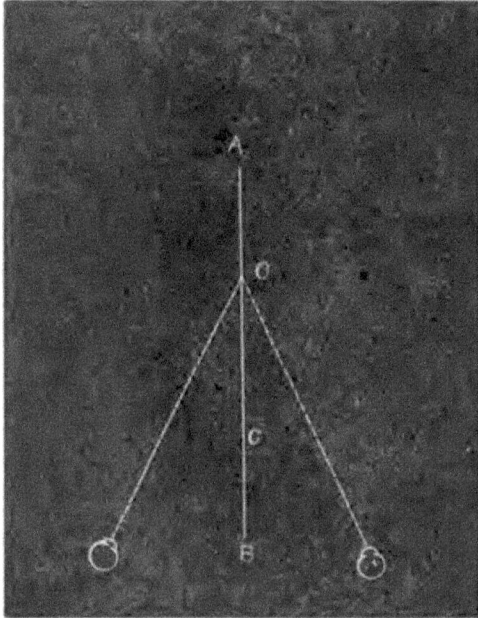

Fig. 27.

at 12″, then a convex glass, viz., about $\frac{1}{14}$, can be used, which will carry the accommodation to 72 inches, while a concave glass, viz., about $-\frac{1}{8}$, will be usable, which will bring the accommodation to 5.33″ (Donders). The former, found by the convex glass, gives the negative side, and the latter, found by the concave glass, gives the positive side of the relative A. With parallel visual lines concave glasses $-\frac{1}{14}$ can be overcome, which bring the object to 11 inches. But if convergence be at 4″, concave glasses can no longer be used ; only convex can be employed, and therefore the relative A is entirely negative. The practical result of these investigations is that for a given amount of convergence there must be a certain ratio of positive A to negative A, else the eyes soon grow weary. Graefe said that the positive side must be about equal to the negative. The eye must have a reserve of A in store for a given angle of convergence, else continued effort is not possible. A considerable range will be found to prevail in practice in this matter, and much is to be allowed for peculiarities of refraction, and of muscular capacity.

Relative A is also modified by keeping A fixed while the visual angle is altered by prisms. For instance, if the object **be at 16** inches, it may be **seen** distinctly when prisms are put before each eye, **either** with their bases **to the** median line or to the temporal side. **If to the** median line, the visual lines will be rendered more nearly **parallel, or** may even become slightly divergent. If the bases be to the **temporal** side, the visual lines are made convergent. The limit of prisms which can thus **be** interposed varies greatly in different persons. But, while the accommodation may be able to focus correctly, despite the influence of the prisms on the convergence, it is not true that it is not itself modified by their action. In fact, with abductive prisms (whose bases are to the nasal side) the accommodation is relaxed, and becomes largely, or at last wholly, negative. **With** adductive prisms (whose bases are to the temporal side), the **accommodation** becomes positive and the eyes feel a strain. In the **former case letters** seem larger, in the latter case they seem smaller. The statement first made by Loring ("Trans. Am. Oph. Soc.," **p.** 57, 1868) is undoubtedly true, **that** "for every increased action of the **ciliary** muscle there is a corresponding tension in the recti interni, **and** *vice versa*, and that this increase, so far as relative accommodation is concerned, is counterbalanced by an opposing muscular force in the recti externi, which maintains the direction of the visual axes."

Mechanism of accommodation.—The essential facts may be repeated again in few **words.** The ciliary muscle is the active agent; its contraction loosens the tension of the zonula of Zinn, and by relaxing the suspensory ligament permits the crystalline to become more convex, which it does in virtue of its own elasticity, because its fibres are arranged in partial spirals, and its surface layers are very pliable; the pupil grows smaller, the periphery of the iris retracts; the ciliary processes swell, **but** do not come in contact with the edge of the lens; the posterior **surface** of the lens scarcely changes its form while the anterior makes the principal increase. The continuous decline through life in the ability to accommodate is less the result **of** loss of power in the ciliary muscle than of **loss of** pliability in the lens, because it grows harder and less yielding.

To determine the limits of A, viz., P and R, the near point is **found** by reading the type 1 Snellen as near to the eye as possible, while the far point will depend on the optical form of the eye. Mauthner remarks, and with this statement **my own** experience concurs, that **to find the** absolute **near point is** by no means easy. Neither Snellen **1** nor **Jaeger** 1 are to **be accepted** absolutely as tests. If the focus be abnormal, the proper glasses **must be** supplied, and the far point be thus taken **to** infinity. If glasses **be not** given, the far point in myopic eyes will be **fixed by** the degree of **myopia. In** emmetropia it will be **at** infinity; in hypermetropia it will **be** beyond infinity, and only to be ascertained by convex glasses. For emmetropia and hypermetropia both near and far points may be found approximately by using convex 12. If with them the near point be at 4″, and the far point at 12″, then $A = \frac{1}{4} - \frac{1}{12} = \frac{1}{6}$. If the near point be at 8″, and the far point at 20″, then $A = \frac{1}{8} - \frac{1}{20} = \frac{3}{40} = \frac{1}{13}$. In the former case the person is emmetropic with good A; in the latter he is hypermetropic with **only** moderate A. For myopia more than $\frac{1}{18}$, no glasses are to be used **in** determining A; **for** degrees less than $\frac{1}{18}$, weak glasses may be employed, as for emmetropia, and the calculation will be the **same.**

DISEASES OF ACCOMMODATION.

We have *paresis, paralysis, spasm,* and *presbyopia.* Enfeebled accommodation is the result of general weakness, and appears after attacks of acute disease, or in the course of chronic affections. It follows typhoid and other fevers, and it is a very common accompaniment of chronic uterine disease and of anæmia. It may also rarely appear as **a** symptom of third **nerve** lesion without diplopia. It may not disqualify a person from reading **fine** print at a short distance, but it makes one unable to continue the effort **for** a long period. Pain and some indistinctness of type is the reason for stopping. It will very likely be found that the extrinsic muscles of the eye are also weak, they being in the same reduced state as the ciliary muscle.

Paralysis of accommodation may be idiopathic, or be a sequel of diphtheria, **or be one of** the **symptoms** of third nerve lesion, or be the result of injury **or of mydriatic medicines.** As an idiopathic condition I have lately seen it **in** the **young** daughter **of** a medical friend, a delicate girl of unusually active mind, who devours books, and who suddenly found her sight give way. There **was** no sign of brain trouble, there was no error of refraction, the pupils were large and responsive to light, but there was no accommodation ; vision $\frac{20}{40}$. She was taken from school and sent to the country. After a few months the faculty was restored. An artist friend, who in his early life painted miniature portraits, completely lost the accommodation of one eye, and its pupil became enlarged and immovable—the result, as he believed, of intense application to fine work. This paralysis continued until **his** death. After diphtheria the lesion is not at all rare, and the pupils may **have** their normal size and contractility, and the impairment of function may or may not be complete. It may come directly after the active period of the disease, **or it may** start up weeks or months afterward. It does not usually last longer **than a** few weeks ; it often coexists with paralysis of the velum palati.

As to affections of **the third** nerve, other parts which it supplies may or may not be involved, and the cause may lie along the trunk of the nerve or in the brain. It has been thought that in the brain there is a separate centre for the accommodation. The cause may be rheumatic or syphilitic. It may be a token of serious central disease, or of only a temporary lesion. A blow on the eye may cause accommodative paralysis, but this is rare. One must look carefully to know whether **the** lens may not have been displaced, and the loss of A be thus accounted **for. In this connection it** is proper **to remark that** when the crystalline is wanting there is no power whatever **of accommodation.** Yet it is true that persons **without a** lens, who have **very small** pupils, may see distant and near objects with the same pair of **glasses.** This is stenopaic vision and not accommodation. It is the same **that is** obtained by looking through a pin-hole in **a card**.

Lastly, paralysis of accommodation is caused by certain medicines, such as atropia, homatropia, daturine, duboisia. The effect may be obtained through the system by ingestion of the medicine, or by its application to the eye. Of the various articles atropia is the one most employed, and all affect both the pupil and the accommodation.

Treatment of impaired accommodation.—For cases of debility of function **the** chief resort is to improvement of general health and removal of the cause, if possible. **As** remedies, strychnia and iron and ergot are directly effective, **while** rest, outdoor life and occupation, and proper hygiene, are the most important. Associated disturbance of other ocular muscles will require careful investigation, and **perhaps** special treatment by **prismatic** glasses or by operation. If syphilis **be** suspected, its treatment **will** naturally follow. For some cases where the **tone of** the system is not so low as to make a local stimulant inappropriate, the sulphate of eserine, gr. ss. ad. ℥ j. (Calabar bean alkaloid), may be dropped into the eye **once** daily, while Dr. Green, of St. Louis, has used pilocarpine, gr. j. ad. ℥ j., and with less inconvenience than is experienced by eserine. Help may be given by weak convex glasses (from $\frac{1}{15}$ to $\frac{1}{7}$), to be used in reading and **near** work. In all cases any error of refraction is to be carefully sought out and corrected. For paralysis of accommodation in both eyes, while it may be **suitable** to resort to sulphate of eserine or to muriate of pilocarpine, it is essential to discover the cause and apply the proper treatment. The use of **strong** convex glasses ($\frac{1}{15}$ to $\frac{1}{7}$) is not advisable, because the eyes will soon tire, although for brief necessities they may be employed. Electricity has but little effect. After diphtheria, preparations of iron and quinine, strychnia, and good food, are to be trusted to hasten convalescence;—and **time.**

Spasm of accommodation is far more frequent **than any** other **disturb**ance of this function. It occurs in normal eyes, **is not** wanting **in** myopia, **is** very frequent in hypermetropia and in astigmatism. It complicates many cases of muscular trouble. Moreover, the supposition of spasm is not to be excluded because of general debility. It also appears, together with other spasmodic affections, such as blepharospasm and nystagmus, and again in cases of hyperæsthesia retinæ. It is often present as one of the factors of so-called sympathetic irritation, such as is caused by an injured globe to its fellow. If it is possible to use the eye, the habit will be to hold print and work too near, while pain, fatigue, weeping and conjunctival irritation soon begin. Sometimes distant sight seems good, sometimes there is apparent myopia. Examination with the ophthalmoscope, with the upright image, will often unmask the condition, and display **a** real hypermetropia or a degree of astigmatism in an eye which seemed normal, having $V = \frac{2.0}{2.0}$ and the patient may have declined to accept any **glass.** But, while great advantage is secured to the skilled observer by the **oph**thalmoscope, the most experienced are liable to be deceived if they **place** absolute trust upon its declarations. In a case of spasm the observer will often note that the refraction of the eye, or, in other words, the clearness of the fundus, will vary as he inspects it. He uses **in the** direct method **the** strongest convex or the weakest concave glass **which** will clearly show **the** granular texture of the fundus or the fine **vessels near** the macula. **He will** often find that the glass he sees with **is not** in harmony with **the examination** of the patient for his far point. **He** may perceive that **the** eye-ground varies in distinctness as he views **it.** Then he infers **spasm.** Another method, and one upon which I place much reliance, is as **follows : to** ascertain by glasses and the ophthalmoscope **as** well as possible the condition of refraction, and if needful **to use** a glass, let the patient put it on ; note the acuity of vision—if this be less than $\frac{2.0}{2.0}$, the fact is to be borne in mind. Then bid him read Sn. 1, through convex glasses and abductive **prisms**, namely, with $+12$, **and** prisms amounting to $15°$, with their **bases inward.** This combination makes effort of accommodation unnecessary, **and has** the greater effect because it makes the visual lines parallel while **the** object is placed at 12″. If the eye has been **found to** be ametropic, **the** required glass will be combined with the $+12$. Now, if Sn. 1, with the above arrangement, cannot be read **at 12″,** push the **print closer until it is** read ; if it have to be moved to within 10″, or 8″, or

7″, this indicates the degree to which spasm exists, and is reckoned by the formula $\frac{1}{12} - \frac{1}{15}$, or $\frac{1}{12} - \frac{1}{8}$, or $\frac{1}{12} - \frac{1}{4}$. Spasm is to be inferred; only due allowance must be made for any amblyopia found by testing the acuity of vision. By these several methods the existence of spasm can be clearly discovered, and subjective symptoms will be pain, tension, possibly headache, and the habit will be to hold work unduly close. Often it will be found **that there is** refractive error, such as hyperopia or astigmatism, or there **is weakness of** muscles, most usually of the interni, sometimes of **the externi, and possibly of those** which turn the globe up or down. It **is of consequence to dwell on this** matter of spasm of accommodation, because **it accompanies so** many functional and refractive disorders of the eye. **It is the immediate cause of the** discomfort of the great body of workers upon small objects who complain of eye troubles.

Treatment.—Sulphate of atropia is to be used in strong solution, and so frequently that the full effects are obtained. There will be great differences in susceptibility. The emmetropic and myopic will generally succumb easily; the hyperopic and astigmatic **are** often difficult to conquer, while some cases resist so stoutly as to be very embarrassing. A solution, grain **iv.** ad. ℥ j. (1 to 120), may be used four times daily—or four times an hour three times daily. In most cases this frequency and strength of dose will suffice. This may be continued for one, three, or seven days. Should the ciliary muscle not yield, a solution, grain ij. ad. ℥ j. (1 to 30) may be employed with more or less frequency. If, as happens, constitutional symptoms of poisoning, viz., quick pulse, prostration, flushed face, etc., forbid the continuance of the drops, apply to the temple the artificial leech, and try the atropine again. If still the spasm defy the remedy, tap the anterior chamber, and then the muscle will be compelled to yield. I have never had to resort **to** the last-named measure. I lay stress on this matter because I know how easily one may be deceived and be content with inadequate atropinization because the pupil is dilated and some relaxation of A is procured. To be sure that the full effect is gotten, the tests by the ophthalmoscope, by the +12, with abductive prisms 15°, must be applied, and refractive error corrected to full acuity of vision. I do not mean that the last is always possible, but it must not be despaired **of until a** thorough effort has been made.

The most remarkable **reversals in the** apparent quality of a patient's refraction are produced by **this** treatment. The findings of **the** spectacle-box, and sometimes those too of the ophthalmoscope, are contradicted or modified. I do not mean to unduly discredit the capability of the ophthalmoscope **in** diagnosis of refraction; on the contrary, I learned its value **from** Jaeger in 1859, **and** sedulously practise and enjoy it. But few **observers** ever acquire control of their accommodation to the necessary **degree for this** use, and while I habitually confide in my own capacity, mistakes have taught me discretion.

After learning what is his real refraction, the patient, if found to be ametropic, must wear the needful glasses, and the rules to be enforced will be indicated under the several kinds of refractive error which we are to consider. **Caution must be** observed about returning to work, about its continuance for long periods of time, about holding objects too near, about maintaining an erect posture of the head and body, about working in dim light, or when one is fatigued.

Presbyopia.—We have learned that the power of focal adjustment continually declines as years increase. This abatement finally makes it laborious or impossible to do fine work or read ordinary type continuously, and at the customary distance. The first token usually is that evening work is fatiguing, the light seems dim, and one gets as close to it as possible, the type is condemned as poor, the book is pushed away to sixteen or eighteen inches, and the head goes back. There is often smarting of the lids and irritability of the eyes, and often the person is alarmed lest some calamity is impending over his sight. These symptoms come when one's birthday cake carries about forty-five candles, perhaps not until forty-eight or fifty are to be counted, if numerical accuracy is by this time considered essential. If the symptoms occur at the age of forty, there is usually cause in enfeebled health, or in unsuspected hypermetropia. Whenever these signs occur, it will be found that the near point has receded to about 11′′, and hence the reserve of accommodation is small. The eye soon wearies at being held up to its highest possible accommodative capacity, and some of the labor must be taken off. The remedy is to wear a convex glass, which shall bring the near point to a comfortable distance. The question is often put, is it not better to defer the use of glasses as long as possible? The answer is, No. When the above symptoms arise, there is urgent call for help, and the help should not be denied. If hypersensitive people choose the alternative of giving up employments which demand close looking, the loss is chiefly theirs, and time and change march on. But it is not right to permit people to wear glasses when no actual need is felt. The result of their premature use is to actually restrict the power of accommodation. When they are required, glasses properly chosen give comfort and economize the power of the eye. While no fixed rules can be given as to the time when glasses should be adopted, or as to the number to be chosen, this fact must be remembered, that there is no absolute reading distance. It varies between ten and twenty inches with different persons, according to their acuity of sight, their stature, the length of their arms, and the habits they have acquired. But the rule is to adopt the glass which makes ordinary type distinct at the distance customary to the person. Practically, the glass first chosen is $+\frac{1}{4.5}$ (.75 D) or $+\frac{1}{3.5}$ (1.D) in most cases. This will be worn for two or three years, and signs of fatigue, as at first experienced, will suggest increase of power in the glasses. It is important not to go too fast in addition, and a weak glass may serve for the day, while a little more help will be needful at night. A decided jump in the loss of A should awaken suspicion as to the possible onset of glaucoma simplex, or of cataract. If the intervals of addition be $\frac{1}{3.5}$, or sometimes $\frac{1}{6.9}$, and the intervals of time be two years, the needs of the eye will be sufficiently met. When, however, $+10$ or stronger glasses become needful, the reading distance is fixed at a particular point, and will not admit of being brought closer or removed farther. It also comes to pass at about sixty—perhaps sooner, perhaps after that age—that sight for distance requires aid from convex glasses, viz., $\frac{1}{4.5}$ or $\frac{1}{3.5}$. This is acquired by hypermetropia, and is a frequent occurrence. It is asserted by Landolt that presbyopia increases one dioptry every five years up to the age of sixty, and from this time on it increases sometimes one and sometimes one-half dioptry in the same period. There are great variations on this point, and actual experience must decide.

Presbyopia, or the natural abatement of accommodation through age, affects eyes of every type, whether normal, or myopic, or hyperopic. Some persons go to advanced age without needing glasses. They are generally exceptionally robust, or they have very small pupils, or they have been slightly myopic. Again, certain old people lay aside the glasses they have been wearing, and read without any, gaining,—as it is said, their "second sight." A number of such persons I have examined, and have found them to have incipient cataract and very small pupils. With incipient cataract a slight, or even marked degree of myopia can occur through swelling of the lens, and, as acuity of vision declines, print must be held nearer. It will be found that such persons usually have vision less than $\frac{2}{2}\frac{0}{0}$ or $\frac{2}{2}\frac{0}{0}$. In addition to incipient cataract there may be haziness of vitreous or cloudiness of the retina or choroid, which continues for a time and afterward clears up. When this better state arrives they find themselves no longer dependent on their glasses, and are proud to boast of their vigorous sight. This may be allowed to them as a harmless bit of comfort, but the test of vision would show that their faculty is much below the normal standard.

ERRORS OF REFRACTION.

HYPERMETROPIA.

Hyperopia—H.—Hypermetropia is the condition in which, with suspended accommodation, a person requires a convex glass to get his best acuity of sight for distance. He may gain $\frac{2}{2}\frac{0}{0}$ or better, or he may not be able with any glass to read XX at twenty feet, but will read a smaller type than he can without its aid. It is essential to put aside the accommodation, because up to a certain age and for small degrees it overcomes and conceals the hyperopia. But it is not to be expected that all cases of high degrees of H will with glasses gain $V=\frac{2}{2}\frac{0}{0}$. Many who need $+10$ (4 D) or stronger glasses, have $V=\frac{4}{5}\frac{0}{0}$ or $\frac{2}{2}\frac{0}{0}$. In fact, the greater number of strongly hyperopic persons do not have normal acuity of sight.

Hyperopia is the optical result of a shortened visual axis or of want of the crystalline. We are indebted to Donders both for the name and for a clear portrait of its character, existence, and symptoms. Both in its theory and in its practical bearings the description of Donders in 1864 has needed little enlargement. It also appears in later life, after sixty, by flattening of the crystalline, being the outcome of advancing presbyopia. Flattening of the cornea may also cause it through distention of the globe, as ensues in glaucoma. The usual cause is shortening of the visual axis. It is a congenital condition. It is asserted by Jaeger that the infantile eye is ordinarily slightly hyperopic, and that emmetropia comes after two or three years. Leaving this aside, an imperfect rotundity of the globe gives H, which becomes a permanent error. A very slight anatomical defect can make an appreciable optical error, as appears in the following table by Loring, showing the length of axis corresponding to degrees of H :

H.	Axis shortened.		H.	Axis shortened.
$\frac{1}{20}$	0.21 mm.		$\frac{1}{10}$	1.00 mm.
$\frac{1}{18}$	0.26 "		$\frac{1}{9}$	1.12 "
$\frac{1}{16}$	0.35 "		$\frac{1}{8}$	1.25 "
$\frac{1}{14}$	0.45 "		$\frac{1}{7}$	1.40 "
$\frac{1}{13}$	0.52 "		$\frac{1}{6}$	1.60 "
$\frac{1}{12}$	0.58 "		$\frac{1}{5}$	1.89 "
$\frac{1}{11}$	0.65 "		$\frac{1}{4}$	2.30 "
$\frac{1}{12}$	0.74 "		$\frac{1}{3}$	2.90 "
$\frac{1}{12}$	0.85 "		$\frac{1}{2}$	3.97 "
$\frac{1}{11}$	0.92 "			

This does not mean the glass required, but simply the anatomical error in the eye.

Symptoms of H.—The objective evidence is found by glasses and by the ophthalmoscope. The test by glasses is made for objects at a distance, i.e., for the test-type at twenty feet, providing the accommodation is eliminated. If A be not neutralized, there may be a discoverable degree of H, which is called the *manifest* hyperopia, Hm. But there will be another portion of the error not brought to view, called the *latent* H, and the two combined make up the *total* H. Donders also makes use of the term facultative H, to state that a person can see distinctly at a distance both with and without convex glasses. The term is not much employed.

The acceptance of convex glasses for distant sight becomes the evidence of H ; but the refusal to accept them does not disprove the presence of H, unless the accommodation has been completely paralyzed. Therefore, the use both of atropia and of glasses is required to demonstrate H. The ophthalmoscope may also be called in with great advantage, but this subject will be left to a future chapter.

Subjective symptoms of H are often entirely wanting, despite its existence in decided degree. Usually the persons are in vigorous health, or do not exact hard work from their eyes, and the H will not often be large ; but no precise statement can be made on this point. Temperament and occupation and acuteness of observation are important factors. It has been shown by Iwanoff that hyperopic eyes have an unusual development and form of the ciliary muscle. This is the product of necessity, because its form compels the eye to maintain an active state of adjustment, both for distant as well as for near objects, and the muscle acquires an extraordinary growth. For this reason we understand why hyperopes may utter no complaint respecting their sight. But the development of the muscle may not be adequate to the work demanded ; a failure of health may reduce its power ; extraordinary demands in work, reading, writing, sewing, drawing, and in various crafts, may exhaust its energy, and then complaints are made. It is a common thing to date the trouble from an attack of sickness, a fever, child-bed, sunstroke, the practice of reading in bed when confined by some protracted illness. Especially does chronic uterine disease unmask such optical errors and reduce the accommodation. A great grief or mental shock may have the same result ; protracted weeping impairs the eye-power by its depressing influence and by the chronic conjunctival hyperæmia which it excites. Moreover, advancing age gives unpleasant revelations, and at thirty or thirty-five, glasses are suddenly found indispensable. Such occurrences are apt to concur with premature gray hair, a marked arcus senilis, atheroma of arteries, and tokens of early senility. The phases under which H betrays itself are manifold, and are by no means exhausted in the above category.

More specific *symptoms* are as follows : use of eyes is possible for a limited time—a few hours, during daylight, for successive days, and then fatigue, aching or sharp pain, is felt in the globes or their vicinity. Generally—and this item is to be noted—the work becomes blurred. Shutting the eyes, or a little rest, makes it again distinct, but the blur returns. With persistent use the eyes water, the lids become red and smart, feel sandy and hot. I have known nausea to be caused, and in many H is the cause of chronic headache. The sufferer seldom attempts to aid himself by pushing the work to a greater distance. On the contrary, if its position is changed, the eye and the work come nearer together. In the higher degrees of H, viz., more than ¼, it is the rule to find the person bringing work so close to the eye as to cause him to be pronounced very myopic. Especially is this true in subjects under twelve years of age.

Stress is to be laid upon this statement because there is liability to **fall** into a most serious mistake, and the possibility of mistake is enhanced **by** the fact that many hyperopes of high degree have very poor sight **for** distance, and may even profess to see better through weak concave glasses. It is evident that the distress which hyperopic persons suffer is because of overstraining of the ciliary muscle and its attendant neuralgic complica- **tions** ; and moreover, the inadequacy of the muscle is exhibited in the want of clear definition which usually supervenes after a certain **degree of** work.   This behavior of the ciliary muscle differs from what we **see when** its power declines by the ordinary progress of time.   In the latter case **the** process has come gradually, and the person by instinct has learned **that** holding print farther off and getting a strong light will assist him.   But the break-down in the muscle which H causes is not so complete.   It can rally for another struggle after a little rest, and it strives against defeat by summoning all its force.   Hence, pain and spasm of accommodation.   Be- cause of the spasm the eye often refuses the aid of convex glasses for dis- tance, and sometimes they are not at first acceptable for near work.   It is **not** rare to see such persons wearing weak concave glasses, notwithstanding they still further tax the accommodation.

The natural association which subsists between the ciliary muscle and the adductor muscles has a two-fold effect: that in hyperopes the recti interni are apt to be hypertrophied, while the externi are weak, and the ciliary muscles become able to exert themselves even with parallel visual lines to an uncommon degree.   This makes the relative Λ in hyperopes very largely, or even entirely negative, and explains, as before stated, why, with convergence to a near point, fatigue must soon be produced.

Diagnosis.—It will be seen from the above statements, that there are two obstacles to the ready detection of hypermetropia.   One is the spasm of accommodation, and the other is a possible obtundity of vision for distance.   This last **may** be due to inherent amblyopia, or to a complicat- ing astigmatism.   The existence of amblyopia may often be suspected **from** ophthalmoscopic inquiry.   If H be high, the optic nerve, besides **being** small—which is the natural result of the inferior optical power **of the** eye—will be deeply red, its edges will be ragged and fringy, **the** adjacent retina will seem thick, the veins very dark, and the substance **of the** nerve opaque.   Dobrowlsky (*Klin. Monatsbl.*, April, 1881, p. 156) suggests that idiopathic retinitis may be promoted by hypermetropia.   It is common to find that while the media are clear the retina lacks perfect brightness, and **the** nerve is in the state **above** described.   Frequently it is anatomically oval in outline.   The ophthalmoscope, by the direct method, gives invalu- able aid in establishing both the existence and the degree of H.   Under the conditions of this examination—the darkness, the want of a fixing-point, and **the** glare in the eye—the accommodation greatly or entirely relaxes, yielding up information which often contradicts the findings of the spectacle-box. If the discrepancy between these two means of investigation, viz., the spec- table-box and the ophthalmoscope, is great, or if the visual acuteness cannot be brought to standard by glasses, and if spasm of accommoda- tion be strenuous, then the vigorous use of atropia must be invoked to settle both the **diagnosis and** the remedy.

Treatment.—Several questions arise.   Must atropia be employed, and how vigorously ?   Must the whole hyperopia be neutralized ?   Must glasses be worn all the time ?   Atropia is demanded if the neuralgic pain is great and accommodative spasm severe ; also, if visual acuity cannot be raised **to** $\frac{2}{2}\frac{0}{0}$.   But if a convex glass is accepted which does not differ much from the ophthalmoscopic finding, and with it $V = \frac{2}{3}\frac{0}{0}$, and if with it Sn. 1. is read easily at 12″, and the larger type at proportionate distances, and if the patient confidently expresses satisfaction in its use, atropia need not be employed. This course will be suitable with H not more than $\frac{1}{16}$, and in persons under thirty (not speaking dogmatically) who have a good degree of A. In them it is proper only to correct the *manifest* H, and they will seldom need at first more than $+\frac{1}{36}$ or $+\frac{1}{24}$. Some *slight* cases of H are attended by so much neuralgic pain, that a glass $+\frac{1}{60}$ or $+\frac{1}{48}$ is of unmistakable value, and to be prescribed. To these highly sensitive persons a light blue tint is sometimes grateful. While most of these patients are women, they do not enjoy a monopoly of being ophthalmically neurasthenic. I have record of many men, students, litterateurs, clergymen, bookkeepers, and others, who find a slight H a heavy burden, and which is lightened by $+\frac{1}{48}$ or $+\frac{1}{36}$. It is often true, of this class of persons in both sexes, that the ocular muscles are weak. There may be a disparity between adduction and abduction, or the whole apparatus may be weak. In another chapter these cases will be referred to again. In deciding upon the propriety of correcting small degrees of H, the whole aspect of the individual case is to be considered. In this regard an oculist must not make a pin-hole diagnosis of his patient's condition (stenopaic inspection), but must take into account character, health, surroundings, occupation, and a variety of circumstances which naturally occur to any wise physician.

If the degree of H be important, $\frac{1}{12}$ or greater, it will very frequently be advisable to use atropia until A is fully at rest, and to give the glass which fully corrects H. Most often it will be better to wear it continuously. Under what circumstances should convex glasses be constantly worn? Some people answer for themselves by finding that they are wholly uncomfortable without them, even though the degree is not strong. The comfort of the individual is of necessity the fundamental reason for constant use, and that only in this way can it be secured is not always to be anticipated. That such use is likely to be needful will be probable of cases of high degrees, viz., $\frac{1}{12}$ and more, whether in young or old subjects; the more advanced in age the person, the more likely is he to require constant use. Again, the same advice is to be given to very sensitive persons whose eyes give them much pain. Under this head will come a large class of semi-invalids and impressible and neuralgic persons. On the other hand, the dull and torpid and unobservant will often be quite indifferent to the aid of glasses for distance, even though they have a marked degree of error. Furthermore, something has to be conceded to the sense of what is becoming to their personal appearance in persons of both sexes, and, while a physician will not modify his deliberate opinions and advice in deference to what may please his patient, there are doubtful cases in which his abstract views must be modified by the patient's experience. For persons with marked H there can be no doubt of the advantage gained by constant use of glasses, because the range of accommodation is brought within the physiological limits and the continued strain on the ciliary muscle is removed. Moreover, it is a frequent observation that in this way acuity of vision decidedly improves in the higher degrees of H. This is not simply the effect of enlargement of retinal images, but of improved capacity and health of the retina.

For persons who have already lost so much accommodation as to be presbyopic, two different glasses must be employed—one for distance and another for near. These may be separate, or may be combined in one frame. Both curves may be ground upon one piece of glass (lenses of double focus), or the glasses may be separate, in which case there will be a split or crack running horizontally across the glass—the weaker being above, the stronger below. To some persons such glasses are very accept-able, to others they are intolerable, and especially because of the displace-ment of objects through the prismatic effect of the glasses. I have noticed that glasses of double focus are more useful if the upper and weaker part

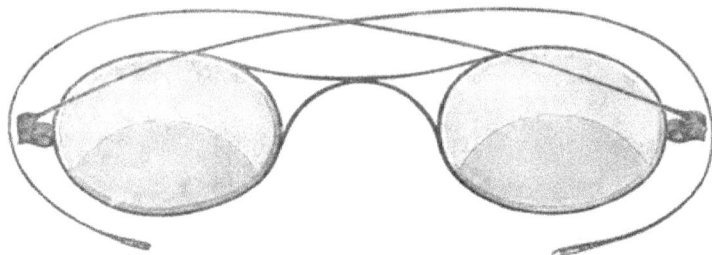

Fig. 28.

shall occupy the greater part of the glass. The separation between the two in this case is always a curved line, and its concavity should be down-ward. This gives greater range and freedom in distant vision than if the curve is oppositely arranged. To artists, and to all who must look quickly from a distant to a near range, this contrivance is valuable. This arrangement was first made by Mr. Hunter, optician.

It often follows that a notable change in the character and behavior of a person takes place when a great hyperopic error is corrected : clear perception, absence of strain, unconciousness of eyes, and entire freedom in their use, work sometimes a change of character which is as delightful as it seems wonderful. The popular objection to glasses is that they often become a permanent necessity, and are inconvenient things to be tied to. Such objections have obvious force, and need not be ignored. But the answer is that nothing else can remove the trouble. We soon accept the disagreeable when it is found to be indispensable.

MYOPIA.

M.—We have in this error the antithesis of that which has just been considered. It has been recognized for a long period, but its full under-standing belongs to the era of the ophthalmoscope. To Jaeger, Donders, Graefe, and many others do we owe our present knowledge about it. The condition depends on elongation of the visual axis. What relation this elongation bears to the degrees of myopia is exhibited in the following table, taken from Loring ("Determination of the Refraction of the Eye with the Ophthalmoscope," Wm. Wood & Co., 1876, p. 30) :

M equals increased axis of	M equals increased axis of
$\frac{1}{50}$ .... 0.22 mm.	$\frac{1}{11}$ .... 1.17 mm.
.... 0.27	.... 1.31
.... 0.37	.... 1.50
.... 0.46	.... 1.97
.... 0.56	.... 2.07
.... 0.63	.... 2.56
.... 0.71	.... 3.34
.... 0.82	.... 4.81
.... 0.97	.... 8.61
.... 1.06	

The slighter degrees of M are attended with insignificant degrees of elongation—for instance, up to $\frac{1}{20}$ amounting to about 0.5 mm.; but the damaging effect on sight is appreciated when it is remembered that M $\frac{1}{20}$ means that the far point, instead of being at the horizon, has been brought to twenty inches. The myope of this degree, by half shutting his lids, and in light which narrows his pupil, will make out familiar objects, but the world is misty, and he rarely has better sight than $V = \frac{20}{200}$, whatever he may do. An emmetropic person can understand the myope's situation by wearing the plus glass which corresponds to the latter's minus glass. But the myope claims advantages in near vision which in many cases are substantial. That he can see small objects and do fine work with special ease, because his near point is close and accommodative effort is little called for, is in many cases true. But we shall find that even this compensation is not enjoyed by a large portion of such subjects. By referring to the table, it is seen that elongation of the globe, amounting to 1 mm., gives rise to M $\frac{1}{10}$, while double this degree, viz., 2 mm., causes M $\frac{1}{5}$, and that above this point the rate of elongation increases alarmingly. Between E and M $\frac{1}{2}$ we have 20 D. In the first 5 D, viz., to $-\frac{1}{8}$, we have 1.59 mm. stretching, in the next 5 D, viz., M $-\frac{1}{4}$, we have 3.34 mm., which is more than double, and in the next 10 D, viz., to M $\frac{1}{2}$ or 20 D, we have 8.61 mm. It is frequent to find decided changes in the tissues of the eye, with less M than $\frac{1}{8}$, and beyond this degree it is the rule that they occur, and their gravity becomes more and more serious with any additional dioptry.

Myopia may be congenital, is often hereditary, but is usually acquired between the ages of five and fifteen. Exceptional instances are found in which M is acquired after twenty, and even beyond middle life. I have notes of such occurrences. When once begun, myopia increases for a term of years. It often is arrested at twenty, but may continue its increase until twenty-five or thirty. In special cases it seems to advance throughout all of life, but less rapidly than during early years. The cause of this change of form is found in the want of resistance of the sclera at the posterior part of the eye. That this part of the structure should most readily give way is favored by the want of direct support at this region, whereas in the other parts of the globe the muscles reinforce it, and are themselves the agents to a large degree of the mischievous pressure. The habit of fixing the eyes upon objects at short range implies strong efforts both of accommodation and convergence. Let this be kept up for a long period and let the sclera be relatively weak, and the globe begins to stretch. Nor is this result limited rigidly to the very juvenile stage of life. I have seen it begun in an apparently robust student of twenty, during his third year in college. Habitual occupation with small objects held near the eye, as when children first take to books, either for pleasure or study, or when learning to draw or to sew, or they are kept at a piano which stands in a dark corner, or the school-room has not the full quota of light which is each child's right, or the benches and desks are so contrived that the child sits crouched or crooked, or a heavy lexicon compels a stooping posture, or the text is badly printed, or is in a language whose characters are intricate and unfamiliar, like Greek, or German, or Hebrew: such are some of the occasions of myopia.

When the ocular change begins it is often unattended by subjective symptoms. It is sometimes a surprise to the person to find that his sight for things once perfectly clear has grown so dim. But, on the other hand, there may be considerable discomfort, such as pain, weariness, sensitiveness to light, conjunctival irritation, hyperæmia of the optic nerve and retina. So long as M is decidedly advancing, the eyes are apt to easily become weary and irritable.

A lack of toughness in the sclera has been indicated as the invitation to myopia, but another element appears in a great number of cases, viz., weakness in the muscular apparatus. This is often a natural concomitant of softness of other tissues; but it has seemed to me that want of proper balance among the muscles, as they combine in their action, provokes unnatural strain and leads, in what seems a paradoxical way, to undue pressure on the globe. The difficulty of steady fixation induces closer approximation of the work to the eye, and thus cause and effect are set going in a vicious circle. It is a fact widely noticed, and on which Graefe laid stress, that in myopes the abductive muscles are very liable to be weak; but I also think the whole muscular apparatus of the eye is in them specially liable to be feeble. Add to the above that many of these persons are lacking in general muscular energy, are fond of books and in-door life, prefer quiet and sedentary pursuits to active out-door work or play, and this temperament and character predetermine both their pursuits and their ocular defect. In defining myopia it may be said that, with correct acuity of sight, the far point and the near point are both brought too close to the eye. How can we practically be assured that there is correct acuity of sight? We trust to the efficacy of concave spherical glasses to demonstrate this. If, therefore, we find a person who without a concave glass has V less than $\frac{20}{20}$, and with a concave glass gains $V = \frac{20}{20}$, and also has a near point corresponding to the degree which the glass implies, he may be counted a myope. But many myopes do not get $V = \frac{20}{20}$ with any glass. They may be amblyopic or astigmatic. Furthermore, spasm of accommodation will locate the far and near point in positions which perfectly simulate myopia, and the patient is able within certain limits to answer the test by a concave glass. Such a condition

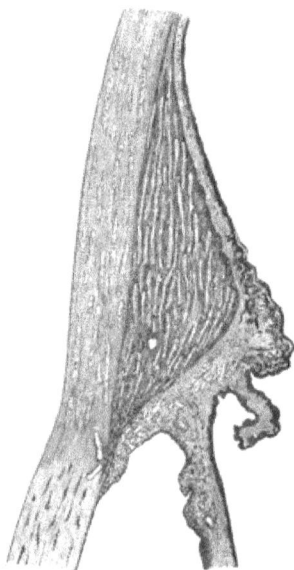

FIG. 29.

may be called fictitious myopia. How are we to escape such errors of diagnosis? By the ophthalmoscope and by using atropia. There is much less annoyance to myopes in atropia than to any other persons, and therefore it should be used in many cases, simply for diagnosis.

It is obviously a consequence of myopia that the accommodation should be little taxed. If the M be more than $\frac{1}{12}$, a book may be read at 6″, and only $\frac{1}{3}$ A will be employed. This makes myopes fond of reading fine print and doing fine work. Otherwise the nearness of the work restricts their range in the extent of surface which they can cover, i.e., the horopter. It is not surprising that for want of use the ciliary muscle should be imperfectly developed, and often the possible A be quite inferior. Iwanoff demonstrated the small and attenuated form of the ciliary muscle in myopes as he did the contrary condition in hyperopes. That many have most inadequate accommodation is proven by the difficulty they have in easily seeing near objects with the glasses which correct them for distant objects.

But the serious tissue-changes of myopic eyes are found by the ophthalmoscope. In a certain proportion no intra-ocular abnormities can be found except the refractive error. This is true chiefly of the slight degrees; but it is verified also, sometimes, in cases of M as high as one-fourth. What we find, however, in most cases is that (concurrent with the stretching of the sclera) there takes place absorption of the choroid on the temporal side of the opening at which the optic nerve pierces it. Necessarily the retina, the choroid, and the sclera are stretched simultaneously and to equal degrees. On the retina we may conceive the effect to be to spread out its microscopic elements, say the bacilli; over a large area on the sclera, that its fibres become pulled apart; while in the choroid the edge of the opening in contact with the optic nerve becomes absorbed because naturally it adheres closely to the inner sheath of the optic nerve as this joins the sclera. In this way a crescent-shaped figure is added to the outline of the disc, which is the inner surface of the sclera thus exposed to view. As a rule, this appears on the temporal side. It less frequently appears above or below and in oblique positions. It almost never primarily occurs in the nasal (medial) side. A decided pigment-deposit usually appears on the temporal side of the crescent. During its early stage the crescent is narrow, like the new moon; but if it advance, the shape changes by extension toward the macula, and also around the disc. (See Fig. 39, which shows a crescent of choroidal atrophy of moderate size. Also see Fig. 31, which exhibits the distention at the back of the sclera, which causes myopia, and figure illustrating choroiditis disseminata in chapter on choroid.)

Because of some confusion in pathological interpretations, this crescent was in the early ophthalmoscopic time called staphyloma posticum, and also

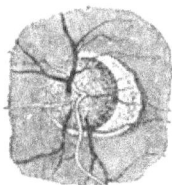 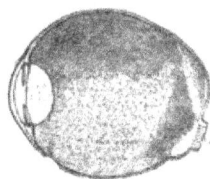

Fig. 30.   Fig. 31.

sclerotico-choroiditis posterior (Graefe). The latter name has been wholly abandoned, and the former is evidently not fitted to the description of the choroidal atrophy alone, but to the whole condition of the back of the globe on which myopia depends, and of which the choroidal lesion is only an element. We shall therefore observe this discrimination. There may be great disproportion in the extent of choroidal atrophy in similar degrees of M. The more extensive it is, the less healthy are the membranes and the lower the acuity of sight. When, as sometimes is seen, the atrophy appears on the nasal side, or surrounds all the nerve, sight is very defective. The erosion of choroid is sometimes incomplete; that is, within the white space are seen isolated choroidal vessels and flecks of pigment, vestiges of tissue not yet removed. It also happens sometimes that the hexagonal

5

epithelium is destroyed for a certain space beyond the white crescent, because for some reason it has been less capable of resistance than the subjacent tissue. As a rule, the whole thickness of the choroid, including the hexagonal epithelium, disappears together. When the atrophic crescent becomes large, there may be an increase in the size of the blind spot, because there is generally partial atrophy of the retina in this locality, *i.e.*, of the bacillary layer (Schnabel). The vessels of the retina are unaltered, and by contrast come out strongly to view. But the case is serious when, as too often happens, the region of the yellow spot becomes affected. Before visible changes appear, the state of vision may indicate the beginning of danger ; but soon the choroidal vessels begin to disclose themselves, then irregular pigment-deposits, and finally an irregular patch of exposed sclera, come to view. A hemorrhage into this spot is not rare, nor are masses of pigment uncommon. Under extreme conditions the eyeball becomes distended in all its diameters, as well as in its axis. Often the optic nerve takes an oval form because seen in perspective, and the long axis may be either perpendicular or oblique, according to the location of greatest bulging. Sometimes excavation extends into the disc, and it may be difficult to define where the edge of the nerve is to be found. Changes also occur in the optic sheath outside the globe. The external becomes separated from the internal sheath. In the anterior part of the eye we find the ciliary muscle undergoes elongation and atrophy, as already remarked, fully accounting for the want of accommodative power. The transparent media undergo changes. The cornea in high myopia may grow flatter, *i.e.*, its radius be longer, although in a rule it is unaffected. The anterior chamber is often deep, because the aqueous is apt to be in large quantity. The vitreous undergoes degeneration by becoming liquid, by being filled with the debris of its structural elements in the form of floating shreds and specks ; it may also be detached from the retina behind, and the space be filled by serum. Its front is also the seat of liquefaction, and hence the lens often becomes tremulous. In the last degree of myopic change the lens becomes partially or wholly cataractous, and, because of the vitreous fluidity, is liable to be luxated backward or downward. As an indirect result of these nutritive degenerations of the vitreous, we find sudden detachment of the retina. According to the most recent opinions, this takes place because of hinderances to osmosis, by reason of alterations in the chemical qualities of the vitreous. Intra-ocular hemorrhages are prone to occur in myopic eyes. In extreme cases the distention of the eye attains the condition called hydrophthalmus, and it may be difficult to say what part of the result is to be ascribed to a genuine irido-choroiditis, and what to the processes of simple myopia. Enucleation may then become a necessity. In external appearance myopic eyes often attract attention by their prominence and their observable ovoid form. Usually the pupils are large and inactive, but in all these particulars the contrary conditions may be true.

It is often the fact that myopes of high degree are unable to maintain binocular vision for the near, and they therefore frequently have diverging strabismus. It is also common to find that they have an unusual capacity for directing the eyes in opposite ways upon a vertical plane, one looking up and the other looking down, as discovered by their ability to overcome prisms with the angle vertical. This signifies that their hold upon binocular vision is, so to speak, insecure.

After this long arraignment of the miseries possible to myopia, there can be little pretence of its being an advantage except for low degrees (less than $\frac{1}{6}$), in that the time for wearing glasses for the near is postponed. To offset this there has been the need of using them for distant sight during preceding life, and to some the time arrives when for near a convex glass is needed, and for distance a concave. This was the plight of Benjamin Franklin, who invented the split glasses.

Treatment.—This is prophylactic and corrective. As means of prevention, all those inducing conditions previously mentioned are to be removed : in the posture of the person, in the habits of eye-work, in the allowance of light, in the arrangement of school-rooms, workshops, and dwellings. The physician must of course address himself to school authorities, to parents, to employers, etc., as well as to the individual. Congenital predisposition may not be combated with much success. The resort to atropine has been much commended ; but it has not, in my experience, served more than a temporary good purpose. It will abate the myopia from $\frac{1}{12}$ to $\frac{1}{6}$, usually about $\frac{1}{18}$; but I have failed to find that it retards the advance of the error. The degree of M returns to what it was before atropia was used, and the gain has been only what comes from entire rest of the eyes.

The corrective treatment is in the selection of glasses. If with unaided eye any letters on the card are read, the degree is not great. If no letters are read, an approximate idea is gotten by noting at what distance the person holds the book in reading Snellen 2. Begin with trying, on one eye at a time, the weaker numbers. If strong glasses are being used, viz., above—6 D, and the vision is nearly corrected, try whether sight is helped by holding the glass nearer or farther away. If the former, it is too weak ; if the latter, it is too strong. Myopes are often sensitive to an interval of $\frac{1}{60}$ or less. With too strong a glass they complain of being dazzled, and shrink from the unwonted brightness of objects. Much respect is to be paid to their impressions ; but, when they have been wearing inadequate glasses, they are liable to mistake the surprising distinctness conferred by a proper glass for a strain of overcorrection. It is often impossible to give them $V = \frac{20}{20}$; but, before admitting this, careful inquiry is to be made for astigmatism, and the state of the fundus minutely explored by the ophthalmoscope—especially the region of the yellow spot. Moreover, in doubtful cases the patient should have the benefit of atropia —using a solution gr. iv. ad. $\frac{\text{Ʒ}}{\text{j}}$. (1 to 120) several times. In seeking the best correction the danger is of getting too strong a glass, which will excite tension of the ciliary muscle. To this, as a rule, myopes are very sensitive. It often is needful to give a weaker glass for reading, because A is not strong and adduction may be feeble. If the person be under twenty, in good health, and M not greater than $\frac{1}{10}$, he can generally be enabled to wear one glass for all purposes. This it is desirable to accomplish, and, for all degrees greater than $\frac{1}{6}$ and less than $\frac{1}{2}$, is often feasible. But for older persons and for the stronger grades, a weaker glass for near work is to be used. This will be such as to bring the work to the point which the individual habit and the state of the converging muscles indicates. It will be from 8″ to 14″. If 10″ be the distance sought and the distance glass is $-\frac{1}{8}$, the reading-glass is found by the formula $-\frac{1}{8} + \frac{1}{10} = -\frac{5+3}{80} = -\frac{1}{15}$; or, by dioptries -7 D $+ 4$ D $= -3$ D. With myopes, the habit of holding things near, is so strong that often an insufficient correction for distance is selected, that it may not for near work embarrass their accommodation.

Another condition to be considered is the power of convergence. It is because of weakness in this regard that glasses for near work are to be advised, and not seldom a prismatic combination must be employed to still further relieve the feeble muscles. This will be done either by decentering the glasses or by adding a prism. Allusion will hereafter be made to this matter. Under these circumstances, tenotomy of one or more muscles may be practised (*vide infra*). To be certain that the glasses selected are proper, ophthalmoscopic examination must confirm the **finding** and give a clear picture of the fundus, especially near the yellow **spot.**

If now there are complications in the case, viz., deep congestion of the optic nerve, considerable attenuation and atrophy of the choroid, vitreous opacities, and symptoms of flashing light, photophobia and pain, treatment should be employed to allay the congestion and irritation. There must be entire rest and abstinence from work, protection by dark glasses, and moderate use of atropine. The bowels and kidneys are to be kept free, and exposure to hot sun and to causes of cerebral congestion are to be avoided. The artificial **leech or two natural leeches are to** be applied to the temples, and afterward the person **is** to be kept **for** twenty-four hours in a dark room. Such a depletion **may be** practised once **or** several times, with advantage. The period of **rest** may extend from a few weeks to months. A moderate use of Vichy or of some **similar** water is good. But the main thing is attention to general **health,** avoidance of near work, and plenty of out-door life.

A nice question is **often** to be **settled**: whether a myope with decided intra-ocular lesions ought to **wear glasses** at all. For the great majority my judgment is that **they should.** Without them they grope **in** a dismal obscurity, which they **are continually** seeking to explore, **and are** provoking irritation of mind **and eyes by** attempting to penetrate. **I have** been abundantly convinced that **it is far** better to permit glasses **than to** forbid them. At the same time **I** would lay great stress on the necessity of searching for and neutralizing astigmatism and muscular **errors,** which **I believe** are much more potential factors in the "miseries of myopes" **than has been** generally comprehended. A curious case presents itself **at this moment,** which serves to show how the exigencies of experience **call for discriminating** judgment which cannot be formulated in rules.

A married lady **with** several young children is, because of a worthless husband, compelled to support herself and family by her pen. **She** has in one eye M $\frac{1}{5}$, in **the other** M $\frac{1}{18}$. In the right eye is commencing choroidal atrophy **at the macula** lutea, with V = $\frac{2}{4}\frac{0}{0}$, and a wide choroidal crescent at the **nerve. The other** eye has V = $\frac{2}{2}\frac{0}{0}$, and the choroidal atrophy at the nerve is not **only large,** but progressive, as indicated by the successive portions which have **been** destroyed, as one crescent has been added to another, **like** terraces, while vestiges of pigment and vessels remain in the outermost terrace. She had four weeks ago an attack of scotoma in each **eye,** causing total blindness over a central spot, and which at ten inches **would** blot out all but the first two and last two letters of a word like "vicissitudes." The same hiatus was noticed in all objects. This lasted **several** hours, was followed by severe headache, and disappeared after a night's sleep. Her medical attendant at once forbade all use of her eyes, and put them under the influence of atropine. She has also dynamic divergent squint, and uses only one eye for near work. I found no opacities in the **vitreous.** She has phosphenes in one eye. She had just had an offer of employment which would call for four hours' daily writing for six weeks. I deemed it safe to permit this because the attack of central scotoma was evidently *migraine* due to cerebral and not to ocular causes, but stipulated that she should have two weeks in the country, and that the atropine effect must pass off and that the work be done with the help of $-11_{\circ}$, and for not more than four hours daily. After six weeks had passed, she reported that she had used her eyes for the time allowed, viz., four hours daily, with

— 11. and broke these at the end of five weeks. Then she wore her full correction, viz., —5½, O.U., and could work with comfort.

To many myopes strong light is a distress, and their glasses may be tinted a light blue. To some the constant observation of objects is a weariness—they prefer to take off glasses and remain in ignorance of what is about them until their eyes are rested. Many are sensitive to the form of the frames, their weight and adjustment, and the eyelashes must not touch the glass. All these points deserve attention. Some persons affect the wearing of a single glass which they have learned to hold in place by nipping it with the brow. If such have two equally good eyes, which generally is not the case, such a practice is no less damaging to the eye, than offensive as a mannerism.

The above description of the possible lesions of myopia is calculated perhaps to make the picture of near-sightedness too gloomy, because so many woful conditions are grouped together. The very large proportion of myopes escape all such disastrous occurrences ; but it is highly important to convey the impression that myopia is more than a mere inconvenience or trifling defect, because it does embrace such sad possibilities.

ASTIGMATISM—As.

When the refraction is such that rays emanating from a single point cannot be brought again to a focus on a point on the retina, this state is astigmatism ; of this there are two kinds, the regular and the irregular. The latter is caused by opacity of the cornea or lens, and does not admit of satisfactory correction, although it can sometimes be mitigated. The former is chiefly dependent on abnormal curve of the cornea or lens, or want of homogeneousness in the lens, and is correctible by cylindric or sphericocylindric glasses. The defect may be acquired or congenital ; irregular astigmatism in the cornea is an acquired error, and some rare cases of regular corneal astigmatism are acquired ; but, as a rule, the regular astigmatism of the cornea is congenital. Acquired astigmatism in the cornea, where no opacity exists, comes from conicity of the membrane, or happens, as I have seen, after tenotomy of muscles, or after wounds of the cornea, such as iridectomy and extraction of cataract. But these cases are a minority of the whole. Of correctible astigmatism the greater portion are congenital cases. Objection is sometimes made to this statement because the error often does not announce itself until middle life. The explanation is that the accommodation can conceal a considerable degree of error until its vigor begins seriously to decline. It is also to be said that a small degree is natural to almost every one, varying from $\frac{1}{120}$ to $\frac{1}{15}$, and because the radius of the vertical meridian of the cornea is shorter than that of the horizontal.

We have occasion now to treat only of regular astigmatism and without regard to its locality in the lens or in the cornea. Consisting as it does in a want of uniformity in the radii of the meridians of the media, this error manifestly may complicate either emmetropia, hypermetropia, or myopia. For this reason we have simple astigmatism, either hyperopic or myopic ; and compound astigmatism, both hyperopic and myopic ; and lastly, there may be mixed astigmatism in which either hyperopia or myopia may predominate. The symbols of these several conditions are as follows, as they have been given to us by Donders. To him we owe the systematic study and development of this subject, which he made with as much completeness as did Helmholtz the theory of the ophthalmoscope. We have. 1st, myopic astigmatism, Am ; and compound myopic astigmatism, M. + Am ; 2d, hyperopic astigmatism, Ah, and compound hyperopic astigmatism, H + Ah ; 3d, mixed astigmatism, with prevalent M, viz., Amh, and with prevalent H Ahm, or both M and H may be alike.

To gain a practical understanding of the phenomena of astigmatism two methods may be adopted. The one is to take a spherical convex glass, of say 8″ focus and form the image of a luminous object, say a 6-inch circular hole in a shutter or a ground-glass globe of a gas-burner. Then add to the 8″ glass a cylindric glass of 16″ focus, and observe the change which takes place in the appearance of the image. Calculation shows that for the plane at right angles to the axis of the cylinder we have a lens which equals the two combined, viz.: $\frac{1}{8} + \frac{1}{16} = \frac{3}{16}$ or $\frac{1}{5\frac{1}{3}}$. Hence, at $5\frac{1}{3}$ inches is the place where the image of the round hole is to be found for rays in this meridian. While at 8″ is the place of focus for rays which lie in the axis of the cylinder, and midway between these points the image of the hole or shade becomes circular, but with exceeding indistinctness of outline. When we hold the spherico-cylinder at 8″ with axis horizontal, we find the globe cast an image elongated vertically with sharp sides and hazy ends. At $6\frac{2}{3}$″ the image is round, but with hazy edges, while at $5\frac{1}{3}$″ the image is elongated transversely, is smallest, has its upper and lower sides sharp and its ends hazy. Between 8″ and $6\frac{2}{3}$″ is a vertical oval with hazy borders, and between $6\frac{2}{3}$″ and $5\frac{1}{3}$″ is a horizontal oval with hazy borders. To understand these phases, remember that when the combination is at 8″ the rays going through the axis of the cylinder, which is horizontal, find their focus at 8″, and in this direction they form sharp points ; but rays going transverse to the axis reach a focus at $5\frac{1}{3}$″, then cross and expand to a wide separation, which has the two-fold effect of stretching out the image in this direction and forming circles of dispersion, which are indicated in the fringed edges above and below. At $6\frac{2}{3}$″ which is the sum of the spherical lens and half the cylindric, none of the rays find a focus, but half are diverging to a degree equal to the angle formed by those which have not yet come to a focus ; of this the effect is a circle with fringed edges. At $5\frac{1}{3}$″ rays transverse to the axis find a focus, while those at 8″ have not yet come to a focus, and, being in a state of convergence, form circles of dispersion which show themselves in a horizontal oval whose upper and lower sides are clean, but whose ends are fuzzy. Of course the image is smaller than when at 8″, because the refracting power is greater.

If now to $+8s, -16c$, with axis horizontal be added, the first focus will be at $\frac{1}{8} - \frac{1}{16} = \frac{1}{16}$″ and then at 16″ we have an image transversely oval with upper and lower sides sharp and ends fuzzy because the cylinder permits rays having 16″ focus to come to a point, and these lie in the direction transverse to the axis, while the opposite rays crossing at 8″ diverge to form circles of dispersion which make fringes on the end of the figure, and necessarily elongate it in the horizontal plane. As we advance to 8″, similar changes take place as in the previous experiment, but in an opposite sense. A +cylinder has its shortest focus for rays which are transverse to its axis, and this is positive, while for planes oblique the focus grows longer until in its axis no focus is possible. A −cylinder has a negative focus which is shortest in a plane transverse to its axis, for oblique planes it grows longer, and the plane of its axis has no focus.

If an emmetropic person put on a plus cylinder with axis transverse, he is made myopic in the vertical plane ; a point at which he looks is stretched out to a vertical line ; a **vertical line** has sharp sides and blurry ends ; **a** horizontal line has ends sharp, **and** top and bottom blurry. If he put on a —cylinder with axis transverse, he is made hyperopic in the vertical plane, and a point is again drawn out in a vertical direction, because rays in this plane touch the retina before crossing, and make circles of dispersion ; a vertical line has fringed **ends and** sharp sides. The above facts will explain some of the symptoms **observed** among astigmatic persons. We have to distinguish always between meridians of greatest and of least refraction, **and** we speak of the axis of greatest **error.** We find this either vertical **or** horizontal, or in any obliquity. **We find** one eye alone or both affected, and we find the axis in the two eyes similar or dissimilar ; usually they are somewhat symmetrical, but may be just **the** reverse. In designating the axis the nomenclature is nearly uniform, but **there** are exceptions. Most oculists accept the method of Javal, who **recommended** beginning the scale at the left side of the circle, and to go **around its** upper half from left to right, **as** the hands of a clock move. The place of beginning will be at nine o'clock, and the terminus at three o'clock. Some (Green and others) begin at twelve o'clock, and count ninety degrees each way to the right and left. I prefer the method of Javal. In writing **a description** of a case one is supposed to be in the place of the astigmatic person, and to be looking at the clock. Trial-frames are marked in degrees according to one or other of the above methods, with intervals **of** five degrees. Those made by Nachet, of Paris, I have found the most serviceable, and have had the scale put upon the front, where it may most readily be read. Concave cylinders have two lines drawn across their face parallel to the axis, and midway between the centre and the edge. **Convex** cylinders have the same lines, but in addition another mark is scratched **on** the edge at the end of the transverse meridian. For both concave **and** convex it would be better to scratch the end **of the** axial meridian **on the** edge, so as to gain in accuracy of notation.

To **make the** practical effect of astigmatism more intelligible, **because I** know **it is** thought to be a difficult subject, the following illustrations **may be** of use. We will **begin** with simple astigmatism : Ah, Fig. 32, supposes the eye to be hyperopic **in** the vertical meridian V, and emmetropic **in the transverse** H, in which **one** dot, H, gives the focus of one meridian **on the retina,** and of the other, V, in front of it. Simple myopic astigmatism, **Am,** Fig. 33, supposes one meridian to be emmetropic, and the other **myopic (where** one dot, V, is in the retina, the other, H, beyond it). **Compound** hyperopic **astigmatism,** H + Ah, Fig. 34, is thus shown: an eyeball **too short,** but with **one focus, H,** in the retina, and another, V, in front of it. **Compound** myopic astigmatism, Fig. 35: **an eyeball** too long, with one **focus in the** retina, H, **and the** other in front of it, V. **Mixed** astigmatism, Fig. 36 : in which one **focus is in** front of the retina, V, **and the** other behind it, H, while the axis **may** be too short or too long.

Among 1,500 cases of refractive **error, gathered in** private practice in **ten years,** I find 800 which have the error **in one or** both eyes. The *symptoms* **are** subjective and objective. **The subjective may** be insignificant and scarcely discoverable, but more **often they consist of** imperfect sight, not fully corrected by spherical glasses, **of an** inclination to hold fine objects too close, often of a disposition to **half** shut the lids for better sight, and of pain on use of the eyes. There is often deficient visual acuteness, but, for most cases, V may be made equal to $\frac{2}{9}$. Objectively, we have certain peculiarities of sight shown by certain tests, and we have the view of the fundus oculi by the ophthalmoscope. Some persons have been made aware

that they could see objects which were in certain positions better than
when in opposite positions, such as the bars of a window-sash or the masts
and spars respectively of a ship. It is common to hold print and work
quite close. It is exceedingly common to have spasm of A. Symptoms of
asthenopia, in other words, are frequent. There may be pain and blur and
fatigue in work.

Diagnosis and treatment.—The first point to be settled is whether the
fault is of the hyperopic or of the myopic variety. One must be watchful not

FIG. 32.—Ah: **Simple** Hyperopic Astigmatism.

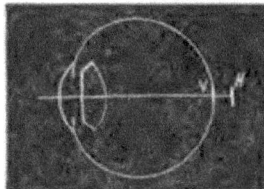

FIG. 33.—Am: Simple Myopic Astigmatism.

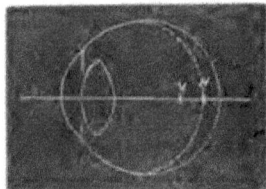

FIG. 34.—H+Ah: Compound **Hyperopic**
Astigmatism.

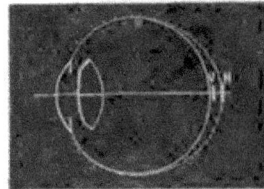

FIG. 35.—M+Am: Compound Myopic
Astigmatism.

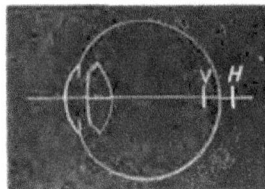

FIG. 36.—Mixed Astigmatism. **H and M equal.**

to be misled by the patient, who will, in many cases, insist on taking concave glasses when convex will be proper. In testing by Snellen's type at
20 feet or 6 m., vision being subnormal, the rule is to begin with convex
spherical glasses and find whether sight improves ; press upon the patient
the strongest convex which he will accept. If concave spherical must be
used, take the weakest, which permit the best vision. If acuity still be deficient, try a test-card for astigmatism. My favorite is one of Green's
series (see Fig. No. 37). A clock-dial is the background of all of his tests.
By means of the hours the situation of conspicuous lines can be described.
The patient wearing the spherical glass is called upon to say whether the
lines are all alike, and the following particulars are to be specified about
them : their color, distinctness, outline, length, and penetration to the
centre. Much deliberation and patience are to be exercised, and pains

taken to know whether the patient understands what is meant. Not having accurate sight, his discrimination is imperfect and his practical acquaintance with terms of precision is very dull. Hence, the same point must be pressed over and over again. Should a patient find some lines more distinct than others, learn the direction in which these lie and the number of them, and whether there be an equal number on both sides of the centre. The direction is readily described. The number of lines, if few, indicates a decided degree of error ; if many, say five to eight lines, are distinct, the error is small ; if the distinct ones are not equal on opposite sides, the astigmatism is irregular and cannot be perfectly corrected. Next call attention to another diagram which has but two lines (see Fig. 38) and place

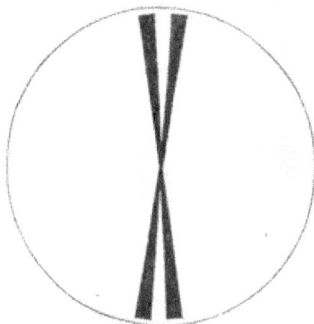

Fig. 37.   Fig. 38.

them in the direction of the lines declared to be most distinct. **Adjust** this with care to hit off the place where they appear best. Now **vary the** strength of the spherical glass to that degree which will confer **the best** sight of these lines. Ask whether they appear very black **and** distinct, whether they run down sharply to the centre, and especially whether the white space between them can be discerned to within one-half inch of the **centre.** This last question is the key to the inquiry, and it **may be** varied in **several** ways so as to excite the patient's attention and **quicken** his perceptions. When the spherical glass is found which gives **the** best view of **the lines** in the first position, next turn them 90°, and ask how **they** appear. **They ought** to be less distinct, and to a degree varying **with the** amount of **astigmatism.** Now increase the power of the spherical **glass** until in their **new** position the lines are as satisfactory as they were **before.** If, for instance, **in the** first position the lines were best seen with $+16$, and now are seen best with $+10$, the difference between the two expresses the astigmatism, viz., $\frac{1}{16} - \frac{1}{16} = 8 - 5 = 2\frac{1}{6}$, and this glass **as a plus** cylinder added to spherical $+16$, will make the card of 30 radii (Fig. 37) both distinct and uniform. It **will** be well to put up a card having couplet lines drawn **at right angles, Fig.** 39, and adding to the spherical the cylinder, let the **patient state whether** both limbs of the cross are alike and distinct. If he so pronounces them, try his ability on the card of test-letters. Visual acuity should rise and ought to reach the normal standard.

Another **test** may be tried if the result obtained is uncertain, viz.,

Pray's test-letters. These **letters** are composed of short strokes which run in different directions in the several letters. If by any correction the patient has already gained $V = \frac{2}{9}$, he will appreciate Pray's letters. **Some** of them will exhibit the strokes strongly and others faintly. This card is good for eliciting the principal meridians **and for** corroborating the findings **of** Green's tests. Dr. Thompson, of **Philadelphia**, has ingeniously adapted the experiment of Scheiner to this **work (see p. 11).** He puts at twenty feet a screen, perforated **by** a small **hole, behind which is a** light

Fic. 30.

—the **room** being **darkened.** Before the patient's eye he **holds** a metal **disc** perforated with **two** fine holes, **4 mm.** apart, **and if the diameter of the pu**pil **is big** enough, **puts the holes at 5** mm. Looking through **this at the dis**tant bright spot, which may **be 4 mm.** in diameter, the ametropic eye **sees** two instead of one spot. The disc is turned until the place is found where the spots appear most widely separated ; now a spherical lens, convex or concave, is put before the disc until the two spots unite to one. Whether to choose a convex or concave glass is decided by putting before one of the minute perforations in the disc a red **glass.** This makes one image red and **the** other white. If images are homogeneous the error is hyperopic and needs **a** convex glass ; if the images are crossed, **the** error is myopic **and** needs **a concave glass. (See** figure of Thomson's disc in " **Trans. Am. Oph.** Soc.," **and in Wells.)**

By a process **of** analysis as above described, **and by one of the three** methods indicated, the quality and degree of the error may be **ascertained.** But seldom is **a case** disposed of at one sitting. The first obstacle **is the** activity and pertinacity of the patient's accommodation ; consistency **is a** jewel which **these** patients have not found. They contradict themselves, and puzzle both themselves and the doctor, despite their anxiety to be accurate and truthful. They cannot help it, because they have always done a large part of their seeing by guess-work, and their sense of exact**ness** has never been developed. Therefore, it is a good rule to examine all **such cases** after getting the full effect of atropine. If for any reason this **is** inadmissible, repeated sittings must be given to verify or correct the **findings.** It often turns out that seeming myopic As becomes hyperopic.

Cases of mixed astigmatism are not excessively rare. They will be **elicited by** strongly insisting on the method by Green's cards, as I have **set forth.** For instance, if the meridian has been found by pushing the strongest convex glass which leaves any **lines** in the card of 30 radii distinct, **the** couplet of radii **has been** correspondingly placed, **and** we find that $+18s$ is the glass **which** makes it most clear, turn the couplet around $90°$, **and** now, instead **of** a convex glass, we need a concave to make the lines clear, and this may **be** $-9s$. (It is supposed that the patient still has on the $+$ spherical which **was at** first needful.) This demonstrates mixed astigmatism, viz., $+\frac{1}{4}$ in one meridian, and $-\frac{1}{8}$ in the opposite. The sum of these, viz. : $\frac{3}{8}$, is **the** measure of the astigmatism. We may correct the error vari**ously :** 1st, we may order a bi-cylindric glass, one surface convex, axis say

at 90°, and the other surface concave, axis **180°** ; or 2d, we may order +
9s with −18c. 180° ; or −9s with +18c. 90°. All of these come to the
same result, yet the bi-cylindric is theoretically the best because it gives a
larger and flatter field. To work out such cases atropine is indispensable.

While for most astigmatics the accommodation is an obstacle to the
examination, it is also the means of compensating for their error ; and while
it remains vigorous they often know no visual defect. To be sure, they do
not define objects accurately, but by using the middle of their focal interval they get on fairly. Moreover, if their error be hyperopic or myopic,
with axis nearly horizontal, in both which cases lines nearly perpendicular
are most distinctly seen, they get on moderately well, because most objects
with which we deal have greater height than breadth. Such is the case
with type (roman letters), and with trees, men, buildings, and the majority
of objects. True, objects are exaggerated in height, but of this the person is unaware. But if the axis is in the contrary direction, or if it be
oblique, or if the two eyes are unsymmetrical, trouble announces itself
early. An attack of illness, chronic uterine disease, excessive eye-work,
great grief, etc., will reveal astigmatic error, previously unsuspected. I
have also seen astigmatism, which **severe** uterine disease had brought to
view, retire into obscurity and **unconsciousness when** by an operation the
uterine lesion was cured. The **glasses which the patient** had with extreme
reluctance consented to, **were put away and satisfactory** use of the eyes regained without them.

Myopes with astigmatism, who **wear only spherical** glasses, sometimes
find that by tilting the glass forward they **see better** ; this increases the
refractive power of the glass in the direction of the tipping and proves
astigmatism. The habit of **myopes to** half **shut their** lids is often due to
astigmatism. I have seen a **case of** simple myopic astigmatism in which
the want of A existed which **belongs to** most myopes, and he was obliged
to use **OD−20c.** axis 95° **OS−24c. axis 90°** for distance, and OD +20c.
axis 185° **OS +** 24c. 170° for reading.

When astigmatics become presbyopic, they are often obliged to **wear**
two glasses, and the cylindric error must be respected in the glasses **they**
wear for near work.

To demonstrate to an audience the phenomena of astigmatism, I have used, with
much satisfaction, a suggestion of my friend Dr. John Green, of St. Louis, viz., to prepare suitable slides and put them in the magic lantern ; then to render the objective
asymmetrical by putting in front of it a weak cylinder. If the screen be at 40 or
more feet distance, a lens of 60 inches focus is strong enough ; **if at 20 feet distance,
36c. may be used.** By rotating the cylinder the phases of the **error may very** beauti-
fully be exhibited.

Besides the tests above detailed, others **have been** contrived by Becker
and by **Snellen,** and Wm. Wood & Co. have published **some** of them. The
stenopaic **slit** is theoretically a means of determining As, **but** in practice
it does not answer. Patients with high **degree of** As **should** be advised
to wear **their glasses** constantly.

Anisometropia.—This term signifies that the refractive state of the eyes
is dissimilar, and sometimes they are even opposed to each other (anti-
metropia). Almost **any** variety can be found of this condition, and all that
needs to be said is as to the correction which is to be adopted. The rule is
to fit each eye by itself, and attempt to make both combine. The attempt
may not succeed. Oftentimes there has never been any true binocular
vision, and the needful mental or cerebral education has never been gained.

To **fit** glasses however perfectly to each eye will not supply this cerebral deficiency. Again, one eye may be highly amblyopic, and while the glasses **may be** tolerated they do little good. Sometimes there is **a** strife or antagonism between the eyes when great refractive difference exists, which renders **a** combination unfeasible. But the only **rule is to** make the **attempt, and if the eyes** do not work harmoniously, to reduce the difference **between them to that degree** which becomes necessary. Of course the eye with least error, and **with** greatest acuity, must be the dominating one, and the less perfect correction **will fall** upon **the poorer.** I have corrected a **great many persons** with **this** dissimilarity, **and** find that a difference **not more than** $\frac{1}{10}$ **between the eyes can** often be easily tolerated in the correcting-glasses.

It may **be of interest to** give an idea how prevalent these refractive anomalies may be. **Of course** no absolute quantity can be **given** representing the proportion in the population, but the following figures are worth something. In my private practice I find **among** 1,500 recorded **cases** of refractive error the following distribution :

Hyperopia	493
Hyperopic astigmatism	387
Mixed astigmatism with prevalent **H**	31
Myopia	207
Myopic astigmatism...............................	354
Mixed astigmatism with prevalent **M**	28
	1,500

Two things **are** no doubt surprising in the above table : that H should so much exceed M, **and that** astigmatism should be so frequent. **The degrees** of As are all above $\frac{1}{8}$, and demanded correction.

It **is** a matter **of common** experience that a very large proportion **of** the troubles of **vision which are** now corrected by glasses were formerly designated by **other** names, and misunderstood. To patients a vast amount of discomfort, **of** disappointment, of hopeless privation, and of penury has thus been abolished ; while vexation, perplexity, and misspent medication are spared to the physician. The subject has become one demanding the best skill and attention of **the** physician and oculist.

DIAGNOSIS **OF AMETROPIA BY THE OPHTHALMOSCOPE.**

Helmholtz clearly perceived **that an** observer is competent to decide **what is the** refraction of an eye by the ophthalmoscope, and gave the theory **for doing it.** It is an advantage which is eminently practical, and of the **highest** value. It to a large degree liberates the observer from dependence **upon the** patient's assertions. Yet to enjoy the highest measure of this capacity **is not the lot of all** ophthalmoscopists, **nor to** the best is it permitted **to boast that for** every case a certain diagnosis can **thus be** made. Admitting some limitations, I nevertheless lay the greatest stress upon the value and trustworthiness of ophthalmoscopic determination of refraction. The observer will not gain his skill without special education of himself, and he must rigidly test his own accuracy before venturing positive opinions. The basis of positive knowledge is, first, that the observer be **able** to completely dispense with all his own accommodation as he inspects the fundus, or, if he exercise a part of his A, that this quantum be uniformly the same, and he knows it ; secondly, the patient, while being inspected, must fully relax his accommodation. This can be brought about by adequate use of atropia ; but without mydriatics it was long since observed that the eye under scrutiny does yield up effort of A to a most surprising degree. The glare of light, the surrounding darkness, the want of any object to be looked at, conspire to cause, in the vast majority of cases, entire abandonment of A. When this is not the case, it may be suspected by the want of perfect clearness in the fundus, or in the variations which are noted to occur under inspection. A remark is to be made as to the observer's competency to declare when the fundus is sharply seen. Every microscopist knows how greatly people differ in their ability to discern a field, and recognize minute details. The same is true of the fundus oculi, and it is hopeless to try to explain what sharp discernment is, except as one shall be able to compare himself with others of admitted accuracy, and upon clearly pronounced cases. It is usually said that the fine retinal twigs of the fourth or fifth divisions are sufficient tests. I find the perfect discernment of the pigment epithelium, which constitutes the texture and grain of the membrane much more correct than the vessels. An illustration may be had in the different look of the fundus, when there remains uncorrected hyperopic astigmatism of $\frac{1}{8}$, with axis 90°, and its clearness when the cylinder is introduced. I quote this as a fine test, and one which I have had reason to put in practice; but larger errors will permit fine vessels to be seen. The observer must not only have his own A perfectly under control, but he must know the state of his refraction, and count this in his estimate. The size of the patient's pupil is important, because, when very reduced, great accuracy is unattainable. For the same reason the perforation in the mirror must not be less than $3\frac{1}{2}$ mm., and the best diameter, as before stated, is 3.75 mm. (Loring).

When illuminated ophthalmoscopically, the emmetropic eye emits rays which are parallel, and cannot form an image ; the hyperopic eye emits rays which are divergent, and form a virtual image which is erect and visible ; the myopic eye emits rays which are convergent, and form a real image which is inverted and visible. The observer using a mirror of about 7″ focus, will first look from about 16″ distance, and the eye will thus be feebly illuminated. In the emmetropic eye nothing but the red glare of the fundus will be seen, no vessels are displayed because the rays do not construct an image, and the field is too narrow. With a hypermetropic eye, vessels are to be seen in a virtual image, and the divergence of the rays is what the degree of H imposes. If H be $\frac{1}{12}$, the observer, by accommodating for twelve inches, sees a retinal vessel, as he moves his head, and he notices that his moving to the right causes it to move to the right ; or, if to the left, it seems to move to the left. For H $\frac{1}{6}$ the observer must more strongly accommodate, and will see several vessels and they will seem smaller. With a myopic eye the observer will see an image of the vessels, if the M is strong enough to give an image so near to the eye as to be within the observer's range of A. For instance, M $\frac{1}{12}$ will give an image at 12″, which will be seen if the observer go far enough away to accommodate for it, say 8′ farther off, making 20″ from the eye. If M be $\frac{1}{6}$, a distinct image of vessels will appear, quite small and readily found, which as the observer moves to the right will go to the left, and vice versa. For aphakial eyes H is about $\frac{1}{4}$, and the divergence is so great as to make distant inspection difficult. For M more than $\frac{1}{6}$, a convex glass may be used to magnify the **image, which will still be inverted** and actual.

But the next step is to approach as **closely to** the eye as possible. To aid **in** doing this the mirror must **be small,** not more than **32** mm. in diameter, while the tilting mirror of Loring (**18 ×** 32 mm.), set in a frame 34 mm. in diameter, is better than a mirror **of the** full width. The lamp **or gas-**burner must be brought to a position near the mirror, and along-side **the** patient's head : the observer's nose **or** forehead will often touch the patient's **face ; one** may usually get to within three-fourths inch of the cornea. (See Fig. **15** page 25.) An allowance **of** one inch from the nodal point must always **be** made, or one-half inch **more than** is required for spectacles. It is important to have **the lenses not less than** 6 mm. in diameter, both to gain light and to obviate stenopaic **effect.** One is now ready to observe the fundus, and both eyes are **to** be kept **open,** and one's visual lines parallel, **objects** seen by the unused eye being disregarded. Absolutely no attempt **to** look at things by accommodating is to be allowed. If objects are **not clear,** and the preliminary view or testing by trial-glasses has indicated whether H or M is present, bring up the proper convex **or concave** glass until **a** good view is **obtained.** The blood-vessels near the **disc** toward the macula **will be** the **first** tests ; but, as **a** closer accuracy is gained, the epithelium must be sought **for.** If one slightly move up and down, **or** across the pupil so as to look through different portions of the crystalline lens, it will sometimes be **noted** that the refraction is not in all parts the same. The high degrees of H and **M** are difficult to study. With H above one-fourth the magnifying power is small ; with M more than one-fourth the magnifying power is large and the illumination feeble, **because** of the great dispersion of light by the correcting lens. **In all cases the** macula is most difficult **to explore, while** it is at **this point especially that** we need to know what is **the refraction.**

Certain important pathological conditions **are revealed by the employ**ment of the ophthalmoscope **as an** optometer. **We are enabled to measure** the depth **or height of an** object in the eye by knowing the number **and** nature **of the glass with** which we can view it. Such, for instance, **is the** depth **of excavation of the nerve** in glaucoma, the height of a tumor, **the** elevation **of a detached retina, the** position of a body floating in **the** vitreous. For instance, we **find the edge of** a glaucomatous cup is to be seen with $+$ 24 ; its bottom requires -16 ; the depth of the pit is $\frac{1}{24} + \frac{1}{16} = \frac{3+2}{48} = \frac{1}{10}$. By referring to the table on **pages** 58 and 62, we find that H $\frac{1}{24} =$ shortening of axis of 0.45 mm., while myopia $\frac{1}{16}$ means lengthening of axis of 1.17. The depth of the cup then equals $0.45 + 1.17 = 1.62$ mm. On the other hand, swelling of the optic nerve in neuritis may permit $+8$ for its summit, and **the** eye is emmetropic. A shortening of the visual axis of $\frac{1}{8} = 1.25$ mm. which measures the amount of swelling. The **same** principle applies to all **other** cases above cited, and by it we are able **to** give precise data in the **facts** and progress of a case. An interesting case **was** one of myopia of -7 **D** $(-\frac{1}{7},)$ which gives elongation of axis of 2.07 mm. In the eye there was detached retina, whose **conspicuous** part or summit was seen by $+7$ D or $+\frac{1}{7}$, which means shortening of axis of 1.6. The true elevation of **the** retina then was 3.67 mm.

In cases of **this kind the** inverted image **has** some value ; **if we** move the objective lens **from side** to side the parts of the object which are highest, and those which are lowest, will not move to an equal degree ; in other words, their parallax will be unlike, and they appear to be displaced unequally. The top of a swollen nerve is nearer than its bottom, and the motion of the objective lens causes its image **to** have less excursion than its bottom. The same thing, to a less degree, can be exhibited in the direct image by moving one's head. So, too, with a glaucomatous nerve, The vessels on the edge of the disc, as the lens is moved up and down, move in front of and faster than do those at the bottom of the nerve. For explanation, see the diagram (Fig. 40) from Abadie. Let *b* be the edge of

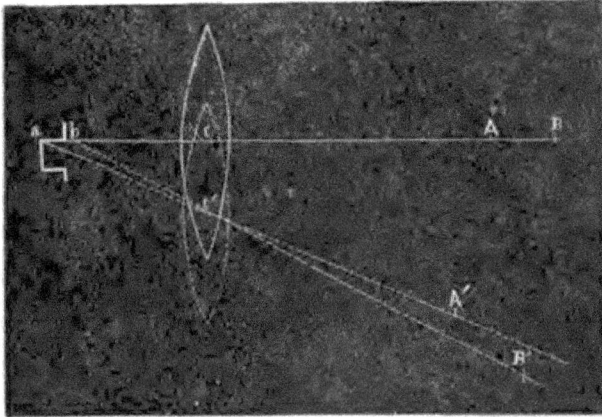

FIG. 40.

the excavation, and *a* be at its bottom, and the image of these points along the axis of the lens will be at B and A respectively; cA is of course shorter than cB. When the lens is moved down, the points A and B are displaced to A' and B' as seen in the figure. They are no longer in the same line, because the surface of the excavation, *a*, *b*, presents itself differently to the lens. The point B' moves faster and farther than the point A', and passes in front of it, because c' A' and c' B' become, as it were, radii of arcs of circles.

Diagnosis of astigmatism by the ophthalmoscope is founded upon the principles already given. The first consideration is to learn the direction of the meridian of greatest ametropia, because in this the retinal vessels will be most distinct. We find if we view Green's 30 radii through a minus cylinder, with axis horizontal that the vertical lines are distinct because in the vertical meridian we are made hyperopic; if we use a convex cylinder, axis horizontal, the vertical lines again are most distinct, because the vertical meridian is myopic. Now, in examining the fundus, if we find that fine vessels in the horizontal meridian need no glass for distinct perception, while fine vessels in the vertical meridian need a plus glass, we have simple hyperopic astigmatism. If we need a plus glass for any vessels and a stronger glass for other vessels, this betokens compound hyperopic astigmatism. In the same way we recognize simple myopic and compound myopic astigmatism. The degree is the difference between the two meridians. The rule in examining hyperopic eyes is to use the strongest convex glass which is available, and, in examining myopic eyes, to use the weakest concave glass. Now, in astigmatic eyes the effect will be that a streaky appearance is produced, and the streaks will run in the axis of the greatest ametropia; of course the least ametropia will be at right angles. Moreover, it will be impossible by any spherical glasses to gain a clear view of the fundus. So noticeable is this fact that one is incited to examine for haziness of the vitreous, or erroneously led to think that the retina is infiltrated with inflammatory effusion. Such an error is obviated by finding that visual acuity by proper correction is satisfactory, and if the proper cylindric glass can be attached to the ophthalmoscope out of the trial-box, obfuscation of the fundus vanishes. I have provided the means of doing this in the ophthalmoscope figured on page 29, and gain the advantage of learning, first, that the deep ocular structures are or are not healthy; and second, that the finding by the trial-glasses is or is not correct. Another feature in astigmatic eyes is that the optic disc is no longer circular; it is elongated in the direction of greatest ametropia, and therefore is oval. The nerve may be misshapen anatomically, presenting a distinct oval, the long axis usually more or less vertical. In such a case the retinal vessels will show no difference of distinctness caused by their various directions.

By the inverted image the streakiness of the fundus can well be seen in high degrees of As, but the lines run in directions opposite to their course when viewed by the upright image. So too the oval of the optic disc is reversed. But the objective lens must be held from the eye at a certain distance. It has already been said that in the upright image the optic disc is elongated in the direction of the meridian of greatest curvature, because the magnifying power is greater. With the inverted image the elongation corresponds to the weakest meridian. With the emmetropic eye the size and form of the optic disc undergo no change in the inverted image when the objective lens is held nearer to or farther from the eye. With the hyperopic eye, when the objective approaches it the optic disc becomes smaller, and grows larger as the objective recedes. With the myopic eye, when the objective approaches it, the optic disc becomes larger and grows smaller as the objective is held farther away. In both H and M the shape of the disc remains round or oval, whatever the distance of the objective. But with astigmatism the location of the objective changes the size and the shape of the disc. If with the lens near to the eye the disc be vertically oval, the disc becomes circular if the lens be held from the eye a distance equal to its focal length, plus the distance of the anterior focus, viz., half an inch. If it be drawn farther away beyond its focal length, the direction of the axis of the oval is reversed. These phenomena have been elaborately studied by Javal and by Giraud-Teulon, and can be utilized in diagnosis, but the upright image is by far the most available and instructive.

In the above remarks it has been assumed that the observer is emmetropic and can perfectly relax his accommodation. If he be ametropic, he may have a personal correction attached by a clip to the ophthalmoscope, or he must calculate it in the lens which he uses for examination. For instance, a myopic observer will, if not corrected, see a hyperopic eye with a less powerful convex glass, but, for a myopic eye, he will need a stronger concave glass than the patient. A hyperopic observer, if not corrected, will need a less concave glass for a myopic eye, and a more convex glass for a hyperopic eye. Another matter must be mentioned, viz., that the correcting lens in the ophthalmoscope is not at the same distance from the eye as it is when in a spectacle frame. In the latter case it is about one-half inch or less; in the ophthalmoscope it may, by skilled observers and with the best instruments, be brought to five-eighths of an inch, but is oftener at seven-eighths of an inch. This increase of distance must be added to the power of the concave glass and subtracted from the convex glass used in testing vision. Quantities so small are not important practically, but if the observer cannot work with his ophthalmoscope nearer than one and one-half or two inches, the difference may require recognition. An observer who cannot wholly relax his accommodation must endeavor to establish a fixity of habit in the amount he uses, and make it a factor in the problem he is studying. Every consideration is in favor of the employment of the upright method in refractive ophthalmoscopy, and it is worth patient and persevering labor to acquire accuracy and facility with it. The indirect method is of comparatively little value for this work.

DISEASES OF THE MUSCLES OF THE EYE.

Anatomy and physiology.—We have for each eye six muscles. They are combined in pairs, and both eyes are co-ordinated in particular ways. In each eye we have the internal and external, the superior and inferior recti, and the superior and inferior obliqui. All the recti muscles take origin from the apex of the orbit around the foramen opticum, and come forward to be inserted into the sclera in front of the equator

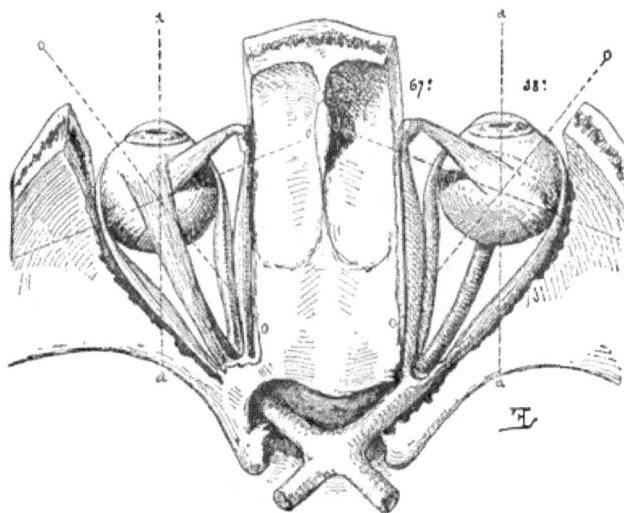

Fig. 41.

oculi, about 7 mm. behind the rim of the cornea, by flat and ribbon-like tendons. The superior oblique also originates at this place ; but, inasmuch as it passes over a pulley at the supero-internal angle of the front of the orbit, this becomes its functional place of origin and assimilates its action to that of the inferior oblique, which anatomically arises from the inner part of the inferior edge of the front of the orbit. Both muscles then pass obliquely outward and backward to wrap around the globe

G

in a thin, fan-like tendon, the superior going over the upper part of the
globe beneath the superior rectus, and the inferior going over the inferior
part of the globe beneath the inferior rectus. The two obliqui hold the
globe, as it were, in a sling, which is entirely to the outer side of the optic
nerve. The recti, combined in action, retract the globe into the orbit;
the obliqui, combined in action, draw it forward. While the recti in
combination have a simple kind of action, the obliqui draw the globe
forward and turn the cornea outward. The rectus internus (called by
Merkel rectus medialis) and the rectus externus, move the globe about an
axis which is vertical. The rectus superior and the rectus inferior move
it about an axis which is transverse to the vertical plane, but which is
also inclined so that its outer end is more posterior, making an angle
of 67° with the antero-posterior axis of the globe. The axis about which
the obliqui rotate the globe passes from before backward and inward
on a horizontal plane at 35° with the antero-posterior axis of the globe
(see Fig. 41). The obliqui thus acquire an action which moves the eyeball so that the rim of the cornea turns like a wheel. Taken singly, the
muscles act as follows: the **rectus internus turns** the cornea inward on
the horizontal plane; the rectus externus turns the cornea outward on the
horizontal plane; the **rectus superior turns** the cornea upward and
slightly inward; the **rectus inferior turns** the cornea downward and
slightly inward; the **obliquus** superior **turns the cornea downward** and
outward, and rotates it from above downward. **The obliquus inferior turns
the cornea upward and** outward, and rotates it **from below upward. In
effecting the movements** of the eyeball all the **muscles co-operate: while
some** predominate, **the** rest antagonize them **to** give **steadiness to the
action.** Moreover, **the** eye requires such precision and delicacy in its movements that the muscular adjustments must be perfectly balanced and exact.
Our judgment of **the** position **of** objects and of our own relations to our
surroundings is founded upon **a** nicety of balance, a harmony and precision
of working which demand the most perfect co-ordination of the ocular
muscles; we rely on their action so implicitly that if there be any disturbance among them we are exceedingly troubled: we cannot walk **or grasp**
objects: **we** become giddy or nauseated, **or** suffer pain.

The following schedule indicates how the muscles co-operate in effecting certain principal movements. For *motion inward, i.e.,* adduction, the
effective muscles **are,** R. interni **and R. sup.** and R. inf., antagonized by R.
externi and Obl. sup. and Obl. **inf.**

Motion outward.—Abductors: R. externi, **Obliq. sup. and Obliq. inf.,**
antagonized **by R.** int., R. sup., and R. inf.

Motion upward.—R. sup. Obl. inf., and when **the cornea** passes a given
point, the upper fibres of R. int. and R. **ext. add to** the effect. The
antagonists are R. inf. and Obl. sup.

Motion *downward.*—R. inf. and Obl. **sup., while,** when **cornea** gets
below a given point, the **lower** fibres of **R. int. and** R. ext. come into
play. **In** motions upon **a** horizontal axis **the R.** sup. and inf. incline the
top of the vertical meridian respectively inward and outward, which tendency is counterbalanced by the Obl. inferior acting with the R. sup. and
by the Obl. sup. acting with the R. inf., which perform the needful rotatory
or wheel movement.

Taking, now, the concomitant action of the eyes, we arrange the muscles
into groups of adductors, which turn the cornea toward the median
plane of the body; **and** of abductors, which turn the cornea away from
the median plane **of the** body. We also **find** the muscles pairing off in

other combinations in turning the eyes to the right side and left side respectively, and in the various diagonal directions. The eyeballs are capable of accomplishing, within a limited range, all possible combinations, but there are certain restrictions which are imposed by the necessity that each must direct its visual line exactly upon the object observed. Hence, for an object near, whether on the median plane or away from it, there must be a slight adduction, as well as an aim suited to the position of the object. For remote objects there will be a degree of abduction, but never to transcend parallelism of the visual lines. This may seem otherwise if the position of the eyes be estimated by the corneæ, because their axes vary according to the value of the angle alpha (see Fig. 42). We are habitually more concerned with objects below the horizontal plane than with those above it; and in the discussion of the movements of the eyes, Meissner has taken an inclination of 45°

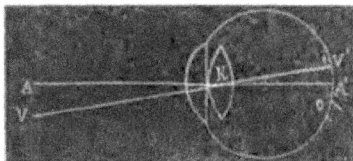

FIG. 42.—Right eye—A, A', the visual axis; V, V', the axis of the cornea; A, K, V is the angle alpha; K is the common nodal point.

below the horizon as the primary position; others, however, assume the horizontal plane as the primary position. The degree of mobility of the emmetropic eye in young persons about a vertical axis is from 42° to 51° inward, and from 44° to 49° outward (Donders). In myopia this is much restricted.

The horopter is a line which represents the curve along which both eyes can join in sight, and it is formed in this way: as the eyes fix upon a given object far to the left side, and move far to the right at the same inclination of the visual lines, they form a triangle whose apex, as it passes from left to right, forms the horopter curve for this plane. If the movement be in any other plane, vertical or oblique, the horopter will be formed in the same way for that plane. In its simplest form, as explained by Johannes Mueller, it is a circle which passes through the centres of rotation of each eye and through the apex of the point of fixation of the visual line. This statement is not strictly correct, but will suffice for our purposes. The subject is a bit of highly complex mathematics.

The fundamental and imperative law which governs the muscles of the eyeballs is that the fovea centralis retinæ of each eye must be fixed upon the object observed. When this is done, all objects lying in the same horopter will form images upon the respective retinæ which will lie at equal distances from the foveæ, and will, therefore, be appreciated as single, giving what is called binocular vision. But objects beyond the horopter or inside the horopter will cast images on parts of the retinæ not equally distant from the foveæ, and will therefore not be appreciated as single, but create the impression of two objects, giving rise to double vision. The maintenance of correct binocular vision is the necessity which dominates the ocular muscles. If the back of the eyeballs be divided into quadrants by vertical and horizontal planes whose intersection shall be at the fovea centralis, and along these lines we mark points one-tenth of a millimetre asunder, and then suppose the two retinæ to be superimposed upon each other so that the vertical and horizontal lines shall exactly coincide, the points which we have imagined will of course also coincide. These coincident points, which are equidistant from the centre, are spoken of as correspondent points of the two retinæ, and there are of course as many of them as there are percipient points, i.e., bacilli in the two retinæ. They are functionally homologated together according to the scheme of arrangement just imagined. By virtue of this arrangement binocular vision is rendered possible. It follows, of course, that the nasal half of the left retina is linked with the temporal half of the right retina; the nasal half of the right retina with the temporal half of the left retina; the superior halves of both retinæ are linked together, and the same is true of the inferior halves. Another name for binocular vision is stereoscopic vision, or the sense of depth. The law of correspondent points of the retinæ was once held to be absolute and invariable, but Wheatstone first showed, when he constructed the stereoscope, that this cannot be true. Further study has resulted in the establishment of the view that while this law is valid in what may be called a coarse sense, it is not maintained with mathematical exactness, and that the two retinæ in individual points are not physiologically identical. On the contrary, the sense of depth is elicited by a slight non-identity of retinal impressions. That is to say, an object appears to us to be solid when it forms upon each retina an image which is slightly unlike the one from the other. This is so in vision of natural objects, because the inter-pupillary distance causes each eye to view objects which are not far away, from different positions, and hence each sees parts of the object which are invisible to the other. For example, a chair stands in the middle of a room: one eye sees more of one of its sides than of the other; the other eye also gets a slightly different view of the chair, each eye having its own picture; two impressions are transmitted to the brain which are the same for the parts of the chair nearest us, but for the more distant parts pictures are formed which are perceptibly dissimilar. This identity in certain parts and dissimilarity of other parts of two pictures received by the brain, causes the mind to interpret the impression as coming from a body which has the three dimensions of extension, and the unlikeness in the two images awakens the idea of depth. Moreover, in viewing the chair we get a view of the floor and of the more remote objects, and of the walls of the room. To each eye the picture is different, the surrounding objects being in part to one eye concealed, and to the other disclosed. This unlikeness in the retinal pictures evokes the sense of depth or distance, and informs the mind of the relation which the chair bears to its surroundings, i.e., gives the idea of distance, correlation, and space. The stereoscope does this for us by artificial contrivance. There are two pictures taken from different points of view, and these are brought to fall upon each eye at corresponding parts, from any close distance, say six to eight inches, by means of convex prisms whose bases are outward and whose convexity produces a focus of six or eight inches. It is true that of the two pictures before us the instrument occasions three to be formed, but two of them are excluded from view by interposing a screen which obscures the unnecessary one from each eye. If one take away the screen, three pictures are seen; the outer ones appear flat, the middle one alone has depth. There is an act of convergence of the visual lines, the accommodation is aided or superseded by the convex lenses, and the concurrent dissimilar images convey to the mind the sense of depth. The same effect is obtained by the reflecting stereoscope of Wheatstone.

So much has been devoted to explanation of the stereoscope because it becomes a means of investigating an important function of sight. It serves not only in the detection of malingering, but it is one means of testing patients in whom the capacity for binocular sight is wanting or partial, or maintained with difficulty. We use the stereoscope in cases of strabismus, in cases of insufficiency of the muscles, and in diplopia. It serves to detect errors and to expose both intentional and unconscious deceptions. In the above remarks it is not intended to intimate that no other factors enter into the perception of sense of depth ; such are in reality perspective, the relative size of objects, the rapid glance from point to point, etc., but they do not belong to our subject.

The limitations of binocular vision evidently exclude those motions by which the visual lines are not directed in proper harmony with each other. That is, one eye may not look up and the other at the same time look down, and the same is true of diagonal movements—they must be harmonious. But in the horizontal, or nearly horizontal plane, the motions of adduction and abduction can be pushed to an extent which shall disharmonize the visual lines. Thus, we may turn the eyes inward so that the visual lines cross by excessive convergence, and we may turn them by abduction so that they fall into divergence. Excessive adduction is possible without artificial aid to a marked degree, excessive abduction is never great, and cannot usually be effected without the aid of prisms. Done by their help, the movement is made to prevent double sight. Thus, if we look at a candle-flame twenty feet away, and put before one eye a prism of five degrees angle, with its base inward, there will for a moment seem to be two candles, but presently they move toward each other and fuse into one. The eyeballs go asunder by a movement of abduction to bring the fovea of one eye inward to the spot on which the prism has deflected the candle-flame. This power of abduction beyond parallelism reaches to a prism of from three degrees to eight degrees in most persons, while adduction for distant objects, say at twenty feet, extends to twenty degrees or to fifty degrees, and for the average of persons who have not cultivated their power it is about twenty-five degrees. If an object come near, adduction increases rapidly, being aided by association with accommodation.

The acquisition of binocular vision belongs to the first months of life. Young infants roll their eyes about in the most inconsequential fashion, and when their visual vagaries give place to binocular fixation an important step has been gained in ocular and cerebral development. In some subjects this function is never acquired, in others it may be lost after having been presumably acquired. All cases of permanent strabismus are instances of suppressed, or of lost, or of undeveloped binocular vision.

Binocular vision is primarily conditioned by the supremacy in acuteness of the fovea centralis above other parts of the retina. But this condition is not the only factor in the function, because experience shows that the brain must possess a certain competency, which sometimes seems to be the quality deficient.

For all objects on which we do not fix the foveæ, and which consequently are not in the horopter, we have diplopia. If, for instance, in one hand a pin is held at sixteen inches, and in the other another is held at eight inches, and upon the same line, when we look at the distant pin the nearer is seen double, and vice versa. We are not disturbed by double vision of objects on which we are not fixing attention ; the mind ignores the impression of the things with which it does not concern itself. This is common experience in shooting, in using the microscope, when the unused eye may be wide open and nothing be known of what it sees. Diplopia follows certain laws. For instance, if the left eye fix on an object, and the axis of the right cross that of the other at a point inside the object—in other words, if there be excessive convergence, the image which in the left falls in the fovea, and whose position in space is projected along the visual line, this image will in the right eye fall to the inner side of its fovea. Now, the position of the object in space is decided by what the left eye sees, and the right eye has an image which, if it were directed aright, would belong to an object situated to the outer side of its visual line, viz., on its right-hand side. This image is recognized, and is mentally located as if it were on the right side of the object seen by the left eye. In other words, if there be excessive convergence of the visual lines, there will be diplopia with correspondent or homonymous images (see Fig. 43). If, now, while the left eye fixes an object,

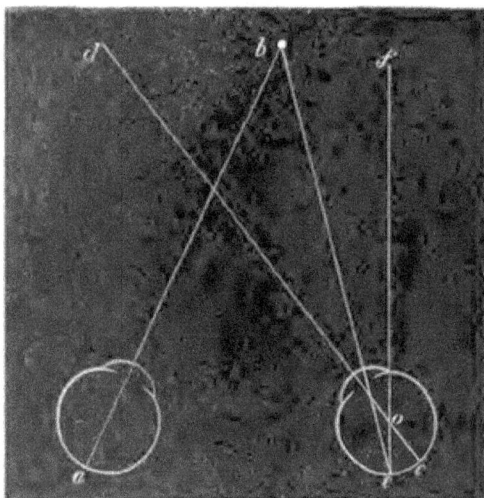

Fig. 43.

the visual line of the right diverge, the image will in the latter fall to the outer side of its fovea, and will be projected mentally as coming from an object on the left side of the visual axis. Hence, for divergence of the visual axis we have crossed or heteronymous double images (see Fig. 44). If the left eye fix an object, and the right eye be directed downward, the image in the right will fall below its fovea, and be mentally projected above the image seen by the other eye. If the left eye fix and the right eye be turned upward, the image will fall above its fovea, and the projection of the image will be downward below the true place of the object. A candle-flame is usually the object chosen, and a red glass is put before the fixing eye so as readily to distinguish the presence and place of double images. If the visual lines form a wide angle with each other, in the deviating eye the image will fall at a great distance from the fovea, and the result will be that the image is less distinctly perceived because it impinges on a less sensitive part of the retina, and it will also be projected to a greater distance from the true place of the object. For these two reasons, the patient is then less likely to be aware of double images. It also happens that persons may fix with either eye and ignore the image of the other. It is also true that in many persons, and by some it is asserted that in all persons, one eye prevails over the other, just as one hand is more depended on than the other.

DISEASES OF THE MUSCLES.

We have weakness or paresis, paralysis, and spasm. One muscle, or several, or all may be affected.

1. *Paresis.*—Weakness or insufficiency of muscles will not ordinarily exhibit any discernible deviation of the visual axes, that is to say, strabismus or squint ; but, because by using certain devices, or under certain conditions the eyes will fix discordantly, their condition was called by von Graefe dynamic (potential) squint. A better term would perhaps be latent squint (Alfred Graefe), as distinguished from the manifest form. In most instances there is no double vision ; or, if there be, the person is unaware of it.

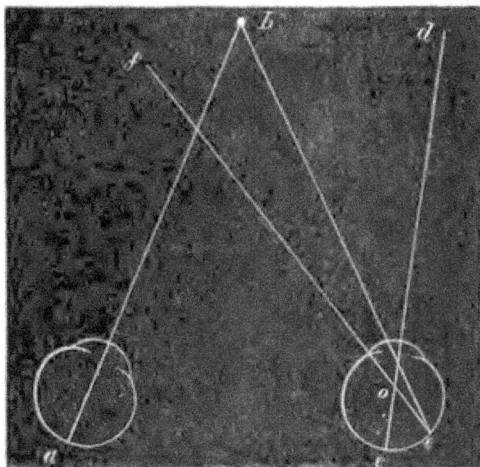

FIG. 44.

The most common forms of muscular insufficiency concern the recti interni, or the recti externi. Much less frequent is insufficiency of the muscles which turn the eyes up or down, while of these the superior oblique is most often singled out. The most frequent cause of these troubles is errors of refraction, and in a general way weakness of adduction goes with myopia, and weakness of abduction goes with hyperopia. Many exceptions, however, are noted. But other frequently operative causes are overtasking the eyes, depreciation of general health by chronic or by acute disease, pressure on nerve-twigs by inflammation or thickening of their sheath, by growths, by injuries, or by congenital disorders ; heredity is not infrequent.

If one or more muscles be of reduced power, it does not show symptoms of debility until its capacity has been exhausted ; hence, such a condition may be present and not be discovered until the period of exhaustion arrives. Hence, too, when the state of enfeeblement has been discovered, its full measure will often not at first be appreciated ; give due opportunity for disclosure, and what at first seemed to be a moderate degree may at length declare itself in much larger proportions. Moreover, the weakness of a muscle is specially, although not exclusively exhibited in those movements which are peculiar to its function. For movements in which it takes no important part its deficiency may have no appreciable influence.

The most important subjective symptom of muscular insufficiency is pain occurring after a shorter or longer period of use. Seldom is there blur of sight, while swimming of letters or work, which is caused by a tendency to and resistance of diplopia, resembles the blur of refractive error, and is sometimes thus spoken of. **The** pain is generally **in the eyeballs and inclines the** patient to press on **them** for relief; but a **most important fact** is that pain is often **temporal, frontal,** or at the vertex. **In truth, not a small** percentage of **obstinate headaches** originate **in disorders of ocular muscles.** Sometimes **dizziness occurs and nausea. It has been asserted** that chorea is caused **by this condition, but my observation has been that** the order is otherwise, and **that it is chorea which gives rise to debility** and irregular action of ocular **muscles as one of its manifestations.** Epilepsy has been asserted **to depend on** this cause. I cannot deny that **in** a few cases such a relation may have been established as an exciting **cause** or occasion, but there has been behind **it a** deeper lesion of **the** general nervous system. This statement is quite consistent with the disappearance of epileptic fits in **case** the muscular eye trouble **is cured.** That chorea can be cured by **relieving** the ocular **trouble** has not **been verified** in my experience, although I have carefully noted several cases.

The objective symptoms **of muscular** debility **are** as follows: while **both eyes** within certain **limits seem to move in** harmony, for **certain extreme** positions to **the right or left, or up or down, they** become tremulous **or one will** deviate; **so, too, in the median line and at** moderate distances **there may** seem **to be no fault, but if an object be** brought very close to **the nose a** deviation may **occur. If no evidence is** thus obtained, repeat the **same** experiments, bidding the patient regard **the** finger as it is carried to **various** extreme positions, and while he fixes upon it, put a card before one eye **and** note whether, **when** the necessity for binocular sight is thus abolished, **the** covered **eye** may not allow itself to deviate from **its** correct position. Especially useful is this test when searching for weakness of adductive muscles, the finger being brought within a few inches of **the** nose and each eye alternately covered by the card. Graefe suggested **the** use of prisms to cause vertical diplopia with separation of images, **so wide** that their **fusion should be** impossible. **It** follows **that in many cases,** perhaps the **majority, the eyes, no longer co-operating, take** the position most natural **to each, and thus disclose insufficiency either of** adduction or of abduction. **His method was to put before one eye, say the** left, a prism of $10°$ with its **base vertically upward, and present as** an **object** a dot with **a fine** line drawn vertically **through it. The** effect is **to** make the patient **who** views this at 14", **or at whatever** may be his usual reading **distance, see** two dots. **If, despite the vertical** diplopia, both eyes maintain the same angle of fixation, **there will appear** two dots and a single, **but elongated** line. But if **one eye deviate, there will** appear two separate **dots and two** separate lines, **both of the** same size. **In** the latter case, if the **dot seen by** the right eye **be not** only higher, **but to the** left of the other, **the** images being crossed, indicate insufficiency of adduction, *i.e.*, of the recti interni. If, however, the image seen by the right eye, being above, is also **to** the right **of the** other, **the** images are homonymous, and indicate weakness of abduction **or of** the recti externi. In this test the presence of the line tends **to instigate** efforts at fusion, and I therefore use the *dot alone,* and prefer **a white dot on** blackened cardboard, which has a slight notch cut in the upper **and** lower edge, vertical with the dot, to mark the perpendicular, but **which can** be no temptation to efforts of fusion. In this test, and in **various other** tests which **are** made respecting the functions of sight at short range, I employ a simple arrangement which saves time and trouble, and is more convenient than the spectacle-frame of the trial-case, which would otherwise be employed. The idea is taken from a stereoscope. There are three cells before each eye, into which glasses may be dropped, and the central stem, which is $20''$ long, is graduated in inches, beginning at the patient's forehead. This does not supersede the trial-frame, when glasses are being chosen for reading, because it places the glass farther from the eye than a spectacle-frame would. Squared prisms are better than circular ones, because their position can be told at a glance. While great advantage is gained by Graefe's test, it is not true that the latent insufficiency is always thus brought to view. The accommodation is undisturbed, and if there be spasm, the full measure of error of the motor muscles will not appear. In cases of myopia this criticism scarcely applies, but it does hold for some cases of emmetropia, and still more for hypermetropia. This test naturally applies only to errors of adduction and of abduction. We find cases in which the weakness obtains only in the vicinity of the eye, and not for distances remote. Especially is this the case with myopia and emmetropia, and the error will be chiefly of adduction. But it may concern remote as well as near vision. It would seem that for deficient abduction the error should be most emphatic at a distance; but this is not true in the sense which would be expected, provided, as is often the fact, there be hypermetropia. The degree of H need not be large, yet with marked debility of abduction the accommodation and the instinct for single vision may so stimulate the externi that a very small or no defect may appear at twenty feet, while at $14''$ a decided homonymous diplopia may be elicited.

But, in testing for muscular errors, it is important to eliminate the accommodation as we do in testing refraction. It is best to moderately darken the room and place a lighted candle at twenty feet. As he fixes on this, put a red glass over one eye—say the patient's right; the flame may remain single and its color be a mixture of red and yellow. Then put over the other eye a prism of $5°$, with base vertically upward. Two flames will be seen; if the red, which belongs to the right eye, stand directly above the white, there may be equilibrium; but if it go to the left, we have insufficiency of adduction; if it go to the right, we have insufficiency of abduction. But if the lights stand vertically, examine further; move the candle several feet to the right or to the left; or, what comes to the same result, bid the patient turn his head far to the right and then to the left, still fixing on the flame; in these lateral positions want of perpendicularity may be detected, and thus the faulty muscle be found out. This applies especially to the discovery of error of the externi. The next step is to measure the capacity of the muscles at twenty feet. First, the abductive power: hold before one eye a prism of $5°$ axis horizontal and base inward. At first two flames will be seen, which quickly merge into one. If they do not, weak externi may be strongly suspected; but persons with less abductive power often complain of no trouble. If the abductive power go above $8°$, it will be unusual, and generally indicates error of adduction. Next put a prism with base outward before one eye; begin with $5°$, then $5°$ before the other; then substitute $10°$ for one $5°$, go on adding to the total until the patient is no longer able to fuse the images. He will require time and deliberation to learn the manœuvre, and will do better at the second sitting than at the first. Many are much fatigued by the process, and some complain of pain.

In connection with these tests, proper regard must be had to the state of refraction, and when an error is present, the correcting-glasses must invariably be used for myopic cases; and for hyperopic, the test is to be taken both with and without them. The same remark applies to cases of astigmatism.

By the methods indicated, applied with care, and with sufficient repetition to enable the patient to understand what is wanted, and to put forth his best efforts, no great trouble will be needed to get the true state of adduction and abduction.

It will happen that a small number of cases exhibit symptoms of muscular asthenopia, which do not betray any notable fault of the externi or of the interni, nor, if there be error of refraction, does its correction remove the symptoms. In such cases search must be made for error in the action of the muscles which move the globe up and down. To find this needs rigorous inquiry. Darken the room and light the candle at twenty feet, apply whatever refractive correction is needed, then use a prism whose base is inward, which shall cause unconquerable double images—it may be of ten degrees—and be very careful that its axis is perfectly horizontal (for this test the advantage of a squared prism is evident). If now the two images are not on a horizontal line, inquire which is the higher. The images are crossed images, because the visual lines are by the abductive prism made divergent, and if the left-hand image is above the place of the real light, the fault lies with the right eye, and vice versa. A few trials of this kind, with variation from one eye to the other, will soon show whether there be a vertical error. If it seem probable, try what power the patient possesses of overcoming prism with vertical axis. Begin with one degree, and go up until his limit is reached. I have found persons who could overcome prisms with vertical axis of from 3° to 8°. All such cases are abnormal, and in this fact will almost certainly be found the cause of the uncorrected asthenopic trouble. For persons who do not habitually and invariably practise binocular vision, as in some myopes and in cases of great anisometropia, etc., this fault of vertical displacement of one visual axis is exceedingly common, and does not occasion asthenopic symptoms. For them it is often difficult to recognize double images, and if they do catch them it may be only for an instant, even when by prisms they are brought into close contiguity. It will, perhaps, be noted in making this test that one candle stands not only higher than the other, but that it also stands obliquely. This indicates fault with one of the oblique, and will be again referred to.

A full investigation of possible muscular errors as well as of the functional muscular capacity of both eyes, it is seen, requires not a little time and patience, and some versatility. In all that has thus far been described, it is also observed that the tests are not of any absolute value, but are simply relative. That is, we cannot ascertain how much actual force each muscle is able to exert, but we merely learn what amount of resistance of its antagonists a certain group is able to overcome. Of course we infer properly that if a patient have a high degree both of abduction and adduction, the muscles are essentially strong; but we have no measure of their strength, and vice versa, if only feeble prisms held in any direction can be overcome, we infer feeble muscles. We have symptoms of muscular asthenopia most often in cases where muscular groups are not properly balanced or proportioned. We may also have it when there is debility of the whole muscular apparatus, without special disproportion among the opposing groups. It is not, however, an invariable rule that general weakness of the eye-muscles of necessity causes asthenopia. A large margin must be allowed for what may be called nervous excitability or activity. Those of quick and eager and vivid perceptions, whose mental processes are always lively and ready, are the persons most liable to complain, and they generally have corresponding rapidity in all bodily movements, unless they are the victims of disease. The more quiet or torpid or deliberate persons, are less often sufferers from muscular asthenopia. Of course these comparisons have no necessary import as to the mental calibre of the patients.

It now is in order to show how frequent may be this disturbance of muscular function, otherwise this long discussion would be unseemly. It is probably more common among the people of this country than among those of other countries. It belongs, moreover, to the more intelligent classes, especially to skilled artisans and persons who deal with books and artistic labor. It follows in the train of chronic diseases, or of that chronic low health which barely suffices to carry the person through the day's task, and leaves no joyous reserve for either recreation or further work. Dyspeptics, the fagged out, the votaries of exhausting pleasures, the men who sacrifice body and mind and family to business, sufferers from chronic uterine disease, the epileptiform and choraic, those who have come through seasons of exhaustion and night-watching, in exposure, in grief and bereavement and disappointment, such as are confined to labor in close rooms and for long hours, or who get an insufficient diet. Such are the persons, and others like them, who have muscular asthenopia. Besides the above causes, refractive errors occasion a very large proportion of cases, as has been already remarked, and such errors may be large ; or, if the persons belong to the classes just pictured, the errors may be both small and mischievous. It also happens that there may be congenital weakness of one or more muscles, or of all.

Among 1,500 cases of refractive, accommodative, muscular, and asthenopic troubles, I find 280 in which muscular abnormities are recorded. It would go beyond the limits of this work to analyze them minutely, but the following items may be stated : The most common error is insufficiency of the recti interni. This is most often associated with myopia, but with nearly equal frequency is found in emmetropia, and sometimes, too, with hyperopia. Insufficiency of the externi alone occurs in about eight per cent. of the cases. Vertical diplopia, which may be accompanied by disorder of other muscles, is recorded in 12 cases (four per cent.) ; insufficiency of both interni and externi is found, and there are many cases in which all the muscles are at fault. The eyes may be emmetropic and insufficiency belong to a single muscle, or to all of them.

There may be accommodative connected with the muscular disorders. For instance : spasm is very common, while a few cases, usually among the middle-aged, exhibit a remarkable failure of accommodation. But overtaxation of the eyes is an important factor, and is brought about by excessive eye-work of any kind, or by dim light ; by reading on railway trains and in carriages ; by reading when lying down, which convalescents and chronic invalids often find out too late ; by attempting difficult work, such as embroidery, sewing on black, fine painting, as decoration on china, etc. ; bending over the work and bringing it too near the eyes ; by the study of languages whose text is intricate, such as Greek, German, Hebrew, etc. But, apart from the local ocular disorders, as of refraction or accommodation, or undue demands on the muscles, we have to take into account remote causes, which are such as concern the general health or some special organs. Want of vigor, whether by congenital conditions of health, by too rapid growth, by malaria, by any debilitating causes, by shock, etc., are to be duly considered, as above stated. Still further, all forms of uterine disease, hemorrhage, fevers, chronic anæmia, instigate muscular asthenopia.

In cases of apparent congestion at the base of the brain, often there is tenderness over the middle and upper cervical vertebræ. Nasal catarrh is sometimes a complication, while chronic conjunctivitis of the lids, or blepharitis, is very frequent. Sometimes there is extreme photophobia ; sometimes any effort of fixation is intolerable ; riding in a carriage, or a railway, unless with gaze averted from passing objects, often causes great pain, and to look at a crowd or at objects in motion is equally unpleasant. Persons addicted to masturbation, or with disorders of the genito-urinary apparatus, are liable to have muscular asthenopia. All forms of so-called neurasthenia are prone to this complication.

Treatment.—This is constitutional and rational, or local and optical, or both combined. Often it is a nice point to be settled which is to predominate. It is of course never wrong to promote the general vigor and remove any organic or functional lesions which are acting as predisposing causes, and may be also the only effective causes. I need not emphasize this side of treatment, because it appeals to the good sense of every physician, and I heartily concur in its value and importance. In due order come all measures to stimulate the general welfare, bodily and mental, viz., exercise in the open air according to the capacity and situation of the person, sufficient sleep, proper diet, friction of the skin, or the Turkish bath, regulated massage, horseback riding, gymnastics under discreet supervision, etc. In many cases we must give general tonics, and especially strychnia, phosphorus, the bitters, iron, quinine, etc., and remedies to correct or improve digestion, etc.

Next, the special conditions of the eye are to be taken in hand. We search out and correct all errors of refraction, and do it in obstinate or severe cases most minutely, and of course with the help of the full effect of atropine. We also include herewith troubles of accommodation. In case there be no refractive or accommodative error, and the muscular fault is not large, we may cure the case by systematic exercise of the eyes. The merit of this suggestion belongs to Dr. E. Dyer, of Pittsburg, Pa. (see "Trans. Am. Oph. Soc.," vol. i., p. 28, 1865). The patient is instructed to begin to read for three, five, or fifteen minutes once or twice daily. To do so after a meal and by good light. Sometimes a weak convex glass, or a weak prism 2° or 3° for each eye, with base inward, or perhaps a prism with a weak convex surface, *i.e.*, +48 or +36, is given. Each day the period of reading is increased by one minute or by two minutes, and the most scrupulous exactness is insisted on. When thirty minutes or sixty minutes are attained, the period is not increased for several days ; then the additions are resumed until a sitting of two hours is accomplished. Such sittings may be held once daily or twice a day, according to circumstances. This method of systematic training has a most excellent moral as well as physical effect, and has served admirably. It gives hope to discouraged patients and actually develops their reading power. In place of reading, other work may be substituted, but the great matter is to regulate and systematize the eye-work. Combined with this proceeding, the galvanic battery, either the constant or uninterrupted current for a few minutes, with one pole to the closed eyes and one pole on the temple, has some value. Stimulating liniments to the forehead and temples of aconite, or of chloral and camphor, etc., are useful when there is neuralgia.

The douche or spray of cold water, or mild lotions to the eyes, viz., borax and camphor-water, are all helpful. In some cases where only slight refractive errors are found, it is best to prescribe the wearing of glasses, especially if convex or cylindric, all the time. The behavior of the muscles, both with and without the glasses, will help to decide this point.

Another mode of invigorating the eye-muscles, and which is especially suited to the cases where all the muscles are feeble, is by using prisms as means of gymnastic training. Dyer's method deals with muscular action and accommodation together; by gymnastic prisms the extrinsic muscles alone are acted upon. The patient is provided with prisms of $2\frac{1}{2}°$, $5°$, two of $10°$, and one of $15°$, with squared outlines. He takes a candle-flame or door-knob at twenty feet for his object, and performs the effort of adduction and abduction by means of these prisms. He begins, say with adduction, and at first holds the prism of $5°$ with base out, in front of one eye, then substitutes the $10°$, then before the other eye places $5°$, making a total of $15°$; then, if practicable, substitutes the other prism of $10°$ for the $5°$, and so climbs up the ladder of adductive prisms by such steps as he can make. If the interval of $5°$ becomes too great, he may take that of $2\frac{1}{2}°$. On the other hand, he will in a similar way train the abductive muscles by putting before one eye with its base inward the prism of $2\frac{1}{2}°$, then that of $5°$, then one before each eye, and finally, may possibly reach the $10°$. To reach an adductive power of $42\frac{1}{2}°$ and an abductive power of $10°$ will require sometimes several weeks, and when attained should be practised once or twice daily. The daily session need not occupy more than ten minutes, and need not be more frequent than twice. A decided gain in comfort and available use of the eyes is obtained by this proceeding, and it is applicable to most forms of muscular debility.

We come now to refractive cases with notable errors : 1st, myopia with insufficiency of the interni may be relieved by wearing the full optical correction continually; or 2d, by using for near work a glass which pushes out the near point to $8''$, to $12''$, or to $14''$, and which may be of about half the power of the full correction ; or, 3d, with the glass just mentioned a prism may be combined, or the glasses may perhaps be given an adequate prismatic quality by having them set in the frame with their centres outside the visual axis. This brings the inner thick edge of the glass into use, whereby it will have a low prismatic effect. With hyperopia similar methods of proceeding may be adopted, but with such adjustments as the kind of muscular error calls for, these being various. With emmetropia one finds less certainty in the helpfulness of prisms in aiding the performance of near work. They sometimes are utterly intolerable, even with decided muscular error. If their angle be more than $2°$, they cause objects to have an unnatural convexity or concavity, according as the bases are inward or outward, and I resort to them only by way of trial.

In deciding how strong the prisms are to be, we first decide the proper working distance, and the correcting-glass, which for this point is required, and with it ascertain the muscular error. To give prisms equal to one-half the amount of error is usually sufficient. If the insufficiency we are to correct amounts to $10°$, we may order the prisms each $3°$, one before each eye. It is only when error is decidedly more on one eye than on the other that the prisms are made unequal.

But we must sometimes order prisms for permanent wear, either with or without a refractive correction, and this is indicated when a muscular error has been detected in testing for the distance of twenty feet. Here the correction of weakness of the externi, or of vertical diplopia, is of the utmost importance, and the full correction must be prescribed. So, too, insufficiency of adduction may need relief, but is less urgent. Necessarily the glasses needed for this purpose are to be constantly worn, and it may be that different ones must be given for near work. It is also required to include correction of astigmatic errors in the formula, and sometimes a complex order results, which can be executed only by a highly skilled optician. It is sometimes extraordinarily gratifying to see what relief is obtained by suitably chosen glasses of this quality. I could adduce many notable and happy instances, did space permit, and my experience has taught me to place much confidence in carefully chosen prismatic and refractive combinations. But, on the other hand, all cases will not be relieved by prisms. For instance, we may have insufficiency of the externi for distance, and of the interni for the near. Such a maladroit mixture is unfortunate, and really means general muscular weakness, for which only the invigorant plan is proper. The true purpose of prisms, for constant wear or for working, is found in those cases in which the difficulty lies in want of balance between opposing groups, by which one set predominates too much over another ; not in those cases where the muscular power as a whole is below standard. Weakness of all the muscles is seen in that tremor or jerkiness which the eyeballs show when fixed in some extreme position, either to the outer side or very near, or in a sudden jump which the globe makes on reaching a certain point, as the eyes follow the finger slowly carried from one side to the other. The jump comes when another combination of muscles is required by the change of fixation, and means that all do not conspire to an equable and smooth action.

Lastly, we were taught by Graefe that tenotomy can be successfully applied to cases of muscular insufficiency. The operation can be so carefully done as that its effect shall not exceed a prism of 5°; but it is safe not to venture on a smaller allowance than one of 8°, and this demands great precision and caution. Schweigger is, and Graefe was, disinclined to operate for less than 15° of error. The most satisfactory cases are those of myopia with weak interni. To take 8°, or so much as needful, from the abduction, confers great benefit on the adduction. It is quite suitable to leave the patient no abductive power for twenty feet, where his M is corrected, because he has no need of it and will almost never suffer from its lack. In operating in cases of E or H, more care must be taken about weakening the abduction for distance, because such patients more readily develop diplopia than do myopes. An excess of 3° or 5° may not be serious, because it can be corrected by prisms ; but it is extremely undesirable, especially with E, to bring about this condition. As a rule, but one eye will be operated on, and that the one in which the error may be greater ; but if this be not so, choose the eye whose sight may be less perfect—in other cases the choice may be indifferent. I do not assert that both eyes may not sometimes demand tenotomy for insufficiency. It is important to avoid free bleeding, because a thrombus disguises and modifies the effect of the operation. The immediate result is always much in excess of the final condition. To apportion it properly, Graefe gave the rule that the eyes must be in equilibrium for a point which shall be about 20° on the side opposite that to which the divided muscle directs the eye, and on a plane 20° below the horizon. If, for example, the left externus is divided, the patient, with a red glass before one eye, should see a candle-flame singly when held to the right side 20° from the median plane of the face, and depressed about 20° below the horizon. If the left internus is divided, there should be single vision of a flame held 20° to the left side of the median plane, and 20° below the horizon. A prism with base vertical is to be put in front of one eye, and the two flames must stand perpendicularly when the candle is put in the place of election above designated. An error of 3° to excess is to be corrected. The effect is controlled by a suture in the wound, which shall include the conjunctiva, and reach, more or less, into the tendon, and be drawn tightly, according to the result required. The operation is performed with just enough ether to quiet the patient, and the examination is to be made when he regains self-control. For an error of 15° or 18°, usually two operations will be needed at different sessions. For greater degrees of error, both eyes may be operated on at once, but with very exact precautions as to effect, and with the control of sutures. At the end of twenty-four hours it is entirely possible to open the wound with forceps and a strabismus-hook, and readjust the muscle, which should be done if an error of 5° exist in the abduction, or if more than 5° in adduction at the place of election. Further detail as to the operation will be spoken of under "Strabismus." The ultimate effect is not reached for several weeks. For operations on the interni with M, the final effect will be generally much less than at the end of two weeks. But operations on the interni with E or H, or on the externi, are likely either to remain the same after two weeks, or to increase in their effect. There is a degree of uncertainty in these cases, which depends on the contractility of the muscles and on the tendencies to binocular vision. An operator must therefore err on the side of prudence. An old lady has lately given me a good maxim, fit for these and for many other cases: "It is better to be sure than sorry."

PARALYSIS OF MUSCLES.

As introductory to the next topic, the diagram (Fig. 45) is introduced to indicate the nerves which supply the ocular muscles, and their relations. Reference should also be made to Fig. 41, which exhibits the axes of the muscles in pairs, and their modes of action. We may have one or more muscles paralyzed, and the cause may be either or, bital or intracranial. If the latter, it may be either along the course of the nerves which animate the muscles, or in the brain. Spinal cord lesions are also associated with paralysis of the eye-muscles through fibres which proceed to the brain. The effective causes are localized periostitis or inflammation of the sheaths of the nerves, basilar meningitis, hemorrhages, tumors of every variety, degenerations of the nerve-structure, or of the cerebral nerve-centres, injuries, etc. Rheumatic and syphilitic processes are more frequent than other agencies. Some opinion can be formed as to the probable seat of the lesions by noting whether the paralysis affects all the branches of a given trunk, or more than one trunk. Thus, the third nerve supplies the rect. internus, superior, inferior, the obliquus inferior, the levator palpebræ, and the sphincter pupillæ. The fourth nerve supplies the obliquus superior. The sixth nerve supplies the rectus externus. If all the muscles animated by the third nerve are affected, the lesion must be at the apex of the orbit or farther back; while if only one or two of the muscles thus animated are concerned, the lesion is probably orbital. Such a case has just come to my notice where paralysis of the levator palpebræ and of the rectus superior point almost certainly to a lesion at the roof of the orbit. The sixth nerve takes a very long course in the skull, coming from the front of the pons, and hence, it is especially exposed to lesions of its trunk. Again, all the nerves are clustered around the cavern-

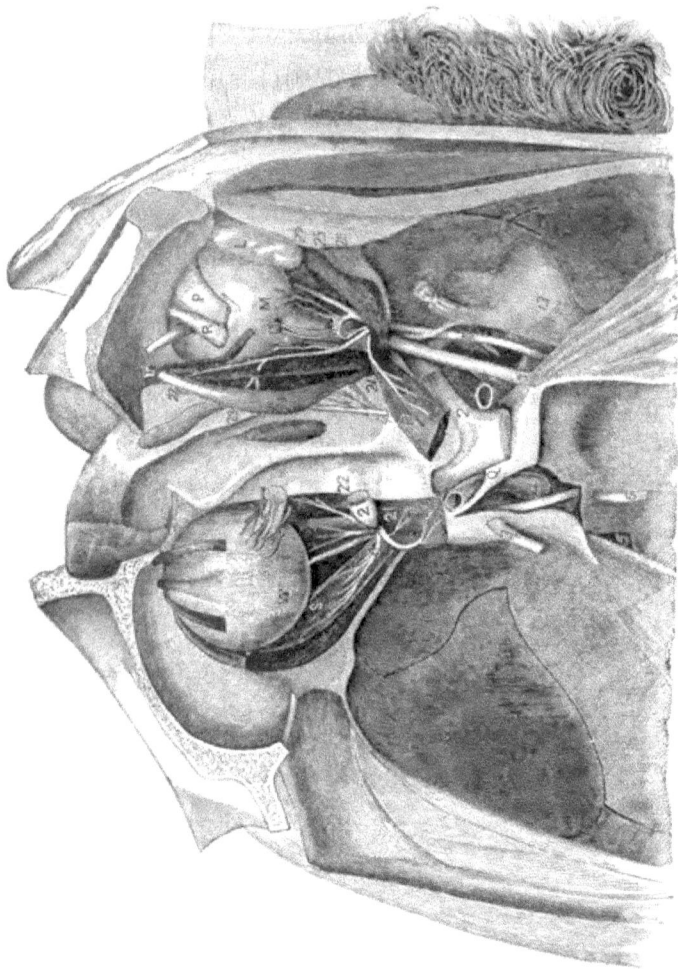

Fig. 46.

L, lachrymal gland.
M, eyeball—right side.
N, superior oblique muscle.
P, levator palpebræ superioris.
R, rectus superior muscle.
S, rectus externus muscle.
T, pulley of superior oblique muscle.
Q, cavernous sinus.
V, inferior rectus muscle.
W, inferior oblique muscle.
X, rectus internus muscle.
a, internal carotid artery.
2, optic nerve or 2d nerve.
3, motor communis oculi or 3d nerve.
4, patheticus, or 4th nerve.
5, trigeminus, or 5th nerve.
6, abducens, or 6th nerve.
13, Gasserian ganglion.
14, ophthalmic branch of 5th nerve.
20, nasal branch of ophthalmic nerve.
21, upper branch of 3d nerve.
22, lower branch of 3d nerve.
23, nasal branch leaving the orbit.
25, lenticular ganglion.
26, long root of lenticular ganglion, from nasal nerve.

ous sinus at the sphenoidal fissure, and may, any or all of them, be here involved in a morbid process. In the determination of diseases of the brain important aid is often furnished by the ocular paralysis which may be present. Thus, a case of double paralysis of the recti externi, which I was permitted by the kindness of Dr. Janeway to observe, was by this circumstance decisively diagnosticated as due to lesion at or about the pons Varolii. So, too, a total paralysis of the third, with other cerebral symptoms, may lead to a localization of a brain-lesion at the peduncles. But caution is needful in urging such deductions stringently, because many obscurities arise, and we must generally be content with probability or reasonable conjecture. In fact, the greater number of ocular paralyses are not of cerebral origin; they are chiefly orbital—and may also be the first premonition of sclerosis of the cord. This occurs with the third, the fourth, and with the sixth nerves. Galezowski and Duchenne have seen cases of bilateral paralysis of the third and of the sixth in spinal disease. Lesion of the sixth is the most common. In spinal cases the paralysis is likely to be incomplete and not to be permanent. No other sign of spinal cord disease may occur for a long time, and this symptom, while unsupported by others, will remain of doubtful significance. The motor nerves of the eye often become implicated at a later stage of the spinal disease, and then the lesion is not transitory, but permanent. The implication of the optic nerve, as will hereafter be mentioned, is similar. Localized or disseminated sclerosis of the brain gives rise frequently to ocular paralysis. But a partial paralysis is more characteristic than is complete paralysis. Irregular and spasmodic movements of the globe, with double images of bizarre character and which often cannot be described or portrayed, are very common. I once observed a case of this kind until death; there were numerous paralyses; but at the autopsy, which was made by expert pathologists, no visible lesions of the brain were found, and a microscopic examination was not made. The weather was very hot and the brain could not well be preserved. In some autopsies (Leube) the trunks of the motores oculorum and of the sixth have been transformed into gray, thick and hard cords. Cases of meningitis over the cerebral cortex as well as at the base, are sometimes accompanied by ocular paralysis. There are many interesting facts as well as speculations in cerebral physiology and pathology connected with paralysis of the ocular nerves. We may instance the relations of the several branches of the third to the medulla oblongata (see Wernicke: "Lehrbuch der Gehirnkrankheiten," p. 354 et seq., 1881; also Bernhardt: "Hirngeschwulste," p. 206 et seq., 1881), by which an explanation is offered of such cases as the following: 1st, an affection of all branches of the third except the pupillary; 2d, one of the iris and ciliary muscle, and of no other muscles; 3d, one of the rectus internus alone, or of the inferior oblique, or of the inferior rectus alone. But the limits of this treatise do not permit us to enter on this field. We have ptosis alone, which may be cortical or central, and may be associated with hemiplegia. Ptosis has also been associated with paralysis of the sixth, because of the existence of an irregular twig of connection. The problems in cerebral pathology are extremely complex and fascinating; but, as already remarked, their solution is not yet perfected, and would not pertain to a treatise like the present. In diagnosticating peripheral from cortical paralysis, electricity is of great value; but in the case of the eye it has not been possible to so isolate the nerve-twigs as to apply the recognized rules of such diagnosis. Double vision often begins with severe headache or with no accompanying symptom. It may

7

happen at any age, may be temporary or permanent. I have lately seen a child five years old who has had partial paralysis of the left third nerve five times. Both eyes may be affected, and the cause is then intracranial and generally basilar.

Symptoms.—They are as follows : a false position of the eye, limitation and irregularity in motion, and double images. Secondary effects are dizziness, nausea, incorrect projection of the field of vision, inability to guide the hands or feet aright ; and if one eye only be involved, the inclination is to close it. Another effect is a peculiar attitude which the head assumes in order to obviate double images. The immediate effects of paralysis are not the same with the more remote. If recovery does not occur, there ensues secondary contraction of opposing or associated muscles. For example, if the left rectus externus be paralyzed, not only will the left rectus internus, by reason of the diminished resistance, turn the eye unduly inward, but the rectus internus of the right (opposite) eye will undergo contraction, and if the left eye look straight forward, the right eye will consequently squint inward. This is because the right rectus internus is associated, in all movements to the left, with the left rectus externus. Laws of association apply to all the muscles and cause complicated effects, which in old cases often make the diagnosis of what muscles were at first damaged very difficult.

Diagnosis.—We meet in practice with the most complex combinations, and sometimes it is more than a puzzle to tell what muscles are at fault. We place most reliance on the character and position of the double images, but to a clear analysis it is necessary to have an intelligent patient with two good eyes, each of which shall be quick to observe the image it receives. To complicate the problem, secondary contractions and involuntary compensations by other muscles, may come in to disturb the regular scheme which ought theoretically to be observed in the behavior of the double images, and we are left in the lurch. But in many recent cases we can tell, without analysis of double images, what muscle is affected. The eye refuses to move to the proper degree in the direction of the action of the impaired muscle, and goes too far to the opposite side ; its movements are often partial and jerky. If many muscles are paralyzed, the situation of the globe in the orbit may be altered, *i.e.,* exophthalmus may occur. False projection arises because the effort to move the eye in the direction for which it is incapacitated is accompanied by a consciousness that such an effort is being made. Because of our habitual reliance on the muscular sense we are deceived into supposing that the effort we make is followed by the effect which we are accustomed to find, and we act accordingly, but find that our assumption of the position of objects in the field of vision is wrong. For instance, if the left rectus externus is paralyzed, especially if only partially paralyzed, and the left eye attempt to see an object to the left side, the effort of movement is so much greater than is usually made, that the mind believes the object to lie much farther to the left than it really does, and the hand, in attempting to seize an object or to put the finger on the point of a pencil, strikes to the left side of the true position, *i.e.,* the projection of the field is too far to the side of the action of the muscle. The inclination of the head, when this occurs, will be such as to favor the lamed muscle, and will be in its line of action and toward its virtual or anatomical origin. For a paralyzed rectus externus of the left eye, the head will turn on a vertical axis to the left. For a paralyzed rectus superior of the left eye, the head will turn on a horizontal axis upward and a little to the right. For the obliquus inferior the tendency will be the same, both being levators of the cornea, and the head thrown back diminishes the effort which falls upon the muscle in looking upward.

Dizziness, nausea, and such cerebral symptoms, are not always present, and when they occur, after a time pass away. They are caused by the confusion of images and by the dissociation which is produced between the conscious effort of the muscles and the instability and falsity of the projected field. Objects are made unsteady, the ground does not seem level, going up and down stairs becomes difficult, movements of the hands are ill directed, and from all these phenomena mental confusion and vertigo result, until further experience corrects the judgment.

In case of partial paralysis it is found that in some persons there will be a singular capacity for correcting the diplopia. This power depends upon the instinct for binocular vision, and is called the capacity for fusion. With the same degree of dynamic deviation, so far as can be estimated by prisms, the extent over which fusion of double images is achieved will be much greater in some persons than in others. Von Graefe declared that, apart from errors of sight in refraction or accommodation or amblyopia, the capacity for fusion is far less in cerebral paralysis than in orbital or basilar paralysis. The reason is that binocular vision is essentially a cerebral function.

To comprehend the value of double images for diagnosis of ocular paralysis, a few graphic illustrations are employed, which are borrowed from Zehender ("Handbuch der Augenheilkunde," 1874, p. 317) and somewhat modified. It has already been stated that two images on the same level, of which the right belongs to the right eye and the left to the left eye, are called homonymous or correspondent. Images on the same level, and of which the right belongs to the left eye and the left to the right eye, are called crossed. The former implies impaired power of abduction, *i.e.*, the eyes are convergent; and the latter implies impaired power of adduction, that is, the axes are divergent. We have also to study differences in height, *i.e.*, vertical diplopia; and the higher image belongs, as before said, to the eye which points too low, and means impaired power of lifting the cornea, *i.e.*, the levators are at fault. The lower image belongs to the other eye. Again, we are to note whether the images are parallel to each other, and for this we must use as a test a long candle or a stick about a foot long. The images may incline at the top or diverge at the top. The cause will lie in deflections of the vertical meridians. We always speak only of the top of the vertical meridian. If now these meridians diverge, the images will incline inward; if they converge, the images will diverge. The obliqui are thus submitted to proof of their condition, while the influence of the recti superiores and recti inferiores may also thus be indicated.

As a condensed and also lucid presentation of the position and situation of the double images, we follow the description of Zehender. They make their appearance on the side of the function of the paralyzed muscle, and this side is shaded in the diagrams. The double images are figured as they are seen by the patient. The white candle denotes the image seen by the sound eye, the dark candle that seen by the paralyzed eye. (In practice it is better to put a red glass over the eye which fixes, and which is generally the sound eye, so that the image seen indirectly and by the paralyzed eye may be relatively more distinct.)

Fig. 46 shows the double images in 1st, paralysis of the rectus externus oculi sinistri, and likewise those in 2d, paralysis of the rectus internus oculi dextri—the one being the counterpart of the other, except that in the former the images are homonymous, in the latter they are crossed.

If the same figure were looked at through the paper from its back side, or were looked at as reflected from a mirror, it would be reversed, and then would represent, as seen in Fig. 47, 3d, paralysis of the rectus externus oculi dextri, or 4th, paralysis of the rectus internus oculi sinistri.   In the third case the images are homonymous ; in the fourth, they are crossed.   The

FIG. 46.—Paralysis of Rectus Externus of Left
Eye, and also of Rectus Internus of Right Eye.

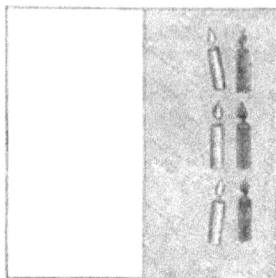

FIG. 47.—Paralysis of Rectus Externus of Right
Eye, and also of Rectus Internus of Left Eye.

images viewed at the middle are vertical and parallel, while the eyes turned up cause them to diverge at the top, and, if turned down, to converge at the top to a slight and physiological degree.   In looking upward the rectus superior predominates and causes the vertical meridian to converge—hence, divergence of the images ; and vice versa, in looking down the action of the rectus inferior causes the images to converge.

Fig. 48 gives the situation and relation of the images in 5th, paralysis of the rectus superior oculi sinistri, and its reverse in Fig. 49 exhibits,

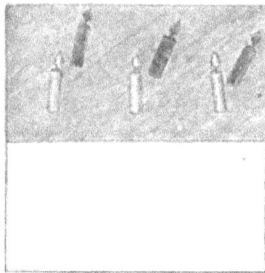

FIG. 48.—Paralysis of Rectus Superior of Left Eye.

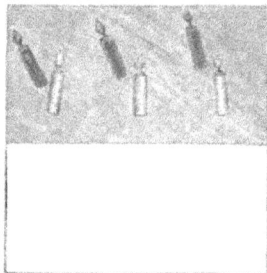

FIG. 49.—Paralysis of Rectus Superior of Right Eye.

6th, paralysis of the rectus superior oculi dextri.   It is noticed that difference in height (Fig. 48) increases toward the left, and obliquity increases toward the right, the reverse occurring in Fig. 49.   In both cases the images are crossed.

In Fig. 50 we have the images seen in 7th, paralysis of the rectus inferior oculi sinistri, and reversed in Fig. 51 of 8th, paralysis of rectus inferior oculi dextri.

In this figure again the images are crossed, and they diverge more widely toward the side of the affected muscle, and the obliquity diminishes toward the same side. The figures give the diplopia only in the extreme upper and lower parts of the field, where the difference in height is greatest; on the median line it will be less, and at the opposite part of the field there will be single vision.

In Fig. 52 are represented the double images found in 9th, paralysis of the obliquus superior oculi sinistri, where they are homonymous; and

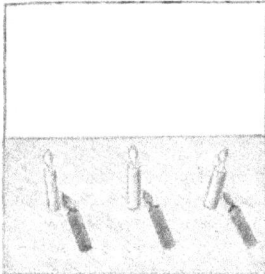

Fig. 50.—Paralysis of Rectus Inferior of Left Eye.

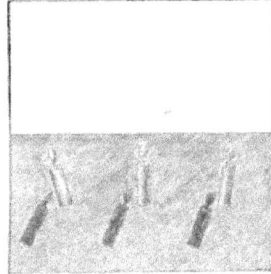

Fig. 51.—Paralysis of Rectus Inferior of Right Eye.

if reversed, as in Fig. 53, we have, 10th, paralysis of obliquus superior oculi dextri.

In these cases the notable thing is that, besides being homonymous, there is difference in height and a remarkable obliquity. The vertical separation increases on the side of the sound eye, while the obliquity increases on the side of the impaired eye. Paralysis of the inferior oblique,

Fig. 52.—Paralysis of Obliquus Superior of Left Eye.

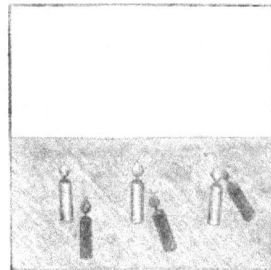

Fig. 53.—Paralysis of Obliquus Superior of Right Eye.

which is rare, gives double images in the *upper* part of the field, and with difference in height as well as lateral displacement, the images being crossed and diverging at the top.

It is not seldom that one image seems to be farther removed than the other. To aid a patient in describing what he sees, it will be well to let him have a stick in each hand, and with them to imitate the position of the images. To determine which is the true and which is the false image is generally easy, because the patient will naturally fix with the sound eye. We shall also be guided by other symptoms in deciding upon the faulty

eye. Moreover, by observing in what direction the least displacement occurs and the line in which the images separate most widely, the erroneous eye will soon be detected. Moreover, while a patient fixes on an object, if the screen be quickly shifted from one eye to the other, the eye which does not remain steady, but makes a slight movement of adjustment, is the affected **one.** Patients can by their own sensations often tell which is **the injured eye.**

Treatment.—Necessarily, we must take into account the probable cause, and when this is doubtful we fall back on general principles of therapeutics. A patient with **double images** will close one eye, or wear over it a screen. It is well after a short time to put the screen temporarily over the sound eye, **to keep the other in practice. Soon after** the lesion there may be headache **or symptoms which** suggest **leeches** or cupping, but not often is depletion proper. Blisters by cantharidal collodion, of small extent, over the temples or forehead, are useful as peripheral stimulants. Iodide of potassium would be given in small doses in non-syphilitic cases, and in large doses in syphilitic cases according **to the** stage and peculiarity of the constitutional disease. Electricity may **be** applied by the faradic current or by the interrupted galvanic current, **the** former preferably—one pole upon the temple or behind the ear, and **the** other by a small sponge **upon** the **globe.**

It cannot be asserted that great **efficiency attaches to** this proceeding, **because** we do **not** see the muscle contract, **even** though, as has been **done, the** pole be applied over the insertion **of the** affected muscle di**rectly to** the conjunctiva. We also use strychnia **in** moderate doses after **a few** weeks have elapsed. In fact, we work in the dark to a great extent **in** such cases, **and** chiefly rely on spontaneous absorption of the mischievous exudation **or** hemorrhage or thickening. We aid the efforts of nature by means which reinforce the natural powers of absorption, and by the special remedies **above** indicated.

When, **however, no** improvement takes place after the lapse of **two or** three months, **we have** little **right** to expect it; but, as a rule, a degree **of** betterment **or entire cure will have** occurred. But, for stationary conditions when **double images are not too** wide apart, we may employ prisms. It is possible **to wear prisms as high as** 8° or 10°, but beyond this they become too clumsy to be **ordinarily tolerated. In** adopting them, this rule is to be remembered : *Put the base of the prism toward the image which is to be influenced.*

The eye which deviates the most, or which **is** weaker in power or in **vision,** will wear the stronger prism, when between the eyes a difference is **to be** made. Oftentimes both **eyes** will wear the same prism, and it is **right to** do this even though **one** only be impaired. It is not to be ex**pected that** the arrangement which is first made in adjusting the prisms **will always** remain good afterward. Frequently the muscles undergo changes, **and** require corresponding alterations of the prisms. For many cases they are to be worn only for a time, especially if they reach high degrees, while permanent prisms are rarely acceptable which have a total of more than 12°. The permanent use of prisms is in fact a rarity, and pertains more especially to cases where there is a degree of vertical diplopia—for example, to those of the fourth nerve involving the superior oblique—the superior rectus sometimes will call for them. Double vision which concerns fixation 10° above and for all the field below the horizontal meridian, or which concerns the median region of fixation, is the most distressing, and calls loudly for aid. The office of prisms is usually confined to these regions, viz., on the median line and for parts on or below the horizon. In fact, to extend their influence over the whole field is impossible, because the relations of the double images become entirely different in its various parts, and it is impracticable to adapt the prism to these changes. It is seen that prisms, like crutches, are greatly acceptable; but they are imperfect substitutes for sound muscles.

When, however, a case has existed for six months or more and is beyond the utility of prisms, and no objection exists in the general condition of the patient, an operation will often serve an admirable purpose. Operative proceedings are two-fold: 1st, simple tenotomy of one or more opposing muscles; 2d, advancement of the impaired muscles.

For such a case, for example, as imperfect paralysis of the sixth, or sometimes when it is wholly paralyzed, a tenotomy of one or both interni may be indicated and give a useful result. Both must generally be divided, because in the opposite eye the internus has undergone secondary contraction by co-ordinated function, and the internus of the impaired eye, by being unopposed, has passed into a similar condition. The greatest stress in such an operation is to be laid on the internus of the sound eye, because undue freedom in loosening the internus of the injured eye will tend to exophthalmus, to sinking of the caruncle, and to render the globe incapable of sufficient movement in any direction. On the injured eye, if any such tendency appear, a suture must at once be deeply entered, and drawn tight to prevent undue slipping back of the tendon. The same methods are to be used in such operations as were spoken of when treating of insufficiency of the muscles. The effect on the muscles is to be measured by using a lighted candle, a red glass and prisms, and single vision must, if possible, be secured to a point within 15° of the normal region of the paralyzed muscle. The ultimate effect will be less than the immediate. If any power remains to the damaged muscle, it gains increase of function by being less seriously overmastered. In fact, this principle has been applied to the advantage of a paralyzed muscle, to prevent both the degeneration of its own tissue and the extreme of secondary contraction in the co-ordinated muscle, by performing tenotomy on a secondarily contracted muscle within a brief time, say two or three weeks after the onset of the paralysis. I have seen Dr. E. G. Loring perform such an operation, and he declared himself satisfied with its effect. I have had no such experience, and do not know that such practice is pursued by any one else. The degree to which the muscle is loosened is very carefully measured and restrained, because the tenotomy is intended to have a preventive effect, and also to aid in the recovery of function.

For cases of marked and permanent limitation of motion, the proper proceeding is advancement of the paralyzed muscle and setting back of one or more of its opponents.

For a correct understanding of this proceeding some remarks on the anatomy of the oculo-orbital fascia are proper. This tissue is also known as the capsule of Tenon. If the upper and lower lids be divided in the middle down to the fornix and the flaps be forcibly drawn back, it will be seen, by lifting the conjunctiva on a probe or a strabismus-hook, that there is a distinct layer of connective tissue going forward under it to the margin of the cornea. It is also noted that the ends of the muscles, as they reach the globe, protrude through it and are clearly displayed. Pressure with the convexity of the hook between the eyeball and the margin of the orbit demonstrates that in this circumocular space something shuts off the parts behind, and forms a layer which adheres on the one side to the globe, and on the other to the margin of the orbit. The structure which is thus demonstrated is the oculo-orbital fascia. If the globe be enucleated, the tendons and the stump of the optic nerve will be seen to stick out through a layer of smooth fibrous membrane, which forms the cup in which the globe rotates, and which is part of the same fascia. The same structure enters into the eyelids and enwraps all the muscles as they advance toward the globe. It thus appears clear that a tendon may be entirely loosened from the globe, and if its lateral and immediate relations with the fascia are not torn up, it still remains in connection with the eye, and can exert an active, although reduced influence upon its movements. If, however, in detaching the tendon, cuts be freely made in the lateral regions, the muscle will lose its control over the eye, because it slips back into the orbit; and if any connection remains, it will be through the medium of some band of tissue which has escaped disruption.

The oculo-orbital fascia does not admit of a clear demonstration as a membrane; it is too complex in its ramifications, and too delicate in structure, besides being perforated by a multitude of organs. It ensheaths to a greater or less degree all the organs, muscles, vessels, nerves, etc., which pass through it. For example, the external sheath of the optic nerve is continuous with it, and it also adheres to the margin of the optic foramen. The periosteum of the orbit is continuous with it, and is sometimes spoken of as its parietal portion. But the analogy of the pleura in its visceral and pulmonary parts cannot be strictly maintained, although it is suggested. For practical purposes we are to bear in mind three facts: 1st, that the fascia serves as a cup, like the acetabulum, in which the globe revolves and makes enucleation possible without opening the deep parts of the orbit; 2d, that it prevents effusions in the orbit from easily finding their way into the lid, and beneath the ocular conjunctiva; 3d, that it constitutes a secondary attachment for the ocular muscles, renders their combined action more perfect, and makes it possible to sever their tendinous insertions without annulling their influence over the globe. A further remark is that the caruncle and semilunar fold are intimately connected with the fascia; and so is the tendon of the muscle of Horner, at the inner canthus, while at the outer canthus the external lateral ligament may be called a process thrown out from the periosteum. Gerlach further calls attention to the check which certain fibres exert over the action of the muscles, and at the inner side of the orbit the figure which he gives shows how firm is the connection between the fascia and the bony wall. It is always somewhat difficult to lift the caruncle in a dissection, and if this is done during life considerable sinking of it is liable to ensue. This has a practical bearing on the operation for converging squint.

STRABISMUS.

In the preceding section we considered a certain class of deviations of one or both eyes, which depend upon paralysis; we are now to consider deviations of the visual axes, in whose production paralysis plays either a very subordinate part, or none at all. The former is sometimes designated as strabismus paralyticus; the latter, which we now take up, is called strabismus concomitans (von Graefe), strabismus muscularis (Alfred Graefe), or simple strabismus or squint. In the typical cases of the former kind, viz., strabismus paralyticus, there is inability to move one or both eyes in certain directions; in the latter cases both eyes are capable of motion to a nearly equal degree, but they cannot fix simultaneously on the same object. In the former cases there may be binocular vision in certain parts of the field, and in other parts there will usually be double vision ; in the latter cases we do not have binocular vision, and double vision is the exception. In the former cases the fault is purely muscular ; in the latter, while the muscles are at fault, there is some kind of visual error, and this is an influential factor of the squint.

It has been already stated that the axis of the cornea does not coincide with the visual line. When the visual lines are parallel the corneæ seem to diverge slightly, because their axes stand outward 5° from the visual lines (angle alpha), causing an apparent divergent squint. This we are accustomed to regard as the normal position when looking at a distance. The angle alpha is not always the same in amount. (See Fig. 54, in which also the places of the cardinal points are designated.)

We have strabismus convergens, strabismus divergens, strabismus sursum (upward), and strabismus deorsum (downward) ; the two latter may be called vertical deviations, and the two former lateral deviations. The

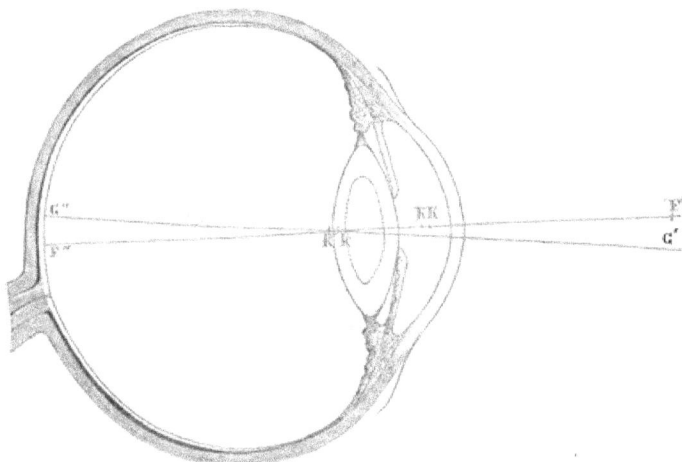

Fig. 54.—Cardinal points of the left eye; F′ F″, visual line; G′ G″, axis of cornea; F′ k′ G″, the angle alpha; h′ h″, the two principal points; k′ k″, the two nodal points; F′, the anterior focus; F^b, the posterior focus.

causes of squint (apart from paralysis) are found chiefly in optical errors, and to a less degree in perceptive errors of the eye, or in both combined. This is proved by statistics. For example, Schweigger ("Klinische Untersuchungen über das Schielen," 1881) gives 446 cases of convergent strabismus, of which 98 were emmetropic, and 348 were ametropic. Among the emmetropic he incorrectly classes some of myopic astigmatism. He gives 183 cases of divergent squint, in which 65 are counted as emmetropic, although one eye may have been ametropic, and 118 were in both eyes ametropic. This fact of ametropia in strabismus is now fully recognized, and was first insisted upon and demonstrated by Donders. It gives a new character to the disease, and removes it entirely from the category of ordinary orthopædic surgery, in which until 1864 it was habitually included. While orthopædic methods must be invoked to remove it, we have learned that the correction of optical errors and acquaintance with the laws of binocular vision **must** co-operate with the rules of technical surgery.

We have *permanent* strabismus in which the axes are constantly crossed, and we have *periodical* strabismus in which the axes sometimes cross and at other times point correctly. The periodic or intermittent squint has also a subdivision. The usual cases are 1st, those in which squint appears only when regarding **near objects** ; and 2d, those much less frequent cases in which, at stated intervals, the squint appears and remains for days or weeks, and then entirely disappears, leaving the eyes normal in action. Schweigger, among his total of 629 cases, sets down 425 as permanent and 204 as periodic. Among the latter he does not isolate, if he observed, the kind last spoken of above. Among other **causes assigned** for squint are an unusual **degree of** divergence in the axes of the orbits, congenital anomalies in the length and capacity of the muscles or in their insertions (Hasner); also anomalies in the interpupillary interval—that this is too small in hyperopia and too large in myopia (Mannhardt). Not denying that importance is to be attached to anomalies in muscles and in anatomical conformation of the skull, it remains true that visual errors are the important and controlling causes of strabismus. Alfred Graefe has in many cases measured the distance from the cornea of the insertion of the tendon of the rectus internus, and found it to vary from 4 to 7 mm.; but he has not found any constant relation between the place of insertion and the degree of squint. A vast catalogue of causes which are in popular belief assigned for squint may be dropped from consideration ; such as an infant's exposure to a window, or wearing a hanging lock of hair, or imitation of a cross-eyed person, etc.; but it is true that squint not seldom follows an attack of illness or of keratitis, especially when an opacity is left on the centre of the cornea. The age at which squint develops is significant, viz., that it appears in early life and at the time when the eyes begin to be employed upon small objects—as in first learning to observe anything closely, or in looking at picture-books, or in learning to read. In the beginning the squint is usually transient, remaining for a few minutes or hours or days, and then going away. Out of 193 cases studied at this period, 176 appeared under four years of age.

Again, if in early life one eye should for a long period be excluded from use, as by any inflammation, this may cause squint. In later life the occurrence of cataract or of marked amblyopia is often followed by squint ; this may be either converging or diverging—more frequently the latter. Causes which impair the power of accommodation induce the error ; marked differences in the refraction of the two eyes favor it. A genuine concomitant squint may ensue after paralysis or paresis of a muscle. The health of the muscle may be restored, but the secondary contraction which befell its fellow may persist, and thus leave a permanent deviation ; or the impaired muscle may not have fully recovered.

Ordinary periodic squint appears only in near vision, and is the expression of extraordinary impulse of accommodation, and this in turn overstimulates the interni so as to cause crossing of the visual lines and abandonment of one eye.

In the early periods of squint, double vision is often noticed, but soon it ceases and may after a time be wholly incapable of being excited. This may be true even in cases where the diseased eye possesses good vision ; as a rule, this eye has vision inferior to that of the other. It may happen that a patient believes one eye to be much worse than the other, and this by testing prove not to be the fact. At the same time he will be unable to use the eye without fatigue. It is a general fact that squinting persons choose one eye for habitual use, and neglect the other. But some will make use of each alternately and without being aware which they employ. In high degrees of squint it is sometimes seen that the head is turned in accordance with the eye which may be used. In divergence the working eye is brought around to the median line of the body, the head turning to the opposite side. In converging squint, if very great and each eye is effective, I have seen the head turned to the right when fixing with the right eye, and this be used for the left side of the field; while if the left eye fix, the head will turn to the left, to command the right side of the field. If now, in established squint, one eye be much below the other in power, it is not hard to comprehend why there should be no confusion of images, both because one of them causes a less decided mental impression, and because in the deviating eye the image falls on a less sensitive, because eccentric part of the retina.

In all high degrees of squint, the eccentric place of the retinal image renders it less likely to be observed, and the faculty of mental abstraction becomes so habitual as to put it wholly out of consciousness. On the supposition that binocular vision is an acquired faculty, and not inherent, it is comprehensible that a squinting person may not have learned the function. The angle of squint varies so constantly that the mutual identity of the retinal images is not established, and if acquired, it may be unlearned. But this explanation cannot be regarded as wholly satisfactory, because in many cases a prism held before the squinting eye, by directing the image to a part of the retina not usually so impressed, will often evoke double vision. Schweigger offered the following experiment many years ago, to prove that the squinting eye can see with its macula while squinting. While the good eye fixes on some object in a dark room, place a lighted candle just behind the head, on the side of the deviating eye; then place between the eye and the nose a slip of plane glass to reflect into it the flame of the candle. One can easily tell, by the position of the image on the centre of the cornea, when the light falls in the visual axis. The moment the image rests on the macula, the person declares that he sees it, and that without changing the position of his eye. This experiment, with another by Alfred Graefe (see G. & S.), unite in proving that the squinting eye is not actively refusing to see.

Because one eye is usually found to be less keen than the other, it has grown to be a general opinion that disuse results in deterioration of the retina (amblyopia ex anopsia). Certain facts antagonize this belief, such as restoration of sight after the existence of cataract for a score of years.

Schweigger (l. c., p. 81 et seq.) attacks this theory of amblyopia from disuse with great energy, and, in my judgment, with much success. His opinion agrees with that of Alfred Graefe, and is opposed to the views of Leber. Out of his 629 cases of squint, he has 177 whose vision was between $\frac{1}{5}$ and $\frac{1}{35}$ or less, and he also collected 98 similar cases of undoubtedly congenital amblyopia in persons who did not squint. His belief is that in squint the amblyopia is congenital. That a real amblyopia results from disuse can be held only in a partial sense, viz., as a regional amblyopia not identical with scotoma, but in the sense that, over the locality upon which the image falls when the eye squints, there is either a dulness of perception or a habitual negation of vision. We find that some patients can recover the function of the squinting eye and combine it with its fellow, and that to other persons this is impossible; and the impossibility may lie either in the incurability of the impaired sight for which no ophthal-moscopic lesion can be discovered, or because of mental incapacity to co-ordinate and combine both eyes. In fact, some persons have what von Graefe called an antipathy to binocular vision, which is unconquerable.

Strabismus Convergens—Concomitans.

We examine in the following way : While the fixing eye looks at the finger six inches distant, the other turns inward. Put a card over the fixing eye, and the squinting one rolls out to look upon the finger. But the eye behind the card will be found now to have taken on the squint. Remove the card, and one of two things may happen : either the axes remain unchanged, or they resume their former position.

In the former case there is power to fix with either eye at pleasure. This power of alternate fixation may be only temporary and imperfect, one eye habitually dominating and the other maintaining direct fixation only for a short time, and with a certain tremor. Sometimes it is found that each eye at will can turn completely to the outer canthus ; again, one or both of them may not fully reach this point, and at the moment of greatest abduction will become tremulous. In looking at a distance, the angle of convergence is least, and suddenly increases as the object nears the eye. This shows how much the accommodation exaggerates the deviation.

We also have *strabismus convergens monolateralis*. While one eye fixes, the other turns in, and when the first is screened, the other comes to the direct position, but the first does not take up the squint. This case is not common, but is properly called monocular converging squint. In the great proportion of these cases both eyes are really involved—one only slightly, the other decidedly.

It is often the fact that, besides turning in, the squinting eye turns a little upward. This will be true of each in alternation. It will also be noted that in looking down the squint is greater than when looking up.

Donders regards converging squint and hypermetropia as standing almost universally for cause and effect. He assigned the undue necessity of accommodation in H, and its association with the interni as both the sufficient and efficient cause of the disruption of or failure to maintain binocular vision.

Large observation has modified these views, and while we find that hyperopia acts the most important part in the production of converging squint, we have many statistics to show that essential muscular defects are also operative.

The latest table is given by Schweigger, who finds in 446 cases the following :

	Permanent strab. conv.	Periodic strab. conv.	
Emmetropia	85	13	
Myopia...............	44	10	
Hyperopia	196	98	
	325	121 446	

Of the hyperopic cases, the higher degrees of error do not produce squint in that proportion which would be looked for were the ciliary mus-cle exclusively the potential agent. The majority of the cases have only moderate degrees of hyperopia. It is found, moreover, that the angle of deviation is not in any demonstrable ratio to the amount of the refractive error. The occurrence of ten per cent. of cases of myopia among this category is a notable fact, and strongly argues for a modification of Professor Donders' views. He estimated the cases of myopia with convergence at only two per cent. Professor Schweigger includes among the emmetropic cases some of simple astigmatism ; but in this I think an error is committed, because these patients are always strongly stimulated to efforts of accommodation, and therefore of the recti interni. In fact, it is a common observation that slight astigmatism is attended with an unusually near working distance for small objects. Primary or congenital anomalies in the muscles are not less frequent than important in this connection, and are not only an invitation to, but are undoubtedly a cause of squint. This view is strengthened by the correlative fact given by Schweigger, that while he collected 219 cases of strabismus convergens with H not less than $_2^1{}_\sigma$, he also had 117 cases of H not less than $_2^1{}_T$ which had no squint whatever. Of course the influence of the ciliary muscles would in both categories be the same, while the influence of the extrinsic muscles must be invoked to explain the difference as to squinting. This fact of the importance of irregularity of the extrinsic muscles becomes manifest in strabismus convergens associated with myopia. In them the affection may not appear in early life, but at a late period. Out of eleven cases in which the beginning was ascertained, Schweigger found that the inception was after the tenth year in eight cases. Irregular action of the muscles will cause squint in choraic and in hysterical subjects, and this may be either permanent, or more usually temporary.

It is important to measure the degree of squint. This can be done accurately only in terms of angles, and then it would be needful to know the exact centre of motion of each eyeball to construct a perfect diagram. But, if each eye has sufficient perception, an approximate measurement in angles can be gotten by the perimeter, provided the nose does not cut off the view of the squinting eye. But, for practical uses, it is customary to measure the squint by noting the place where a plumb-line from the centre of the pupil falls upon the lower lid. Instruments have been devised for this purpose : one, an ivory plate laid off in lines by Mr. Lawrence, and a more correct instrument by Galezowski. In it the action of a screw places a stem over the centre of each pupil, and the distance of each from the median plane of the face is read in millimetres ; the difference gives the amount of squint. Without an instrument the deviation can be laid out upon the lower lid by marking with ink the place of the pupil for the eye which fixes, and the spot to which it moves when the squint occurs. In all cases, some mode of measurement, with tolerable accuracy, must be adopted. The facility with which the eye moves to the outer canthus, and whether this excursion is attended with jerks, is to be noted. By this we gain an idea of the relative strength of the externi as they oppose the contraction of the interni muscles.

Prognosis.—Strabismus, after becoming permanent, usually remains about the same. There are, however, notable variations from this statement. It can last for several years, and at ten or fifteen years of age, or even so late as thirty-five or forty, disappear. In the very young it is quite frequent for strabismus to disappear. This would seem to be the result of increase in power of the externi, but on this point there are as yet no sufficient data to warrant a theory. In cases where the degree is six or more millimetres, and the recti interni are evidently greatly superior to the externi, as shown by the limited abduction and by the jerky movement in the endeavor to reach the outer canthus, the case may be reckoned as permanent. Should the deviation decrease or disappear of itself, binocular vision is not necessarily established. It is understood that spontaneous correction of squint can occur in persons with hypermetropia who do not use glasses. When, however, squint has lasted for one or two years, if the subject is over five years of age it is likely to remain.

STRABISMUS DIVERGENS.

In explanation of this condition we have to assume either that the externi are relatively stronger than the interni, or that one eye taking no part in vision, because of its great deficiency, is allowed to deviate passively. In these cases the accommodation cannot participate. That optical errors are, however, to be kept in mind, is exhibited by a table of Schweigger, viz. :

	Permanent strab. div.	Periodic strab. div.
Emmetropia	37	28
Hypermetropia	4	5
Myopia	59	50
	100	83 183

It appears now that myopia comes to the front as a cause in sixty per cent. of the cases, and in this is justified the statement that while for convergent squint hypermetropia is the chief predisposing cause, for divergent squint myopia takes this part. In the latter case, it is seen that we have both periodic as well as permanent strabismus, and as in convergent squint, the application of the eyes to near objects develops the deviation, while for distance the eyes are parallel. For myopic cases, when not corrected by glasses, the parallelism for distance is intelligible, because there is no incentive to active use of the eyes, by reason of the imperfect sight. If glasses be worn continually, they remove the near point and stimulate the active use of the motor muscles and of the accommodation, so that in mild cases the squint may disappear. For emmetropia with divergence we have to deal with pronounced insufficiency of the interni, and we may or may not have asthenopic or painful symptoms such as were described under a previous heading. If the strabismus be permanent, there is no effort at binocular vision, and no pain, while the person may either use one eye habitually or each alternately. A periodic and manifest divergence may be attended by pain. As to recognition of double images, the same facts exist as in cases of convergence.

It is frequently the fact in divergence that there is an emphatic difference in the refraction of the eyes. One may have E, the other M. Sometimes the normal eye entirely excludes the use of the other ; sometimes the normal is used for distance, and the myopic for near. So, too, one eye may have decided astigmatism, and the other little or none. The better eye is the one depended on ; while, if both be properly corrected and vision reach a sufficient degree, it will often be impossible by any operation or other means to change the patient's habits of seeing. Instead of combining the eyes, there will be suppression of one, and often it is remarkable to see how quickly a patient accustomed to the separate use of each eye will continue to turn first one and then the other to the object, even though the quality of sight has been corrected by glasses. Oftentimes, besides the lateral deviation there is vertical deviation, and this has a powerful influence in preventing binocular vision. It is specially characteristic of divergent squint that there should be an extraordinary capacity for overcoming prisms held vertically. This of course implies the capacity for recognizing double images. It is the fewer number who can do this, and that only by the help of strong abductive prisms or after an operation. To excite double images in divergence is even more difficult than in convergence. It is also to be noted that many cases of divergence ensue after the loss of sight in one eye. This happens after deep opacities of the cornea after occlusion of the pupil, after cataract, or detachment of the retina, etc. Under these circumstances the eye wanders away, apparently not because of weakness of the interni, but usually because it is thrown out of use, and perhaps the configuration of the orbit, or the mode of insertion of the muscles favors it. Divergence is more often the effect of monocular blindness than is convergence. It follows that divergent strabismus is not so uniformly confined to the early periods of life, but may begin at any age, and as a rule does belong to an age later than is characteristic of convergence.

Cases of vertical squint (sursum or deorsum vergens) are the result of paralysis of muscles acting about the horizontal axis, and do not need special description except under the head of treatment.

Treatment.—The object to be attained is: 1st, to remove a deformity; and 2d, when possible, to gain the power of correct binocular vision. The latter is the special object of the surgeon, while the former is the motive which the patient has in view, generally being unaware that an improvement of sight is to be desired or is possible.

Treatment of a gymnastic kind may be undertaken during the early stages of squint, and when the disease is not fully established. If the subject is very young, viz., under three years, the chief possibilities are to avoid employment upon very small objects, and to compel the child to use each eye by screening the one most used for two or three hours daily; also to place it in such relations to the window or lights as to induce the eye which squints to look in that direction. The object simply is to promote the equable use of all the muscles of both eyes. A squinting child should not be encouraged to learn to read at an early age, and all causes of irritation should be completely withheld. The eyes should be examined by the ophthalmoscope, and usually under atropine, to learn what the refraction may be. It will, however, not be expedient to order glasses for a child under six years old and usually not sooner than seven or eight. The risk of accidents in breaking them is considerable, and if they are found to be indispensable, they will be given for reading and sedentary occupations. If it be found that glasses obviate the squint, it will be proper to let the child wear them for a certain period daily to keep the muscles in equable action. The use of prisms is not practicable. It is sometimes the case that atropia will break up the squint; but it cannot be continuously employed, and is simply palliative. For older subjects, when intelligent enough to co-operate with the surgeon, atropia and glasses will sometimes effect a cure, especially in the periodic type. It is, however, needful to bestow great pains in the matter by teaching the child to recognize and fuse double images. Daily practice with prisms in producing and overcoming double images, and in addition the correction of any refractive errors, will for some time be required. A considerable difference in visual acuity will not prevent binocular co-ordination.

In some cases we resort to the stereoscope. Du Bois Reymond, and more lately Javal, have strongly commended it. One must choose a card which has on one field the half of a square and the other half on the other field, or on one a dot and on the other vertical lines; or spots of different colors, or at different heights, on the respective fields. By such tests the patient is made conscious of the use of each eye, and when he may be bringing the fields together. There is no guarantee with the ordinary pictures that he really uses both eyes; the usual stereoscope will often not be suitable, but one in which prisms and special correcting-glasses, or prisms alone, will be made from the trial case of glasses to suit the demands of the case. Cases of periodic squint may in this way often be cured, and permanent squint is sometimes dealt with successfully. But, for the former, and especially for the latter cases, the undertaking is extremely tiresome. For patients with periodic squint able to appreciate and fuse double images, and not with too great disparity in visual acuity of their eyes, the attempt may be promising. But even with them we have a lack of balance in the muscular apparatus which may demand mechanical correction. It therefore results that an operation is to be advised in the great majority of cases. I consider it not proper to operate until six or seven years of age, when a child has acquired a proper understanding of test-types and other incidents of an examination. There is no harm to sight from waiting, and the observance of hints as to the exercise of the muscles will be all that can be usefully done. An earlier operation is very likely not to be as advantageous, both because the muscles are not fully developed and because of the difficulty of applying optical treatment. The operation is strictly a tenotomy—not myotomy, as in the days when first performed. The tendon is severed from the globe, and by the elasticity of the muscle is drawn back to acquire adhesion in a new spot. Of course, the farther in front its insertion lies, the greater rotation it will be possible for the muscle to effect. As a fact, the insertion of the rectus internus, on an average, reaches forward 7 mm. beyond the spot where it first touches the sclera (Volkman), the eye looking straight forward. For the rectus externus the overlapping part is 13 mm. Now, if the degree of strabismus convergens be 5 mm., the internus is shortened by 5 mm., while its opponent is stretched 5 mm. If we dissect up the insertion of the internus so that the externus may turn the eye around 5 mm., we have gained the desired purpose, although it does not require that the internus shall retract 5 mm. from its original insertion. It does retract, but the activity of the externus must bring the globe to position.

The operation for convergence is performed as follows:

The patient is anæsthetized, but not to the most profound degree; the wire speculum separates the lids. With a spring-catch fixation forceps, seize the conjunctiva near the outer border of the cornea on the horizon, and roll the eye to the outer canthus. It may be let fall as seen in Fig. 55, and will hold the eye steady. Take up the conjunctiva by toothed thumb-forceps, midway between the inner margin of the cornea and the caruncle, so as to lift a horizontal fold; a long-bladed pair of scissors with probed points snips the fold, and, gliding beneath the conjunctiva, loosens the connective tissue overlying the tendon. Through the wound introduce the forceps and seize the end of the tendon; have it clearly in view, and with the scissors make a small opening, and through this aperture enter a strabismus-hook. The upper half of the tendon is cut away from the globe, the hook next slipped below the remaining part of the tendon, and this similarly cut away. To bring the hook in place after it has entered the perforation through the tendon, it is laid flat and the point is directed away from the part to be severed ; then turning it over the convexity of the eye, it is rotated under the bit of tendon to be cut. All futile fishing is thus avoided, and when in position the handle of the hook is laid over the nose, and its convexity pressing upon the insertion, there is no difficulty in making a close division between the hook and the sclera. Care is to be taken to divide all fibres adhering to the barbed and probe-pointed tip of the hook. The blunt barb (a suggestion of Dr. Theobald) prevents the hook from slipping away, and makes it easier to secure all the tissue. When all is supposed to be severed, the hook is to be swept around to catch any remaining fibres ; but no cuts are to be made into the capsule of Tenon on the sides of the tendon. At the close of the operation the conjunctival wound will have been stretched somewhat, but at first it is to

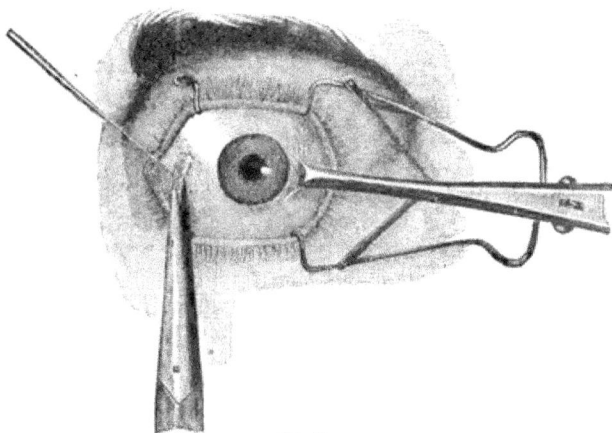

Fig. 55.

be as small as will permit the scissors to pass through, and great care is to be observed not to disturb it unnecessarily. Usually no important hemorrhage occurs until after the tendon is cut, but there will be very variable experiences. If a large thrombus form under the conjunctiva, it will increase the effect of the tenotomy. It is useless to try to get it out. If bleeding occur during the operation, an assistant will remove it by bits of rag, which I prefer to sponges. Only when extreme, is irrigation with iced water needful, or pressure on the closed lid with a sponge wrung out of iced water. The great point is to make the wounds as small as possible, to limit the interference strictly to the insertion of the tendon, and to avoid all unnecessary dissection. A tenotomy thus performed will not always be attended by the same degree of effect, but it will usually cause a correction of about 5 mm.

Shall one eye, or both, be operated on ? For deviations of 6 mm. or less, but one eye should be touched. After two weeks, if notable squint remains, the second eye may be dealt with, and very circumspectly. For strabismus of higher degrees, both eyes are to be operated on ; but if not greater than 10 mm., it is prudent to take them at different times, giving

8

at least two weeks or more interval. For myself, in such cases I most frequently deal with both eyes at once. Much regard is to be given to the amount of elasticity of the rectus externus, and how well the eye rolls to the outer angle. The best form of scissors is shown in the cut, as well as Theobald's strabismus hook.

In estimating the effect produced, the patient must be allowed to fully recover from the anæsthetic. During anæsthesia the eyes will often widely

Fig. 56.

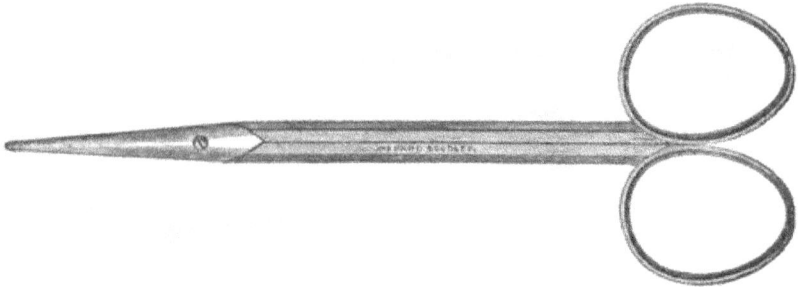

Fig. 57.

diverge, but to this little value attaches. If convergence of some millimetres remains after one tenotomy while the patient is still insensible, I do a second tenotomy on the other eye, carefully attempting to proportion it to the degree desired.

When all is done that seems proper, the patient must be able to perform all the associated movements of the eye, and to converge up to five inches easily. Should a slight squint of say 2 mm. remain, this may be safely disregarded. Should a divergence of 2 mm. be found at near fixation, one's conduct will depend on the degree of hypermetropia, on the visual acuity of both eyes, and on the power of recognizing double images. If the patient have H $\frac{1}{12}$, or greater, and both eyes have good sight and capacity for double vision, a slight divergence will be spontaneously rectified. If the refractive error be small and vision in one eye be poor, no divergence must be allowed to remain. The means of controlling the result after doing the tenotomy is found in a suitable introduction of sutures. In cases where the tenotomy gives an immediately satisfactory effect, a suture is used to draw together the edges of the conjunctival wound; this procures quick union, and prevents sprouting granulations as an after-result. The stitch of finest silk (black, so as to be easily seen afterward) is to be put in from above downward, by a curved needle and proper holder (Sands' is by far the best). The caruncle is drawn upon, and its tendency to sink is counteracted. But the stitch can be made to restrain the effect of the tenotomy by taking in its bite a larger quantity of tissue. The needle must be short and have a sharp curve. If sunken deep into the tendon, and also well under the conjunctiva near the cornea, a decided reduction in the effect, follows its being tightened. By a little care, very accurate results can thus be secured. Some time is consumed, but the patient will submit then much better than later. Even after twenty-four hours if it is seen that a decided over-correction has been made, the error can be rectified by loosening the recent attachment of the muscle with a small strabismus-hook, and putting in a stitch deeply and firmly enough to draw the eye into its proper position.

On the other hand, we may find that the tenotomy has not been adequate. There may have been a single or a double operation. If, after a single operation, only two or four millimetres remain uncorrected, a little more of the subconjunctival tissue may be divided, or a little cut ventured into the lateral connections of the tendon ; but such measures must be very cautiously undertaken. A better and safer thing is to aid the ineffectual contraction of the externus by an intensifying suture. This may be put into the conjunctiva over the outer side of the sclera, gathering up a roll of the membrane, which, like a tuck, shall draw the eye-ball around (v. Graefe). Another way is to run a thread, which shall have a needle at each end, vertically into the conjunctiva at the outer side of the cornea, pass it through the skin at the outer canthus, and tie it over a small roll of sticking-plaster, or bit of match (Knapp). The suture may be cut in twenty-four hours and removed.

Where, however, we have a paralyzed muscle, or one which, by having been too freely divided, has lost control of the eye, and the deviation is large, say more than ten millimetres, something more than tenotomy must be done. For such cases we perform *advancement of a muscle*. There are several ways of doing this. In all cases the tendon of the impaired muscle is loosened, and by means of sutures drawn forward to a place in front of its original insertion. The simplest proceeding is to divide the opponent, then take up the tendon of the weak muscle and put sutures through it and the overlying conjunctiva, and attach them to the strip of conjunctiva remaining at the border of the cornea. Some of the tissue must be cut out, and the effect obtainable is small, because the sutures will not bear much strain.

Another proceeding is much aided by using a clamp strabismus-hook made by Wecker (see Fig. 58). We suppose the rectus internus to be operated on. The opponent is divided ; the tendon is brought to view by a free opening of the conjunctiva, and loosened down to the caruncle, lifting the tissues freely. Then slip one blade of Wecker's hook under the tendon and push it down as far as it will go, so as to get a deep bite. Then shove down the other blade, and don't let the muscle slip. Its insertion is now severed. It is lifted up and fully loosened, while accessory cuts are made into the adjacent tissues on the sides of the muscle. A thread with a needle on each end is passed transversely across the muscle, quilting it in and out, behind the Wecker hook. Each needle is then thrust beneath the conjunctiva, in the direction of the insertion of the rectus superior and rectus inferior, respectively. The globe is turned strongly inward by fixation-forceps ; the superfluous conjunctiva is snipped away, so as to avoid a wrinkle at the inner side of the globe, and the ends of the thread are tied. They pull up the conjunctiva so much that the thread

FIG. 58.

does not touch the cornea. The edges of the conjunctival wound are su-
tured. The opponent of the other eye, as is usually necessary, is now
divided, and sometimes both the opponents are freely dissected up. It
requires two assistants for this operation. I have mentioned the use of a
single thread. Usually two principal sutures are used, and in adjusting
them care must be taken not to twist the globe in its vertical meridian.
Free incisions are made in this operation, and parts have to be extensively detached. Hence, pretty sharp reaction occurs, and the conjunctiva
becomes much infiltrated.

Another mode of performing this operation has just been proposed by
Dr. Prince, of Jacksonville, Ill. (see *St. Louis Medical and Surgical Journal,*
June, 1881). I have practised it once in a case of paralysis of the third
nerve, and have been pleased with the method (see Fig. 59).

I quote his description :

"*Operation.*—The patient being asleep and a speculum introduced, a
fold of conjunctiva over the insertion of the tendon of the muscle to be ad-

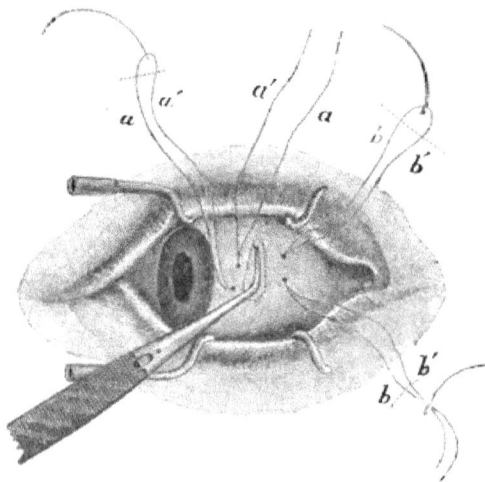

Fig. 59.

vanced is grasped with the fixation-forceps and elevated. A needle armed
with a silk suture is passed through the elevated conjunctiva, parallel to
and about two or three millimetres from the corneal margin, after which
the needles are cut off, making two loop sutures, as represented in the
Fig. (a) and (a').

"A small opening through the conjunctiva and Tenon's capsule below,
and opposite the insertion of the tendon to be advanced, is then made in
the usual manner, to admit of the introduction of one branch of Wecker's
double hook, or appropriate forceps,* which is passed underneath the ten-

---

* For forceps, address Aloe, Hernstein & Co., 300 North Fourth Street, St. Louis,
Mo.

don and drawn tense when the remaining branch is lowered upon the conjunctiva, including tendon and cellular tissue.

"This done, the tendon with the conjunctiva is separated from the ball at its insertion, by the introduction of one blade of the scissors through the opening previously made.

"Lifting the detached tendon from the ball by means of the forceps, the needles, carrying the *double-loop* *suture*, are introduced from within outward through muscle and conjunctiva, as indicated (*b*) (*b'*), the position of the points of puncture depending on the effect desired. This step is facilitated by arranging the needle-holder to carry both needles at once, which is important when no anæsthetic is employed, for the requisite time is thereby reduced to about two minutes.

"Upon the introduction of the sutures, the forceps is to be liberated by separating the contused tendon and conjunctiva with scissors along the dotted line to the right in the figure, the needles being cut off as indicated.

"A subconjunctival division of the opposite tendon having been made at the commencement when necessary, the advancement is accomplished by isolating and twisting sutures (*a*) and (*b*) which form secure loops, respectively through the conjunctiva and muscle. The parts being cleared of blood, a knot is formed and drawn until the tendon is deemed sufficiently advanced, when it is secured by a simple bow. The conjunctival gap will have simultaneously been closed. After recovery from the anæsthetic, and upon the return of the muscular tonicity, an examination is made to ascertain the success of the operation. The patient being directed to fix some object, there should be no motion upon alternately covering either eye.

"If this be the case, the remaining sutures (*a'*) and (*b'*), after some hours, should be withdrawn. Should the effect not have been sufficient, the sutures (*a*) and (*b*) are to be still further tightened—as much as may be necessary to give parallelism to the eyes—and the knot secured, some time after which the remaining sutures (*a'*) and (*b'*) are to be removed as before. Should, on the contrary, the effect of the operation have been too great, the stitch may be loosened or cut with the scissors and removed, the tendon allowed to retract, and the remaining reserve sutures (*a*) and (*b*) brought into requisition, twisted, formed into a knot, and sufficiently drawn to make the effect of the operation perfect.

"The stitch may be allowed to remain until it becomes loose, when it can be removed without pain."

After-treatment.—Simple tenotomy is never attended by serious reaction; in two or three days the patient will be going about, and the subconjunctival ecchymosis will last ten or fifteen days. The bandage may be kept on for twelve hours, and afterward a compress wet in cold water used as the patient likes. Once I had severe reaction after tenotomy of the externus, causing orbital cellulitis, and calling for active treatment during a week. The same result has been recorded by others. In another case, after tenotomy of both interni, true diphtheritic inflammation attacked the wounds and threatened the loss of both eyes. The sight was preserved to each, but an extreme strabismus divergens was substituted for the pre-existing convergence. The boy was about twelve years old, and seemed healthy, but lived in some squalid home. I have operated on many hundred cases by tenotomy, and remember only these two serious events. Usually the patients, if treated at the public clinic, are allowed to go home after the operation and return on the third day. Then the conjunctival suture is taken out, or, if let alone, it drops in about a week. Advancement of the tendon is a more important operation, and in direct ratio **to the amount of** surgical interference.   The patient must be kept in bed; the bandage left on for twenty-four hours, and then cold water, iced or not, as the sensations of the patient indicate, kept on the eye continually, night and day, until the reaction abates.   There will be some chemosis and swelling of the lids; often at the operation the cornea will be overlapped by the conjunctiva. In two cases I have had suppuration of the cornea and panophthalmitis. In one the man had undergone twelve tenotomies at the hands, first, of an ignorant practitioner in New York, and afterward at various European clinics.   He was still distressed by diplopia, and had divergence.   In my search for the retracted and much adherent internus, extensive dissection was unavoidable, and the resulting reaction was more than the eye could endure.   At the expense of his eye the patient was content to be freed from diplopia.   In another case of congenital paralysis of the third nerve, with extreme upward strabismus and a very myopic eye, in a girl fifteen years old, I had already made an ineffectual attempt to sever the inferior oblique at its origin.   Four weeks later I performed advancement of the inferior rectus and tenotomy of the superior rectus of the damaged eye, **and** this was followed by ulceration of the cornea, perforation, staphyloma, and panophthalmitis.   In both these cases of failure the complicating circumstances seem to account for the disaster.   In all my other cases of advancement, which have been not less than twenty, the requisite result has been attained.

Ultimate results of **strabismus** *operations.*—A perfect cure, by which must be included establishment of true binocular vision, is not gained in more than twenty per cent. of the cases.   To **reckon this** result so high as fifty per cent., as Knapp does, is to confound an apparent binocular fixation, in which double images cannot be evoked by prisms, with true binocular vision.   The removal of the deformity and restoration of a comely and harmonious appearance, is a result whose frequency will vary among operators.   Inasmuch, however, as convergence is far less offensive in appearance than divergence, it is always better to leave one or two millimetres in this direction than in the other.   It may be added that excessive effect in cases of H can be diminished by avoiding glasses, and by resorting to adductive prisms and the stereoscope; while the effect, when too small, can be increased by atropine, which permits sight only for distance.   Moreover, turning the eyes toward or away from the erroneous muscle will **increase** or diminish the effect within the first week.   If a given effect is to be carefully maintained, the patient must wear a bandage over both eyes for several days.   This applies only to critical cases.   It therefore follows, that after an operation a patient is to be watched for several weeks, and perhaps kept under treatment by prisms, stereoscope, atropine, glasses, etc., to enable him to acquire command of his eyes.   But when, as often happens, one eye is persistently left out of play, we then have to depend solely upon the correction of the muscular adjustment, and must be sure not to fall into the error of incipient divergence.   Strabismic patients often claim **that** by the operation they gain better sight.   This will be true in a proper sense, when full binocular vision is gained.   It will also be true, in that increased usefulness of the squinting eye is gained by enabling it to fix properly, and perform accommodation as it could not before.   But the actual gain in visual acuity is generally only moderate.   Many anomalies and varieties offer themselves among strabismic patients, which must be left to each one's experience to trace out, and, if possible, to classify.

STRABISMUS DEORSUM VERGENS, OR SURSUM **VERGENS,**

is to be treated by operating on the rectus inferior, **or** on the rectus superior. The obliqui are not suitable for interference. **As a** rule, only tenotomy is to be done ; and it is to be observed that, because of the greater delicacy of the oculo-orbital fascia about these tendons, the muscle will slip back more freely than happens with the lateral muscles. It is also **to be** remembered that division of the superior rectus acts by association on the levator palpebræ superioris, and is followed, not only by a depression of the globe, but by lifting of the upper lid, by which an unusual amount **of** sclera will be exposed above the cornea. This fact may be utilized, if **there** be partial ptosis, both to aid the levator of the lid as well as the depressor of the globe. v. Graefe correctly taught that, if, for instance, the left eye squint upward to a slight degree, its rectus superior is to be cut so as to give proper correction when looking down. Division of the opposite (right) rectus inferior, would cause this eye to go up toward its fellow, and induce insufficiency for both inferiores in looking down. The degree of defect in motility, on the part of the faulty muscle, will decide which alternative is to be chosen. In bad cases, both the rectus superior of one, and the rectus inferior of the other, will **have to be** weakened. For strabismus convergens or divergens, with upward **or** downward deviation, it is proper **to cut** the adjacent tissues rather **freely on that** side of the contracted muscle to which it most deviates.

NYSTAGMUS.

An oscillatory movement of both eyes quick and jerky, greater **at** some times **and in some positions than** in others, is the characteristic of this disease. **In very rare** cases one eye alone is affected. **The con**dition is usually congenital. It is almost always associated **with am**blyopia, while it of necessity much impairs the available acuity **of** sight. We often find it with congenital cataract, both partial and total, also after ophthalmia neonatorum with central opacity of the corneæ ; it is almost invariable in albinoes, and we see it in cases of extreme hyperopia, and sometimes with congenital choroiditis at the macula. Frequently there is convergent strabismus. The movement may be lateral, vertical, or rotatory, or all combined. The cases are almost always congenital ; yet I saw, **by** the kindness of Dr. H. W. Williams, of Boston, a man who had ac**quired** the power of voluntary nystagmus after having been for some eye**trouble** confined for several weeks in a dark room. A form of nystagmus, lately noticed, affects individuals among high mountains, and especially those who work in mines. It comes in adult life, is most noticeable toward **night, is** periodic or paroxysmal, **is** induced by looking in certain directions, **and** apt to be attended with **vertigo.** Nystagmus among the English **miners** has been described **by Oglesby, and** is attributed to their unhealthy surroundings and **the downward** posture of the head in **their** work. Some by refraining from work would seem to get well, **only to re**lapse on returning to the mines. Commonly cases of **nystagmus are not** aware of the movement of the eyes, except by its effect upon sight. **Some** persons, despite this trouble, have highly useful vision. They are apt to be myopic, and distant vision is below the standard, but near work may be prosecuted with **great success.** In New York I have known two notable cases—one a distinguished musical composer and teacher, and the other a well-known practical chemist. Both of them were albinoes.

This condition is sometimes dependent on brain-lesions of recent occurrence. For example, it has been seen to follow blows on the head, also apoplexies, but with no definite localization, and in softening as well as in hemorrhagic pachymeningitis. In some chronic brain diseases it has been noted, and the matter has been summed up by Robin (" Des Troubles Oculaires dans les Malad. de l'Encéphale," 1880). "Nystagmus, unilateral or double, permanent or temporary, exhibiting itself with other convulsive or with paralytic symptoms, indicates an encephalic lesion. In general this will be at the base or on the convexity behind the fissure of Sylvius (region of the angular gyrus). In the former case it will often be complicated with paralysis of the motor nerves of the eye or of the optic ; in the latter case (when on the convexity) there will be epileptic attacks, hemiplegia, etc., but we cannot venture on any exact localization." Irritation of the peduncles has caused this symptom in experiments by Schiff. It occurs among the insane and the neurotic. It is very frequent in disseminated sclerosis of the brain and cord. With locomotor ataxy it is very rare. A not infrequent picture in a case of brain disease of the kind now noted is the concurrence of rotation of the head, conjugate deviation of the eyes, and nystagmus ; what the connection between them may be is yet undetermined. Nystagmus may be seen in cases of aphasia and of labio-glosso-laryngeal paralysis.

It thus becomes evident that, while most cases exhibit a complex causation, consisting both of defective sight and of irregular innervation of the muscles, other cases depend alone upon lesion of innervation, and this of central origin. As to the former class of cases, it cannot be doubted that the irregular movements are in very many due simply to the want of motive for correct binocular fixation, from lack of predominance of the macula lutea.

Treatment of these cases is of little service. For some the correction of optical errors, so far as it can be accomplished under the difficulties of the examination, is valuable. For those with strabismus convergens, tenotomy of one or both interni is advisable. I have done tenotomy of the interni when no strabismus existed, but when the lateral movements were excessive, and found benefit ensue. The degree of tremor was abated ; but, as a rule, an operation is not fitting. Albinotic patients wear dark glasses, and especially those with side-pieces to cut off the glare of light ; and all nystagmic patients hold fine objects close, and have some choice position of the head in which their trouble is less annoying. Excitement greatly aggravates the tremor, and it usually remains unaltered through life. For an exhaustive study of nystagmus, see an article by Raehlman : *Arch. für Oph.*, xxiv., 4, pp. 237–317. His conclusions tend to locate the cause of the disease in the brain, but at what region is undetermined. Another elaborate article is by R. P. Oglesby : *Brain*, vol. ii., July, 1880.

CONJUGATE DEVIATION OF THE EYES.

The symptom thus designated has been noted in literature, in isolated instances, by some of the early authors of the present century ; but Achille in 1858, Vulpian in 1866, and Prevost in 1868, systematically described it, and attempted its explanation.

Many have since written upon it, and the facts we now recognize are as follows :

In certain cases of cerebral apoplexy with hemiplegia, both eyes will be found turned strongly in a lateral direction, and the head rotated toward the same side. Again, the same deviation of the eyes and head will appear while the patient is in convulsions. In cases of paralysis, the deviation is toward the non-paralyzed side, and the eyes will not move in the opposite direction farther than the median line. This is usually a transient symptom, enduring but a few hours or days. When, however, the symptoms are irritative and cause contractions, the deviation is toward the side of the contracting members. In the first case, the axes of the eyes point to the side of the brain where the lesion has occurred ; in the second case, they point to the side of the brain opposite the lesion. In the second case it has been several times proved that the bleeding has been into one lateral ventricle (Wernicke : "Lehrbuch der Gehirn-Krank.," p. 352, 1881).

The same author says that if the deviation be permanent, and consist in true paralysis of the abducens and of the other internus, we are to look for the cause in the pons, where the sixth nerve has its nucleus of origin, and where are found some fibres of the third nerve decussating with it (Bernhardt : "Hirnegeschwulste," 1881, p. 206, quoting experiments of Graux and Duval upon dogs and apes). Graux lays down the rule that peripheral lesions of the sixth nerve will be attended by strabismus convergens, with secondary deviation of the sound eye inward, while central (pontine) lesion of the sixth will be followed by a secondary deviation of the sound eye outward. This latter condition of homonymous deviation is never so decided as that which ensues after hemiplegia. It is not possible to specify what may be the cerebral lesion, or exactly where it is located, by the symptom of conjugate deviation.

There are other forms of symmetrical impairment in the movement of the eyes, different from those previously described and dependent on some central cause, whereby different nerves are simultaneously acted upon. For example, Priestley Smith, in "Ophthalmic Hospital Reports," vol. ix., part 3., Dec. 1879, p. 428, alludes to several which he has seen. Mr. Hutchison has written upon ophthalmoplegia interna, i.e., impairment of both irides and both ciliary muscles. These matters specially belong to the subject of cerebral diseases, and about them we as yet have only imperfect knowledge. The subject is enticing, but does not pertain to the subject of our study. Adamük has shown that the anterior pair of the tubercula quadrigemina preside over the co-ordination of the eye-muscles.

PART II.

CHAPTER I.

DISEASES OF THE LACHRYMAL APPARATUS.

Anatomy.—We have to do with the secretory and with the excretory parts of the apparatus. The former, which supplies the tears, consists of a series of small follicles situated in the superior conjunctival cul-de-sac, and the lachrymal gland, while the conjunctiva itself also secretes moisture which may be counted part of the lachrymal fluid.

The lachrymal gland is lodged in a fossa at the upper and outer angle of the orbit, and may be felt by the finger indistinctly, as it is overhung by its rim. It is an acinous gland like the parotid, subdivided into a smaller and a larger lobule, which are separated by a septum of fascia. The smaller **is** sometimes called an accessory gland. There are numerous isolated acini lying near the principal masses. The size of the chief gland is variable, but may be stated at 20 mm. in length, 11 to 12 mm. from before backward (breadth), and 5 mm. in thickness. It is concavo-convex, and lies against the periosteum. Numerous ducts, whose orifices are from ten to twelve in number, give exit to the secretion at the temporal side of the superior fornix. The tears contain 1.25 per cent. of sodium chloride and .5 per cent. of albumen.

The excretory apparatus begins as minute openings (the puncta), about 6 mm. from the inner angle of the lids, which lead into small canals (canaliculi), and they unite to empty by a common orifice into the side of the lachrymal sac. The sac rises a little above the place of entrance of the canaliculi, and is continuous below with the lachrymo-nasal duct, which **empties** into the inferior nasal fossa, behind the tip of the inferior **turbinated bone.** The total length of the sac and duct is about one inch (25 **mm**). **Its** section is ovoidal, with the long axis from before and outward, **backward** and inward. Its calibre varies greatly, and its shape may also vary. **In** the same skull, from which the **soft** parts have been cleared, I have seen the duct on one side **to be round,** and not more than 3 mm. in diameter, and on the **other to be oval** in section, and in its major axis **6 mm.** long. The membrane **lining the duct and** sac is like that of the nostrils, being both **a periosteal and a** mucous membrane. It is highly vascular, thick, **and** covered by cylindric epithelium, lying on several layers of spheroidal cells. The cylindric cells are by some declared to be ciliated. Next the bone **the membrane** is spongy and erectile. It is thrown into folds at two or three points, viz., at the junction of the sac and duct, which corresponds with the beginning of the bony portion of **the** tube in the ascending process of the superior maxillary bone, and also at the lowermost part, where it communicates with the nostril. There is also, sometimes, a less distinct fold at its middle. The lining membrane of the canaliculi is thin and pale, and the puncta are a little whiter than the neighboring membrane. They point toward and rest in contact with the globe. Muscular fibres surround these openings like sphincters, and they are held in apposition with the eye by the action of the orbicularis and tensor tarsi muscles.

The latter lies behind the lachrymal sac, and the tendon of the former crosses in front of it and is sometimes called the tendo oculi. It is brought into relief by pulling upon the lids at the outer canthus. The orbicularis has additional insertion into the lachrymal bone by bundles of fibres which go to it directly. The tears are forced into the excretory passages by the action of the muscles just mentioned, aided by a kind of suction caused by the muscular fibres of the puncta and canaliculi (Klein). Unless the puncta are kept in tonic contact with the eye, the tears cannot enter. The quantity of fluid is usually so small that evaporation and secretion balance, and nothing passes down to the nose. With any irritation of the eye, a larger flux occurs, and frequently the capacity of the tubes is overtaxed and tears brim over the lids (epiphora). Usually the follicles in the superior fornix and the conjunctiva furnish all the needed moisture, but on unusual demand the lachrymal gland comes into play.

DISEASES OF THE LACHRYMAL GLAND.

Acute inflammation occurs in rare instances. I have in one case seen both glands inflamed at the same time. The symptoms are, swelling, by which the border of the gland is pushed down out of its fossa and can be recognized on turning up the lid by its uncommon prominence and by the orifices of its emunctory ducts; there is œdema of the lid, and tenderness of the gland and of the neighboring edge, together with dull pain. The inflammation usually resolves without suppuration. The treatment consists in warm fomentations and anodyne lotions. Constitutional treatment is not often needed, although the possibility of a syphilitic cause is not to be ignored.

The gland may be the seat of neoplasms, such as sarcoma and other tumors, and of cystoid degeneration, and it is liable to chronic hypertrophy; but these conditions need no special consideration. Its extirpation to cure epiphora was practised by Mr. Lawrence, but is not now approved.

DISEASES OF THE EXCRETORY APPARATUS.

We have eversion and stoppage of the puncta, occlusion of the canaliculi, catarrh of the sac, and obstruction of the duct. We also have acute dacryocystitis, chronic distention of the sac, and fistula lachrymalis. Sometimes there are two canaliculi in each lid. Eversion of the puncta is the consequence of chronic blepharitis marginalis or of chronic conjunctivitis, or it follows from paralysis of the orbicularis muscle in lesions of the facial nerve, and necessarily accompanies ectropium. In the first class of cases the orifice is apt to be made smaller; in the paralytic cases the punctum may be uncommonly prominent as a papilla, and while the lower one sags down, the upper also fails to lie upon the globe.

The canaliculi are sometimes the seat of stricture, and in a few cases chalky concretions have been found in them, or a vegetable growth—the leptothrix.

Catarrh of the sac and duct is a lesion which is not often presented to us at an early stage, because people are apt to avoid the surgeon until the disease has lasted so long that simple catarrh has become complicated with obstruction. There is practically no real separation to be made between these conditions. In dacryocystitis we have swelling of the mucous membrane, hypertrophy of its epithelium, and papillary growth—sometimes a state precisely like granular conjunctivitis, and with this a mucopurulent, glairy, somewhat tenacious secretion, which fills the cavity and is there retained. The calibre of the nasal portion of the passage speedily becomes choked, and the morbid secretion, after a time, cannot find outlet ; hence, the sac-wall undergoes **distention.** The three factors of thickening of the mucous membrane, excess of secretion, **and** distention of the sac, gradually conspire to bring about a more or less aggravated condition, **in** which the lachrymal tumor becomes larger and the stricture smaller. The skin, after a long period, becomes thin, and may even get to be translucent. It may in very old cases happen that the lachrymal bone becomes diseased. The constant and annoying effect of this state of things, at almost any period of its existence, whether early or late, **is** to cause an undue quantity **of tears** to be formed ; they overflow the lid or stand ready to drip over. **On** exposure to wind or to cold air, **the eye** waters uncomfortably, and the **fluid** sweeping **over the** cornea makes **vision** misty, and continuous use of **the** eye is sometimes, and more especially **at** night, greatly embarrassed. The tears which thus flow too liberally are called forth, it is true, by a hypersecretion of the lachrymal gland ; but they are likewise mingled with the products of the irritated conjunctiva and its glands. The universal concomitant of dacryocystitis is palpebral conjunctivitis, sometimes severe, and not infrequently blepharitis marginalis coexists. The caruncle and semilunar fold are swollen and injected, and aid in hindering the entrance of fluid into the puncta. The patient is constantly using his handkerchief, and thus materially aggravates his troubles. But he may learn, and this should be taught by the physician, to keep the sac empty by squeezing its contents into the nose, if the passage be permeable or it gushes out of the puncta upon the eye. Wherever it goes, keeping the sac empty affords some relief. But when the disease has lasted long, the secretion acquires irritating qualities, especially if it be permitted to stay long unexpelled from the sac. Then its contact with the eye sets up decided conjunctivitis, and the fluid may even have an offensive odor. In **cases** which have lasted long, the fluid is sticky and unpleasant ; especially **is it mischievous** if the eye has been submitted to an operation. The pus **appears to** have an infectious quality, and is extremely apt to cause suppuration in a corneal wound. It follows that cases of cataract, or cases which require iridectomy, should be first relieved from any lachrymal trouble.

The disease is one of slow progress, and often goes for a long time without causing great annoyance. Even after a tumor appears at the **inner canthus,** the swollen sac may not cause any great discomfort. But, **if it be** impracticable to empty it by pressure, the stricture is evidently close and the condition obstinate. But a most unpleasant complication in the progress of the disease is the occurrence of acute phlegmonous inflammation and abscess. This is severely painful, and may cause an extreme degree of swelling of the lids and neighboring parts. The tumor will be red, shiny, and tense. If not large, it will be very tender to the touch, and the conjunctiva will be hyperæmic. If the process be left to itself, the matter finally escapes by ulceration, and in this case a fistula lachrymalis is quite liable to ensue. The opening will be below the tendon of the orbicular, and may be large or small. I have seen one case of congenital lachrymal fistula affecting both eyes. It was reported by Dr. Agnew.

In other cases which have been long neglected, the subjacent bone may become carious, and a passage may even take place into the superior nasal fossa, or into the cells of the ethmoid. In general, the disease will either remain stationary or grow worse—it does not get well. It may be tolerated for years with slight discomfort, or it may prove unpleasantly exasperating.

Before entering upon the consideration of treatment, a few words may be given to a condition which causes epiphora, and is apparently not associated with the morbid lesions above described.

I have seen a few patients who were annoyed by an accumulation of tears, in whom I could find no swelling of the sac, nor tenderness over it, nor could I elicit any discharge. At the inner canthus there was swelling of the semilunar fold, and turgescence of the caruncle; the puncta were prominent, but not everted nor choked, neither was there obstruction of the canaliculi. The cause of the epiphora seemed to be the swollen state of the caruncle and of adjacent parts; this irritation excited hypersecretion of tears, while the prominence at the canthus served to obstruct entrance of the fluid into the puncta.

This rare condition has been noted by Graefe, and I have seen it a few times.

Diagnosis.—We have epiphora and a swelling over the lachrymal sac. The tumor will be effaced by pressure of the finger, and its contents will either flow over the eye through the puncta, or else pass through the nose. If by pressure the tumor do not wholly subside, the sac-wall may be very thick, or the stricture be very tight. If very large and the walls thin, its bluish color may suggest a cyst; but the history of epiphora will settle the doubt. The caruncle is red and apt to be swollen, and the puncta also to be swollen and red, and of unusual size. In some very quiescent cases no tumor appears, but pressure will force fluid into the nose. These varieties depend on the duration of the malady, and on the amount of secretion and the degree of obstruction.

Treatment.—We have the palliative and curative. A considerable number of persons are not greatly disturbed by their lachrymal trouble; another portion are too timid to submit to surgical proceedings, and others are unwilling to spare the time which effectual treatment demands. For these patients, only palliative proceedings can be used, and they are as follows:

To keep the sac empty by pressing on it with the tip of the finger from above, down, and backward, so as to force the fluid, if possible, into the nose, and do this with firm, slow pressure. A certain knack is often acquired by the patient which the physician cannot imitate. If the fluid must be disgorged on the eye, the handkerchief must be in hand to absorb the fluid at once without needless rubbing of the lids, and at all times the eye should be gently pressed, and not wiped. The sac must never be allowed to approach distention.

The use of astringent drops or of a lotion upon the lids, or occasional applications to the palpebral conjunctiva, as this surface may be more congested, will do good service. Moreover, the state of the nasal cavity must be inspected, and duly dealt with. Washing out the nostrils with warm, salt water by a syringe, the application of depurating and astringent fluids in spray by an atomizing apparatus, or by the blowing of powders into the nostrils in the manner called for in the treatment of nasal catarrh, will be well worth doing.   For spray I use the following formulæ:

R. Sodæ biboratis.............................. ℥ iv.
     Glycerini.   .... ℥ j.
     Sodæ carbonatis........................... ℥ ss.
     Acid. carbolic............................. ℥ ss.
     Aquæ..................................... ℥ vj.
M.

R. **Ammoniæ et ferri sulphat**.................... ℈ ij.
     Aquæ................................... ℥ iv.
M.

R. Liq. Ferri persulphat......................... ℈ j.
     Aquæ.................................... ℥ iv.
M.

For powder, among many which may be chosen are the following:

R. Bismuthi trisnitrat.,
     Gum acaciæ ....āā ℥ j.
     Pulv. cubebæ............................. gr. x.
M.

R. Acid. boracic. pulv. q. s.

Under such management, some persons get along fairly well and are satisfied.   Many do nothing more than keep the **sac** empty, and expect when they get a coryza to have more trouble—and so they do.

The curative treatment involves a careful discrimination of the state of the sac and duct, and the suitable adaptation of means, and it requires weeks and months for its realization.   It must also be stated that a considerable number of patients will not obtain a full restoration to soundness, but a sort of half-cure, which is less than they would like, but far more than without treatment they would have.   Premising this statement, and also that the introduction of probes is always unpleasant, and to some persons really painful, a patient is prepared to screw up his courage to go through with the business.

If the lachrymal tumor is easily emptied into the nose—and this implies **that the case** is recent—external applications may suffice.   In children of **a** strumous quality it is impracticable to use probes, and often the cleansing **of** the nostrils by a camel's hair pencil, and the use of cod-liver oil, iodide of iron, etc., will bring about recovery.   Carefully wipe out the nostrils with cotton on a holder, and to them apply vaseline twice daily, and a solution of nitrate of silver, gr. x. ad. ℥ j., twice or thrice weekly, or the powdered boracic acid once daily.

But the common run of cases call for treatment of stricture of the nasal duct.   The first step **is to** slit the canaliculus, which Mr. Bowman taught **to be** the best mode of approaching the **sac**.   My preference is for the lower one.   I also **choose** a beaked knife, with a blade wider than is generally used, and set in a long and stiff, but malleable shank.   For a case of no long duration it may be needful to do no more than slit the canaliculus. The surgeon, if operating on the right eye, will stand behind the patient. holding the head against his own body, use the left hand to draw the lower lid out and keep it tight, and insert the beak of the knife perpendicularly into the lower punctum. (See page 129.) Sometimes this is partially occluded. The point of a pin will usually open it for the tip of the instrument. When well engaged bring the hand to the horizontal position, and push the blade with cutting edge inclined inward and upward into the sac until the tip is felt to strike the lachrymal bone ; keeping the point firmly against the bone, raise the handle up, and also lift up the blade so as to incise as freely as possible the conjunctival wall of the sac. Many surgeons stop at this point, and let the patient apply cold water as may be comforting, and on the next day attempt to introduce a probe. In the greater number of cases I do not follow this mode of proceeding, but at the first operation carry the knife down into the nasal duct and divide the stricture. I make two or three incisions upon different sides of the duct, to gain the greatest enlargement. I am always glad to see blood issue from the nostrils, as proof that the passage has been opened. When the stricture is divided, as Stilling recommended, a larger instrument can be inserted, viz., the larger end of Weber's probe, and afterward the probes of larger sizes. My conviction is that the stricture should be expanded to the fullest degree, much beyond No. 6 of Bowman, and I have a size No. 10, which measures 11 mm. in circumference. Cases must be dealt with according to the calibre which is normal to each, and the fullest possible expansion obtained. The probe may be used three times weekly, and be left in place from ten to thirty minutes. Progress can, in some cases, be made rapidly ; others will permit only a gradual increase. The amount of reaction after probing will regulate the frequency of introduction and the rate of enlargement. In passing the probe, carry it horizontally into the sac, and when its point impinges on the bony wall, bring it to a perpendicular and attempt to follow the axis of the duct. The direction is downward, outward and backward, toward the wing of the nostril. The aim must be to get behind the edge of the opening into the superior maxilla, and until this is gained the probe must be handled with delicacy, and in the exercise of a nice sense of touch. Caution at this point is indispensable, and a moderate degree of it will avoid making a false passage. After this opening is gained the probe may be firmly sent down until it reaches the nasal fossa. It should be left in place for ten minutes, and then withdrawn. This exploration will indicate what kind of stricture we have in hand, and what instrument will best dilate it.

I have, during a year or more, made use of Theobald's probes, and find them exceedingly satisfactory. They go up to large sizes, No. 16 being the maximum. Dr. T. has advocated the use of large probes in a paper in *Arch. of Ophth.*, vi., and in "Trans. Am. Oph. Soc., 1879," and was not aware that Dr. E. Williams, of Cincinnati, myself, and others had, for many years, sought to secure the fullest dilation which the anatomical and pathological conditions make possible. Dr. H. W. Williams, of Boston, has introduced probes with bulbous tips and elastic necks, which, while stiff enough to handle easily, find their way around projecting obstacles or through sinuous passages better than straight instruments. I have often had occasion to be pleased with their qualities. But my ultimate resort is to a large instrument, smooth, with conical point, which must press its way through the inflammatory deposit— not with violence, but with some force ; and this is to be left in situ from ten to thirty minutes, but not long if its pressure be extremely painful. Making haste slowly is the password to success with these cases, but I am convinced that the gate must be opened widely and made to stay open, to get full relief. Dr. Theobald's probes are of the following sizes : beginning with the diameter of $\frac{1}{4}$ mm., advancing by increase in circumference of 1 mm. from No. 1 to 16, the last being 4 mm. in diameter. I have found them so well contrived that I have adopted them almost to the exclusion of all others. (See page 129.)

As the result of probing, abatement of the catarrhal secretion is soon manifest. In most cases, nothing more than probing and treatment of palpebral conjunctivitis is needful. In a certain number, secretion is copious, and does not measurably diminish. The syringe must then be employed with a weak solution of argent. nitratis, gr. v. ad. ℥j., or gr. x. ad. ℥j. Perforated probes have been devised for this object, but a small, hard-rubber dental syringe can be readily adapted to the purpose by bending its nozzle in hot water to an obtuse angle. After the probe has been withdrawn, the syringe may be used. It will not require protracted employment.

In cases where persons cannot spend the weeks or months with the surgeon which treatment requires, the plan may be adopted of putting in a leaden wire about size No. 6 or 4 Bowman, which shall lie in the duct with its upper end properly bent downward and inward at the inner canthus. This style may be worn as in old times Scarpa's nail was worn, for two months, more or less, and it will then be found to have brought about absorption of the stricture. It excites considerable secretion, is not agreeable to wear, but answers fairly well. Granulations are liable to spring up at the entrance into the sac, and when the style is taken out, the opening soon contracts and is difficult to find.

I have another suggestion to make in this matter. Some cases permit dilation of the stricture with reasonable rapidity and to a satisfactory degree, but the annoying epiphora does not stop, and the patient does not find the pain of the treatment compensated by good results. It must be remembered that there may be another stricture at the bottom of the duct where it enters the nose. Here I have many times found a nodular projection from the side of the canal, or a decided narrowing of its calibre. To overcome this stricture the common probe is futile. I have had a form made which is a repetition of a very old instrument, with a bulbous tip and of

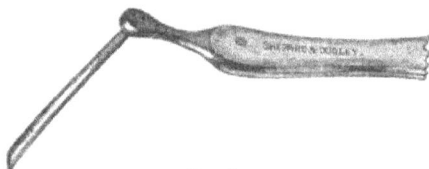

FIG. 60.

unusual length. It is carried down to the lower end of the canal in the ordinary way, and then, to get it into the nose, the flat handle must be rotated toward the temple so as to turn the point backward, and then push it onward. It will go down almost an inch farther, and it may so far penetrate the nostril as to touch the place of junction of the hard and soft palate. Some obstinate cases of epiphora have been cured by ascertaining the presence of this hidden stricture, and resorting to the instrument thus described. In cases where the obstruction at the bottom is osteoid, I have used a narrow gouge with a cutting end, and have bored a way into the nose (see Fig. 60). Afterward steady probing would be needed to prevent the return of the obstruction.

I have found Theobald's probes able to cope fairly with the cases just cited, because of their well-fashioned tips and greater length. But they deserve special notice, and the probe I have devised will sometimes be necessary.

There remains another class of cases in which the passage cannot be restored to its normal state : either because of excessive thickening of the lachrymal sac, or the duct is almost occluded by osteoid growth, and is practically impermeable, or there may be caries. The older writers proposed opening into the nasal fossa by perforating the lachrymal bone. The

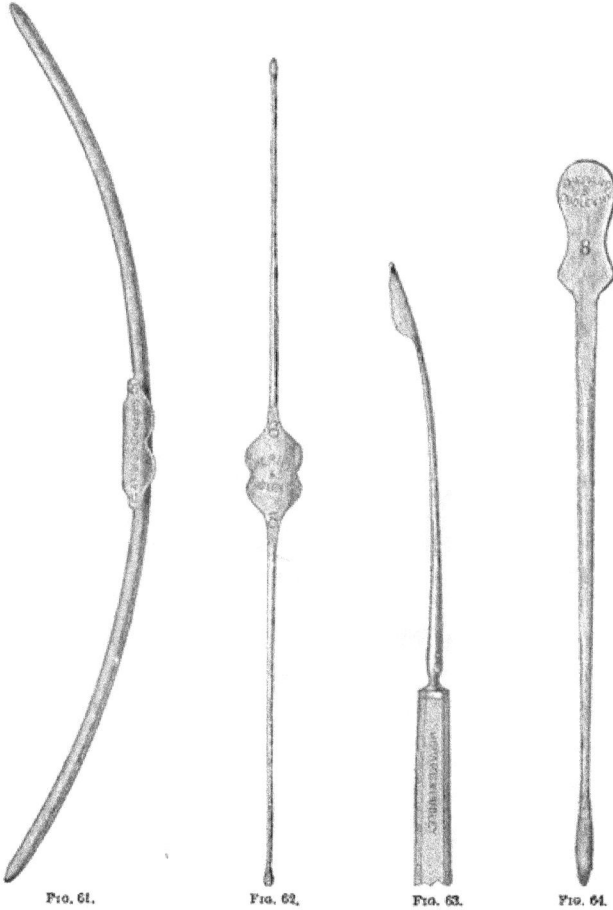

Fig. 61.   Fig. 62.   Fig. 63.   Fig. 64.

modern treatment is the obliteration of the lachrymal sac and duct. This is done by dissecting out the hypertrophied sac, or by destroying it either by the actual or potential cautery. Excision of the sac may be combined with the cautery. After dissecting out with scissors as much of the sac as can be removed, the beak of the heated iron is thrust into the duct. Usually, fuming nitric acid is the agent employed. The sac is freely ex-

9

posed by an incision in the skin, and when the bleeding stops, a bit of wood—the untipped end of a match—with some fibres of cotton on it, is charged with it and freely applied to the mucous surface. Care must be taken to protect the eye, and the edges of the wound must be held asunder by sharp hooks. This operation has been done by Dr. Agnew through an incision upon the mucous side of the sac, with simultaneous division of the canaliculi, and he reports good results. I have not followed his method ; and although, as commonly performed, a scar is left upon the skin, I have not found it a conspicuous thing or a deformity. Still another mode of destroying the sac is by putting into it pieces of nitrate of silver. This causes prolonged pain, and is less effectual than the red-hot iron or the nitric acid. After cauterization, the sac is stuffed with lint, and cold-water dressings applied. It will take two or three weeks for the wound to close by granulation. When the cavity is obliterated, the success which follows in relieving the epiphora depends on the fact that there is no longer an irritation in the sac to stimulate a superabundant flow. The obstruction of the excretory passage causes no inconvenience, except when some special occasion for weeping arises, such as keen winds or mental emotion. In fact, however, it is not easy to perfectly obliterate the sac and duct, and hence this treatment does not give uniform results ; but it is a great amelioration of the previous condition. In very young children a probe may be passed by the help of chloroform. I have seen lachrymal abscess, with stricture on both sides, in a child six months old, and treated it successfully by the usual method. I have sometimes instructed a patient to use the probe for himself, when he had reached the proper size, and simply needed to maintain the enlargement.

Added to the above suggestions for local treatment, the possibility of syphilitic infection must be borne in mind and the suitable medical treatment adopted. The iodide of potassium and corrosive sublimate will do the same service as in any case of specific periosteal inflammation. In all cases where nasal catarrh shows decided symptoms, this must receive attention.

PHLEGMONOUS INFLAMMATION OF THE LACHRYMAL SAC.

This takes place as an incident during the progress of a chronic dacryo-cystitis. The attack is always painful, may be ushered in by a chill, and varies greatly in severity. Swelling, tenderness, and hardening of the sac are always present, while sometimes the lids become puffy, especially the lower lid along the furrow which lies below it, and in a few cases the œdema of surrounding parts has simulated severe orbital cellulitis. Even though the swelling be small and circumscribed, the patient commonly suffers much pain, and the reason is the same as in the case of any sub-periosteal inflammation, viz., the effusions are compressed by dense membranes and the nerves are numerous.

Treatment.—It is rarely of any use to do anything else than to make an incision into the sac. If the case be seen early, this may be done by way of the canaliculus and slitting freely the sides of the sac, thus preparing for the probe at a future period, when treatment of the original stricture shall be in order. But, if much swelling have taken place, the knife should be put perpendicularly upon the skin over the middle of the sac and thrust through it to the bone, and then with one sweep carry the incision down for half an inch, more or less, according to the extent of the tumor. The best surgery is an early and a free incision. By doing this the occurrence of fistula is almost certain to be avoided, while it is very likely to be the disagreeable consequence of permitting the abscess to "break of itself." After opening the abscess, warm-water dressings and poultices will be applied until the attack subsides. The cut will be kept open by a bit of lint.

It is not denied that sometimes, when phlegmonous inflammation begins, resolution may occur, and this is best promoted by the continuous use of hot poultices, of which the ground slippery-elm bark is the most eligible.

Lachrymal fistula is occasioned by the imperfect healing of an abscess, and implies the existence of a permanent stricture. This lesion is not seen as frequently as once it was, nor do lachrymal diseases often attain the extremity which the older writers describe. Surgical aid is better and more ready to be instituted than in the elder day. Hence, a bad case of caries or of fistula does not often get an opportunity for production. It is needless to describe the condition—it declares itself; and if dead bone be present, the probe will soon discover it, if the foul odor and discharge do not betray it.

For a bad case, cleansing by the syringe through the fistula may sometimes be proper, together with attempts to restore the calibre of the duct. If there be dead bone, this may be left to gradual elimination or be removed by a small gouge. For such cases destruction of the sac will generally be a necessity. In general, it is better for those which are less severe to slit the canaliculus and deal with them as if there were no fistula. So soon as a route can be established for the secretions to make their way into the nostril, the fistula will heal. In case it prove sluggish, the process of closure may be hastened by stimulating it with a pointed crayon of nitrate of silver (Squibb's caustic points). Cure of the stricture carries with it cure of the fistula. If the stricture be incurable, the obliteration of the sac is the alternative, in the manner above described.

It may be remarked, in summing up the whole matter of lachrymal troubles, that the larger number may be completely cured, another proportion are relieved of special annoyance, and the remainder gain some benefit, but still have trouble. That a perfect cure should always ensue, it would be unreasonable to expect; that palliation is better than no relief, is evident, while patient continuance and careful discrimination of the precise lesion are indispensable to success. Moreover, in no cases more than these is the tactile address of the surgeon an element of value to win confidence and spare needless pain, and thereby contribute to success.

A case reported by Dr. Bull (*Am. Jour. Med. Sci.*, July, 1880) is worth remembering, where caries of the ethmoid bone caused a prelachrymal abscess, and on opening it the lachrymal sac was not involved, nor its cavity entered. In such cases there will be no epiphora. Treatment will simply be to provide for the escape of the discharge by washing out the cavity with antiseptic and slightly stimulating solutions.

Another condition may occur, and at first give the impression of being a case of lachrymal sac distention. It is a prelachrymal tumor, either a lipoma or a fibroma, or other neoplasm, which may form in the skin of this region. Its mobility, that it can to some extent be grasped by the thumb and finger, and the absence of epiphora, will guard against a false diagnosis.

CHAPTER II.

DISEASES OF THE EYELIDS.

Anatomy and physiology.—The eyelids have, as their framework, the tarsi (formerly called tarsal cartilages), upon which we have externally the orbicularis muscle and the skin, and internally the conjunctiva. The orbicularis muscle shuts the lids; and attached to the upper edge of the tarsus is the broad tendon of the levator palpebræ superioris, to open the lids. In the substance of the tarsi are found the Meibomian follicles for the secretion of an unctuous matter to grease their borders, and follicles to produce the lashes or cilia. There are also other glands at the border of

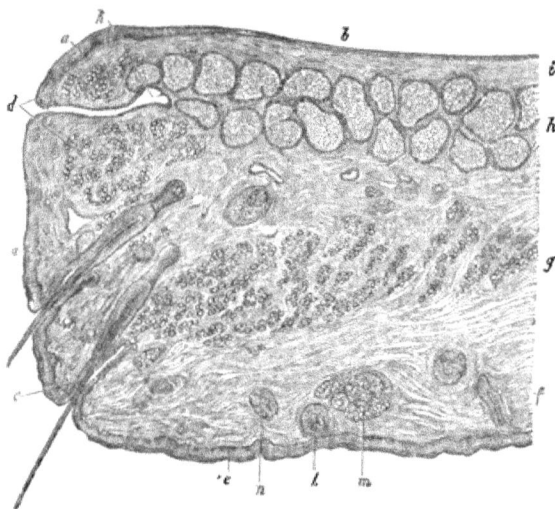

FIG. 65.—Vertical section of upper lid of a new-born child: *a*, epithelium of the free border; *b*, epithelium of the palpebral conjunctiva; *c*, epidermis of the skin; *d*, muscular fibres of Riolan; *e*, cilia; *f*, skin; *g*, orbicularis muscle; *h*, Meibomian gland; *i*, conjunctival tissue between the Meibomian gland and the epithelium; *k*, duct of the Meibomian gland; *l*, hair-follicles of skin; *m*, sebaceous follicles; *n*, sweat-follicles.

the tarsi, which lie close to the hair-follicles. For a fuller explanation, see Fig. 65. The lids are concave posteriorly, are held in close contact with the globe, and slip over it without conscious sensation or friction. Their edges are about two or three millimetres broad, and meet upon a line which is slightly curved and inclines downward to the inner canthus. The outer canthus is sharp, while the inner is of horse-shoe shape, and presents a little space called the lacus lachrymalis. The cleft or fissure of the lids varies considerably in length, and the lids are tight or loose, long or short, in different persons and in different ages. They are held in place at their ends, by bands of fibrous tissue, the external and internal canthal ligaments, which attach the tarsi to the outer and inner borders of the orbit. The internal ligament is really the tendon of insertion of the orbicularis and muscle.

The skin of the lids is thin, and set with very minute hairs and a few sweat-glands, and is destitute of subcutaneous fat. Between it and the subjacent muscle is loose connective tissue. The prominence or recession of the globes modifies essentially the appearance of the lids. It is curious to note how, in cases of exophthalmus, the eyelids will keep pace with the advance of the eyeball, and actually increase in extent. Besides the muscles alluded to, fibres of the occipito-frontalis enter the integument of the eyebrow, and can indirectly lift the upper lid.

The orbicularis muscle reaches, in curved lines, all around the palpebral fissure, extending upon the side of the nose, upon the temple, and for a distance of about one and one-half inches above and below the palpebral slit. Its fibres near the free margin are pale and thin, and certain special bundles which are near the conjunctival surface at the inner edge have been designated as a distinct muscle (muscle of Riolan).

Other muscular fibres, which are of the non-striped variety, belong to the lids and are attached to the peripheral edges of the tarsi, and mingle with the bundles of the levator in the upper lid ; in the lower lid they are fewer and less regular. They were first noted by H. Müller. They reach into the orbit a short distance, and are under control of the sympathetic nerves. The levator palpebræ superioris originates in the apex of the orbit, and by a broad tendon is inserted into the posterior edge of the upper tarsus. It is in close contact with the roof of the orbit. The tensor tarsi muscle has been before referred to. A muscle, similar to and smaller than this, is sometimes found at the outer canthus, and was discovered by Dr. Mosely, of New York. The lids are connected with the upper and lower margins of the orbit by a sheet of fibrous tissue—the tarso-orbital fascia. This shuts off communication between the connective-tissue space of the lids and the orbital cavity, and has been described on page 103, as an offshoot of the capsule of Tenon. The tarsus of the upper lid is about twelve millimetres wide, and that of the lower lid is about five millimetres wide.

BLEPHARITIS MARGINALIS—OPHTHALMIA TARSI—BLEPHARO-ADENITIS.

The disease appears in varying degrees, either at the angles of the lids or in isolated patches, or along the entire border of the lids. It may consist in maceration of the epidermis, which produces a moist, red surface ; or the lashes may be stuck together in bundles, by crusts which cover ulcerations of the tarsal edge ; or the whole border may be glazed with a red, thickened and ulcerated surface, from which almost all the cilia have fallen. It is characteristic of the affection that the hair-follicles atrophy, and the lashes drop out. The conjunctiva will also exhibit chronic inflammation. In very chronic cases, in the aged or the uncleanly, the edges of the lids become everted (ectropium) and the lachrymal puncta are both occluded and displaced. The disease affects chiefly young persons of delicate skin and strumous habit, and in a considerable proportion of cases it is associated with some refractive or muscular error which renders the ordinary treatment of little value. This point deserves recognition.

The ailment is apt to be chronic, but will yield without much difficulty if suitably treated. For all cases, soothing lotions, warm water, or warm water and milk, are to be preferred, and when the crusts are softened, and as much as possible removed, the following ointments may be used : two grains of hydrarg. oxid. flavæ to one drachm of vaseline or amylo-glycerine ; or, ung. citrini, gr. x. vel xx., vaselini, ʒ j., to be applied night and morning, or, in bad cases, more frequently. In a large number of cases, the best method is to pick off the crusts with fine forceps or the finger-nails, and cauterize the exposed ulcers with a fine point of nitrate of silver. It often bleeds, and the caustic hurts. In cases of extensive incrustation, and especially in young children, the lashes may be cut off with scissors to facilitate the denudation and cauterization of the ulcers. The subsequent use of stimulating salve will then control the disease. But if the person be the subject of **error** of refraction, or of other error which causes eye-strain, the removal of the blepharitis will not only demand the **usual local** treatment, but also that the error be corrected. (See Part I. **of this treatise).**

HORDEOLUM OR STYE.

This affection is a phlegmonous inflammation at the tarsal edge, which forms a small and generally painful lump. It is apt to be associated with chronic blepharitis or conjunctivitis, and often depends on general debility. In **its** inception it may sometimes be checked by applying a bit of ice wrapped **in muslin for** a few minutes repeatedly, or by pulling the cilium which passes through it. One is apt to follow another in succession. **When** suppuration is unavoidable, a poultice of ground slippery-elm bark (ulmus flava) is most **comforting,** and a puncture should be made at an early period. General tonics and mild astringents are the proper remedies to prevent their recurrence ; but it is important also to investigate the state of refraction, because what causes eye-strain will provoke styes. Another frequent concomitant **and** favoring condition is nasal catarrh, which will also need attention.

CHALAZION, OR CYSTIC TUMORS OF THE LIDS.

Obstruction and distention **of** some of the follicles **of** the tarsus, more frequently of the Meibomian, **are** the origin of these tumors. They are painless, imbedded in the tarsus, and the skin is freely movable over them. They **vary** in size, and are apt to come in crops. The sac-wall is usually thin, **and as** the tumor enlarges it causes a reddish **or** yellowish projection **on** the conjunctival surface, and sometimes presents granulations. The contents are a glairy, mucilaginous fluid. Fluctuation is never felt, and I have sometimes found a solid, fibrous **tumor** when I expected to meet a cyst, and at times the cyst-wall has been extremely thickened. When small, the tumors are not troublesome, and they occasionally disappear. If they reach a size to be annoying, they must be excised. I prefer to do this on the skin surface, through a wound parallel to the lid-border, and no perceptible scar is left. The cyst may be opened on the inside surface if it project notably in this direction. Special forceps have been contrived by Desmarres, Snellen, Prout, and Knapp, to enclose the tumor in a clamp which shall prevent bleeding during the dissection. A flat spatula, or the operator's forefinger slipped under it, will hold the lid tense until the sac is exposed, and will fully check bleeding, when the tumor should be caught with a sharp hook and removed by the scissors. I have made use of the clamp, figured in the (see Fig. 66) text, which is to be preferred,

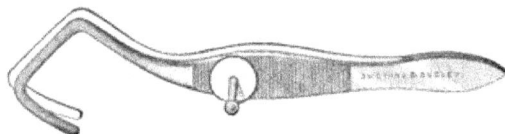

Fig. 66.

because it gives access to both surfaces of the lid, and permits removal from the conjunctival surface when deemed advisable. The posterior wall is often thin and the lid may be perforated. Often the whole sac cannot be excised, and then a point of lunar caustic is to be thrust into the wound. Suppuration ensues, a little, hard nodule remains for a time and then disappears. To guard against recurrences, remove chronic palpebral conjunctivitis, and avoid eye-strain.

PHLEGMON OF THE LID.

If suppuration occur in the connective tissue of the lid, as may happen after debilitating disease, or in strumous children, or without recognizable cause, there will be great swelling, and fluctuation will be detected early. It may come with very little pain and but little redness. It is also important to remember, that a general inflammation of the lids may occur in delicate children, and not result in suppuration : there will be great œdema and slight redness, and the whole may disappear by resolution. On the other hand, the tissue may become gangrenous in cachectic subjects. If suppuration occur, the pus must have vent early by a free incision, parallel to the border of the lid. The best knife is a Beers cataract-knife, or a very narrow, sharp-pointed and curved bistoury. Stand behind the patient, pierce the skin, and run the point along with a quick, steady thrust. The earlier the incision is made, the less will be the likelihood of deformity after the abscess heals.

In cases of erysipelas of the face, if there be much induration of the lids, care must be taken to watch for suppuration. It is very liable to occur, and considerable destruction of tissue may take place, which early incision would obviate. In the severe forms of the disease, it sometimes becomes needful to make deep incisions when there is no evidence of pus, to save sphacelation of the tissue.

SYPHILITIC ULCERATIONS

are sometimes found upon the lids—they may be chancres or secondary ulcerations; although the latter are more likely to appear on the mucous surface. Another lesion which is quite rare is syphilitic tarsitis. It is hardly necessary to say anything about the recognition and treatment of these conditions. They only need to be mentioned (see paper by Dr. Bull, "Trans. Am. Ophth. Soc.," p. 408. 1878).

EPITHELIAL CANCER AND LUPOID GROWTHS

are quite often situated upon and near the eyelids. A discrimination between them is hardly needful for practical purposes. If a nodular, irregular elevation appears on the lid border, or on the skin, and is covered by a dark crust which, when picked off, exposes a bleeding surface, and if **this continue** for months or years, sometimes healing and again breaking out, but never going entirely away, this neoplasm, although quite painless, had better be excised. The true epithelioma is more rapid in development than a lupoid growth, and both may result in ulceration. In either case the neighboring lymphatic glands are not likely to be enlarged, except at a late date. The gland which we look for is that in front of the tragus —the pre-auricular gland. Growths such as we are now considering occur during and after middle age, and usually remain unheeded for a long time. Sometimes soothing lotions will procure healing of the ulcer. I have known **the solution** of chlorinated soda (Labarraque's), diluted with five parts **of water, to** be followed by perfect cicatrization of **a** suspicious **ulceration. The** sore did not break out again for many months.

The **disease will not get** well of itself, and the only treatment **is** to remove it. For **almost all cases,** the best method is by an operation, and not by caustics. **If the latter** be applied to the lids, some deformity will follow which **will necessitate** an operation, while if the knife be resorted to, the deformity may be at once remedied by a suitable plastic proceeding. I am disposed to allow a limited applicability of caustic to cases in which the disease is not very near the eye, as, for example, the temple, or on the forehead where the skin is not elastic. The process is painful, because the caustic paste must be left in contact with the disease for twenty-four hours, and sometimes longer. Its action may extend beyond the region intended. It is therefore a method to be resorted to in only a few peculiar cases. Among "cancer doctors" it is the habitual treatment, and lauded because, as they allege, no operation is done—and this is a word of fearful sound to many people.

But the very large proportion of cases are to be submitted to careful **and** complete excision by the knife. In most cases a reparative operation can be immediately done which more or less perfectly obviates deformity. These operations will be described hereafter.

I have seen one case of *amyloid tumor* upon the border of the lid. The patient was a young woman and under the care of Dr. Prout, of Brooklyn.

Papillomata or *warts* are not uncommon on the border of the lids. They may be snipped off, or accurately touched with nitric acid applied by a platinum probe, or by a small and pointed stick, like a lucifer match.

Horny growths have been known to occur on the lids. One instance I have had in my own practice.

Xanthelasma or *Xanthoma* is a fatty degeneration of the connective tissue of the skin, which seems to have a predilection **for** the lids, although it occurs elsewhere. It appears as yellow or straw-colored patches, usually upon the inner or medial side of the lids. They may have a little elevation, and be smooth or slightly nodular. They are usually symmetrically placed upon both sides. In a case recently exhibited to the New York Ophthalmological Society, both eyelids of both eyes were occupied by **these** patches in prominent welts over their entire surface. The conspicuousness of the marks makes them objectionable, and the only means of disposing of them is by an operation—a proposal to which I have not as yet found a patient willing to consent. They seem to appear oftener in women than in men, and not in early life.

All the varieties of diseases of the skin may appear upon the lids, and it is not worth while to enumerate them. Mention may be made of a dusky discoloration which sometimes occurs, often associated with uterine disease. It is likely to be permanent. Eczema very frequently occupies this locality, and is commonly best treated by soothing remedies. In the moist variety, by absorbent powders, and in the desquamating stage by the oxide of zinc ointment or diluted citrine ointment. If the face be attacked by eczema of the acute form, the lids will share in the inflammation and the conjunctiva will be very probably involved. The latter complication adds not a little to the patient's discomfort, and is to be dealt with as subordinate to the eczema. The stress of the treatment must be for the latter, while remedies usually effective for acute conjunctivitis will be found of small value. Soothing applications which the patient's feelings approve are to be preferred, viz., warm fomentations with fluid extract of opium ; or, if water be irritating, as it often is, coat the lids and their edges with vaseline, and take pains to wash or wipe away all secretion with a small brush. Acute eczema of the face, accompanied by acute conjunctivitis, is a disagreeable combination. For details of such cases I refer to treatises on diseases of the skin (*e.g.*, Tilbury Fox, p. 192, 1873).

HERPES ZOSTER OPHTHALMICUS

is an affection which deserves attention because it exhibits conspicuous and important features. It is called by the French *zona ophthalmique*, and has been extensively described by Hybord, and previously by Mr. Hutchinson. It is, in truth, a neuropathic affection having its cause in degeneration of the ganglion of Gasser, or of the branches of the trigeminus, or of both. Any of the branches of the fifth pair may be thus affected, and the eruption is localized by the distribution of the diseased nerve-twigs. It therefore happens that vesicles may occur on the eyeball as well as upon the skin, and both ulceration of the cornea, acute conjunctivitis and acute iritis may take place. It is also said that small abscesses have been found in the ocular muscles. The mode of occurrence, as illustrated by a case very recently seen in a boy ten years of age, was as follows : the supra-orbital nerve was the one affected. The initial symptom was intense pain along this nerve at the supra-orbital notch, around the lachrymal sac and side of the nose, upon the forehead, and up to the vertex. In a few hours the skin of the forehead became red and swollen, tender to touch, and a few vesicles appeared above the inner end of the brow. While the right half of the forehead, red and swollen, presented the look of erysipelas, the left half remained natural.

The hair could not be combed because the scalp was tender, and a few vesicles could be discovered. The eyelids swelled, a slight conjunctivitis appeared, chiefly affecting the palpebral surfaces, and there was great photophobia. The pulse was quickened ; it reached ninety, and some **fe**brile reaction occurred. The urgent symptom was the pain, which **con**tinued day and night. A few vesicles appeared on the side of the nose ; none whatever showed themselves across the median line.

The treatment consisted in keeping the boy in bed and dropping into the eye every two hours a solution of sulph. atropia, gr. ij., ad. $\bar{z}$ j., to abate the pain, using upon the forehead hot fomentations without intermission, and giving full doses of morphia and quinia sulphate three times daily. By the fourth day there was decided mitigation of the symptoms, but it was not until the twenty-fourth day that the patient could go out. No lesion of the cornea took place. In case the latter should occur it would be much longer before the patient would be well. I have noticed that when there has been an eruption on the cornea its surface is markedly anæsthetic. This suggests a reason for the long continuance of the affection in some cases, and also the need of keeping the eye bound up so long as the irritation continues. The special treatment suitable to cases of ophthalmic shingles, in which the cornea or iris may be involved, will be found under the chapters which treat of these troubles respectively. I have seen one case in which, while one eye was destroyed by the direct mischief of the disease, the other was also lost through sympathetic irido-choroiditis. I have notes of a case in which both sides of the forehead were attacked. Permanent scars remain, which may be recognized by their rounded form, and by a slight depression of the surface.

The disease may take place at any age, and it is most hurtful to the aged and feeble. It is very apt to be regarded as simple erysipelas, but from this it may be discriminated by the intense neuralgic pain following certain nerve-twigs, by the strict localization of the skin trouble, and by the vesicles. The lesions may go down to the tip of the nose, or upon any part of the distribution of the trigeminus. The treatment, as above specified, should **be both** local and constitutional, the latter being such as may control neuralgia, the former to soothe the local inflammation. It is said that when the vesicles appear on the nose the cornea is most likely **to** be involved. I cannot support this statement, because I have found the corneal affection both with and without implication of the nasal twigs.

Vascular tumors appear on the lids with some frequency. They may be simple red patches, level with the surface, or little, flattened elevations, or deeper seated masses of inosculating vessels. The last are cavernous tumors, the former are telangiectasiæ or nævi. The last may be either con**genital** or acquired ; the former are congenital. These growths need no special description, and the only remarks to be made about treatment are that the choice of method depends upon their size. If not large, they may be destroyed by piercing them with red-hot needles ; I have found shoemakers' awls very serviceable, and the dental blast-lamp is convenient, although **a** large alcohol-lamp will suffice. As many punctures may be made at a time as the size of the growth will call for. Reaction is usually slight, and several sittings will be needed. The galvano-caustic is the most convenient appliance.

Excision is sometimes practicable, and for small growths hemorrhage is prevented by a clamp-forceps (Snellen's). Injection of perchloride of iron or other fluids is not to be commended. Some cases are too extensive to be safely dealt with, and it is always necessary to bear in mind that the tissues will shrink by repeated burnings and that deformity of the lids must be avoided. I have seen a case in an infant for which the common carotid artery had to be tied. The tumor disappeared. Threads may be run through a mass if thought proper, and allowed to cause suppuration. Small tumors and even large ones have been known to disappear of themselves.

Minute vascular growths, like little red warts, sometimes appear on the border of the lid. They may be easily tied off with fine silk.

Moles or brown patches may occur as congenital diseases on the lids or in the neighborhood. They should be excised, and afterward a proper plastic operation performed. I was called upon to do this for a young lady, whom I saw again after several years, and found that similar pigment-nodules had appeared upon the neighboring skin which had previously been healthy. The primary growth was congenital, was set with stiff hairs and seemed to be innocuous, although a decided blemish. The subsequent pigmentation showed no malignant or ulcerating tendency.

THE EYELASHES

are subject to a variety of disorders. They are liable to fall out *(tylosis)* *(madarosis)*, as the result of chronic blepharitis tarsalis, and in consequence of secondary syphilis. The loss is either partial or total. When the result of syphilis, it will be noticed that the eyebrows also fall out.

I have seen partial discoloration of the eyelashes *(canities)* take place in two instances, and one was in a boy twelve years old. It presented a white patch among a row of dark lashes, and seemed to be extending.

Distichiasis, which means a double row of lashes, occasionally occurs, and the inner row is apt to touch the eye. More than two displaced rows are said to have been observed. The condition may be congenital or acquired.

Trichiasis is the inversion of a few lashes so that they touch the eye. It is caused either by marginal blepharitis, or more commonly by chronic trachoma. The offending hairs may be so atrophied and fine as to be difficult to discover, and patients complain very much sometimes that the " short hairs" are overlooked. Frequently there is spasm of the orbicularis muscle which aggravates the condition. Trichiasis passes into entropium by imperceptible degrees, and the treatment of it will vary according to its extent. When but a few hairs are misplaced, the patient sometimes prefers to have them pulled out as often as they become troublesome. This must be repeated every week or two, and frequently is learned and practised by some member of the family.

A method commended by Snellen consists in passing through the border of the lid, alongside of the errant hair, a very fine thread, both ends of which have been put through the eye of a fine needle ; as the loop comes to the base of the hair, the latter is caught by forceps and put into it. The loop is then drawn through the skin, dragging the cilium into the substance of the lid. Its follicle is by this means said to be compelled to change its direction and the hair falls from the eye. Another and effective proceeding is to pierce the edge of the lid with a glover's needle heated red-hot and pushed up alongside the hair. The lid must be held by a clamp.

The operation most generally approved for trichiasis not accompanied by marked shrinking and incurvation of the tarsus is that known as the Arlt-

Jaesche. It consists in transplantation of the ciliary border upward. The patient should be etherized, an assistant lifts and holds the lid tense on a horn spatula. A thin and narrow scalpel is pushed into the border of the lid at the space between the lashes and the mouths of the Meibomian follicles, and splits it from one end to the other into two layers for a depth of six to eight millimetres, or one-fourth to one-third of an inch. By this incision the front layer which carries the cilia is made movable, but is not separated from its connection at either end nor above. Sometimes it is separated at its upper edge and then is freely movable, but more risk of sloughing attends this proceeding. A narrow strip of skin from one-fourth to one-half an inch wide, and the subjacent muscular fibres, are now taken out of the lid at or below its middle, which leaves upon the ciliary border about one-fourth of an inch of skin, and this piece is united by sutures with the cut edge above. Careful manipulations and a just adaptation of the parts are necessary, and the case must not be too far advanced in tarsal deformity. The wound at the edge of the lid is left to heal by granulation. When the layer of skin which bears the cilia is completely dissected up, except at its extremities, there is, as has been said, some danger of sloughing. This I have had take place. Arlt warns against the proceeding. His own practice is to put two or three narrow strips of isinglass or court-plaster lengthwise of the lid, without sutures, and then to close the eye with a cotton pad and bandage. I have had many successes with this operation and think it is to be resorted to ; but it is sometimes difficult to say whether it or another proceeding by Dr. John Green, which will be described under the head of entropium, should be chosen. If there be serious doubt, the latter should be preferred.

In former years, the abscision of the whole ciliary border was frequently practised ("scalping the lids"), and in the extremest cases it may sometimes be necessary, but the occasions for it are rare. In doing it the edge of the lid is split into two layers, as in the Arlt-Jaesche method, and the lifted piece cut away. No sutures are used, and cold-water dressing is applied.

ENTROPIUM.

An operation for entropium has been proposed, within two years, by Dr. John Green, of St. Louis, which I have practised with much satisfaction. It may or may not be preceded by enlargement of the palpebral slit by canthoplasty. The clamp figured on page 135 seizes the lid and everts it so as to expose the ciliary border. A Beers cataract-knife is run along the edge so as to slit it into two leaves, the front one containing the cilia, the rear one the Meibomian orifices. A cleft is made three to four millimetres in depth, to fully liberate the cilia and allow them to turn out. To extend the cleft to the inner and outer ends of the lid, the clamp is removed and the lid controlled by the fingers or by forceps. Next a cut is made in the skin about six millimetres above the ciliary margin along its whole length. The fibres of the orbicularis muscles are dissected out with scissors as far as they can be reached, and the outer surface of the tarsus cleanly exposed. It is optional whether a strip of skin from four to six millimetres wide shall now be excised. If the skin be abundant this will be done, otherwise none will be removed. Sutures are passed through the ciliary border by small, sharply curved needles,

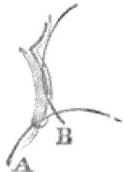

B

A

FIG. 67.

and over the forefinger of the other hand placed beneath the lid to stretch it. The thread is drawn through and the needle pierces the *upper edge of the tarsus and tendon of the levator*, and having been drawn through, next perforates the upper edge of the skin. Not less than three such deep sutures, which gather up the ciliary border and the tarsus and skin together, are used, and other sutures as many as need be through the skin alone. The deep sutures, by their hold on the tarsus, turn out the ciliary edge most effectively. They are left in situ several days. The efficacy of this proceeding consists in the deep insertion of the sutures, and this feature gives it most satisfactory success (see Fig. 67). A method which includes this idea was published by Anagnostakis ten years ago, and revived by Dr. Hotz, of Chicago, two years ago. In this we have a decided advance on former modes of operating. The incision into the posterior surface of the tarsal edge is the valuable suggestion which makes Dr. Green's method admirable.

Ectropium

is due to chronic thickening of the conjunctiva, to contraction of the skin by cicatrices, and to injuries. The excision of a strip of conjunctiva along the whole length of the lid, and of suitable depth and width, will often cure the first variety. The second, due to cicatrices, very often needs the introduction of additional skin, and constitutes a true case of blepharo-

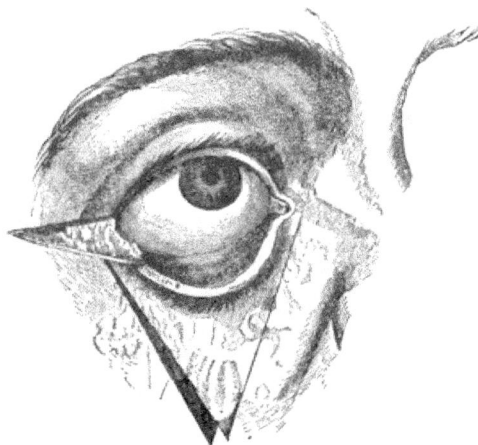

FIG. 68.

plasty. But a proceeding by Wharton Jones is sometimes useful in mild cases. He intended it for the upper, but it serves still better on the lower lid. For a mild case the proceeding will be as follows: a triangular flap with the base toward the ciliary border, and of length according to the degree of deformity, is dissected up, and the dissection not carried too far toward the conjunctiva, else the flap may slough. The lid is pulled up to place and the sides of the wound brought together to push the flap upward, and the lines of the cicatrix will form a letter Y in place of the letter V, which was first made.   For a more serious case a more extensive method may be used, which I have done with advantage (see Figs. 68 and 69).   The figures are from Ruete, and the essential modification consists, in the removal of a horizontal triangle of skin at the outer canthus, which allows a greater lifting of the lid and also provides for shortening its excessive length.   The figures explain themselves.   There are many other ways of dealing with ectropium by excising portions of the lid at the middle or at the outer

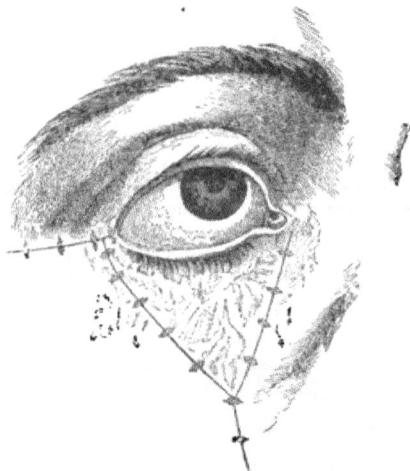

Fig. 69.

canthus.   One must have some inventive faculty to meet the various contingencies of practice, and space does not admit introduction of other cuts to illustrate various methods.   Ectropium from injuries or after removal of growths or tumors will be mentioned under the head of blepharoplasty.

Lice sometimes find lodgement in the cilia.   They are the pediculi pubis.   Their eggs will be found as minute globules strung like beads on the lashes, and a little searching will set the animals in motion.   The latter will be disposed of by fine forceps, and the ova destroyed by a few applications of mercurial ointment made soft with vaseline (about three parts to one).

AFFECTIONS OF THE MUSCLES OF THE LIDS—PTOSIS.

We have partial or complete paralysis of the levator palpebræ superioris. It may concur with impairment of other muscles supplied by the third nerve, or be isolated.   If complete, the upper lid falls over nearly all the cornea, and is raised only by extreme contraction of the occipito-frontalis, which lifts the eyebrow, and by traction on the skin pulls the lid up enough to enable the patient to peep under it when he throws his head backward.   The attitude of the head, when patients attempt to use their defective eye, is highly characteristic. The causes of the disease are peripheral or central ; the most frequent is syphilis. It may be well to remark that a little drooping of the lids may be congenital, and that an inability to lift the lid to its full height is always present when there is chronic conjunctivitis or trachoma. The true action of the levator is conspicuously suggested when the only muscle able to act on the lid is the occipito-frontalis. The latter simply stretches the skin, and if it be lax can exert but little effect while the levator pulls from the cavity of the orbit and rolls the lid over the convexity of the globe into its opening, and at the same time causes a fold in the skin.

The *treatment* of ptosis consists, first, in combating the cause if this be ascertained ; second, in stimulating the muscle by the faradic current of electricity ; thirdly, in operating. As to the first indication, we give iodide of potassium in small doses, gr. v. ter in die for supposed rheumatic cases, and in larger doses with mercurials in syphilitic cases. This treatment should be held to for four or six weeks. After the first week or two the battery may be used for a few minutes, once daily, or as often as practicable. With one pole behind the ear, the other is placed on the lid, and the current should be only of moderate strength.

When time enough has elapsed to prove the impossibility of recovery, say three or four months, we may resort to an operation. It will consist in removing a narrow bit of skin, and a much wider strip of the orbicularis muscle undermining the skin above and below for this purpose. Then unite the wound with sutures which go through the muscular fibres and deep fibrous membrane, as well as through the skin. Another mode lately suggested is to carry a thread subcutaneously through the eyebrow and the tissue of the lid to the tarsal edge—to draw it tight and let it cut its way out. The result is said to be a sufficient drawing up of the lid by the cicatrix. For cases in which other branches of the third nerve are involved and the movements of the eyeball are impaired, it may be inexpedient to operate for ptosis, because this would subject the patient to the serious inconveniences of double vision.

For that form of ptosis which depends on redundancy of skin or its hypertrophy, the operation will not concern the orbicularis muscle, but will include only a portion of the skin large enough to lift the tarsal edge to the level of the pupil when the sutures are in place.

The proposal to advance the tendon of the levator palpebræ has not served a good purpose. I once tried it unsuccessfully. In all these cases the relief is partial, and results only in enabling the patient to use the eye, without remedying the disfigurement. An attempt has been made to enable a patient to open the eye by using a cord of india-rubber (Von Bibber), one end of which was fastened by plaster to the lid, and the other to the forehead. I saw some patients wear these comical appendages for a time, but they cannot be expected to keep up such inconvenient amusement perpetually.

An interesting, but slight and rare form of ptosis, has been noted by Horner, in which the lid droops very little, the pupil is contracted, and the face turgid. All these symptoms point to a paretic affection of the sympathetic nerve in the neck. Their counterpart is seen in exophthalmic goitre, where, by stimulation of these fibres, the upper lid is retracted through spasm of the organic muscular fibres (Müller's muscle) and the pupil dilated.

Paralysis of the Orbicularis Muscle, or Lagophthalmus,

causes more annoyance than ptosis. The eye **is** fretted by external irritation and overflows with tears which cannot be directed into the lachrymal puncta. Chronic conjunctivitis, and even inflammation of the cornea are to be expected. The cause lies in lesion of the seventh or facial nerve. It is not uncommon for the orbicularis to escape when other muscles supplied by the inferior branches of the facial plexus are involved ; but, if the orbicularis is paralyzed, all other muscles are also apt to suffer. The causes of facial paralysis are peripheral, or lie along the track of the nerve, or are **in** the brain. From the crookedness of its course, and the variety of **tis**sues which it traverses, the nerve is greatly **exposed to injury,** and **it** may be wholly or partially affected. One need only remember that diseases of the ear, and of the lymphatic glands, and of the parotid, are all liable to do mischief to the facial nerve. As to the cases of cerebral disease, Eulenberg says that facial paralysis, originating from lesions of the pons, involves the orbicularis ; while, if it proceed from the cerebral peduncles, or from the central ganglia, or from progressive paralysis of cranial nerves, or from spinal cord affections, the orbicularis is likely to escape. Peripheral paralysis of the seventh nerve usually includes the orbicularis. The cases of partial impairment are most common. I have seen instances of total pa**ralysis** caused by wounds in the space between the angle of the jaw and the mastoid process. In one case the wound was inflicted by a mason's trowel.

Treatment will be governed by the supposed cause of the lesion. The **remedies** to be used will suggest themselves. If the cornea be much exposed, it may be necessary to wear a bandage, or to partly close the lid by a strip of plaster near the outer canthus. During sleep the lid will drop a little from its own weight, because the levator is relaxed and the cornea turns up so as to be covered, even when paralysis is total. One of the aggravations of the **trouble** comes from the frequent wiping of the eye to get rid **of** the tears, and the lower lid is likely to be dragged down to an additional degree, and may pass into a permanent ectropium. When this state arrives, relief will be afforded by paring the edges of **the lids for** ten **to** fifteen millimetres at the outer canthus, and uniting them by sutures to shorten the palpebral opening. It is needful to pare away only the inner angle of the tarsal edges and the cilia are left untouched. The sutures should be left in situ from four to six days. I have lately seen a man for whom I did this operation seven years ago, and the relief it gave, continued until within a few months. The continued paralysis and the drag of the lax tissues finally brought on troublesome ectropium of the lower lid at the inner can**thus,** with a return of the former epiphora. I performed the same opera**tion** of tarsoraphy at the inner canthus, which had been previously done at **the** outer canthus. I dissected **up** a parallelogram of skin above and below **the canaliculi, for a** space **which** reached from the commissure to 3 mm. beyond the puncta. I turned the raw surfaces of the little flaps, raised from the respective lids, against each other and stitched through them. The puncta were thus turned inward and out of sight. The edges united, and the palpebral slit was left as a narrow oval through which the pupil could peep, and the annoyance of the epiphora was removed.

SPASM OF THE ORBICULARIS

occurs as an independent disease without co-existing trouble of the eye. It is infrequent, and of variable intensity. It appears as a twitching of a few fibres, which does not continue long, and is simply a trifling annoyance. It also occurs as a paroxysmal and painful spasm participated in by other facial muscles, and constituting the disease known as tic. I have seen several examples of this affection. The face is thrown into ludicrous and painful grimaces, and the spasm is excited by very slight irritations. In the case of a car-driver who had two attacks of it under my observation, it once seemed to be produced by the severe cold of the winter wind. Only one side of the face was affected, and when the attack came it would be thrown into extreme convulsions it would become red, his mouth be drawn up, the lids tightly shut, and he would complain of severe pain. Such turns would happen many times a day, and they recurred during several weeks. He appeared to derive benefit from full doses of bromide of potassium. I have seen a clergyman whose whole face was thus contorted in interrupted paroxysms, and for whom remedies seemed of little value ; but after several years I was happy to find that his affection had become almost imperceptible. For further particulars as to this topic, reference may be made to treatises on diseases of the nervous system. A less serious spasm, which is a kind of nictitation, may occur as an unconscious habit or trick. It may be confined entirely to the eyelids, and be a congenital and life-long peculiarity. A distinguished sculptor of my acquaintance is thus affected, and has a little impediment of speech. A lady friend of mine has had it for seventy years, and her powers of speech are not impaired. Sometimes there will be a point of tenderness over the supra- or infra-orbital nerves, or on the temple, and perhaps at the back of the neck, or the cause of reflex irritation may be a defective tooth. Should such tender spots be found, they suggest indications of treatment, such as neurotomy, local anæsthesia, counter-irritation, hypodermic injection of morphia at the place of tenderness, etc. In many cases no treatment avails.

In the trifling cases, in which an almost imperceptible twitch occurs in a few bundles of the muscle, search should be made for conjunctival irritation, and in default of this, general tonics will generally be in order.

Blepharospasm which is not idiopathic, but associated with some ocular trouble, must be treated, in the first instance, by attempting to remove its cause. This may be severe conjunctival inflammation, a foreign body in the eye—either on the cornea or on the lid, perhaps buried in the fornix ; there may be hysterical photophobia, and Donders says he has seen blepharospasm as an evidence of sympathetic neurosis, caused by irido-choroiditis and atrophy of the fellow eye. In scrofulous children it is the most distressing symptom of conjunctival and corneal disease.

In all the above cases, suitable proceedings will suggest themselves according to the cause, but certain special suggestions may be made which are to be addressed to the spasm. In the hysterical cases, administration of ether or chloroform will often give more than temporary relief, and be a valuable adjuvant to other treatment. Galvanism may also be tried. We have a valuable remedy to combat this condition in the fluid extract of conium maculatum. It is potent and perilous. Its effects must be watched with great care, and the doses graduated to the tolerance of the patient. For an adult, one may begin with ten drops of Squibb's fluid extract three

10

times or six times daily, and notice the first signs of intoxication. There will be dizziness, inco-ordination of muscles, a little thickness of speech, and as these tokens appear the eyelids will open. There may be considerable toleration of the remedy, and the needful quantity cannot be anticipated; one must keep on until evidences of constitutional disturbance are discovered. When they do appear, great caution must be exercised not to let the medicine go too far. But, in watchful and **prudent** hands, much good can be gotten **out of conium.** Severe cases of **strumous** conjunctivitis and keratitis, **both in children and in adults, are subjects** for this treatment, and sometimes **cases of cataract operation, which** make a slow recovery because **of iritis or other inflammation, arrive at a stage** where blepharospasm is severe, and I have **seen the use of conium followed by** happy results. It may **be** tried in hysterical **cases, but their great** susceptibility to drugs must **be** borne in mind.

Another proceeding which we have **at** command **is** the division **of** the external canthus by a straight incision horizontally outward, either by scissors **or** by scalpel; another suggestion in vogue in the Brooklyn Eye and **Ear** Hospital is to forcibly stretch the lids asunder, by the fingers or by two elevators, after an anæsthetic has been administered. Before doing any operation, or even before giving conium, take a look at the state of the skin at the outer canthus. It will probably be ulcerated, or perhaps deeply fissured; in either case, touching either lightly or severely with nitrate of silver will help **to** heal the abraded skin, and materially aid the treatment. Frequent use of **vaseline, or of** simple cerate at this **spot, is** important.

Still another **remedy, suited** to children **with** strumous inflammation, or adults, is **to** drop **iced water** upon the exposed eye every fifteen to thirty minutes by a dropping-tube. It acts as a **local** anæsthetic, and is a suggestion of Dr. Oppenheimer, when house surgeon **of** the New York Eye and Ear Infirmary. **It at** first is unpleasant, but soon becomes tolerated, and has a good influence. **It was** the practice of Graefe to dip children's faces into a basin of cold **water** and hold them in it for several seconds, and this would **sometimes be an** effective eye-opener. The dropping ice-cold water into the eye is less troublesome and can be kept up a long time.

I have already alluded to the operation of canthotomy; another kind of operation, known as canthoplasty, is sometimes done, but **I do** not easily incline to resort to it under these circumstances. It belongs more, I think, to cases of chronic trachoma with organic narrowing of the palpebral slit. The operation will be described under its proper head.

BLEPHAROPHIMOSIS.

This term indicates a permanent shrinking in the palpebral opening, caused by alterations in the form of the **lids,** more especially of the **tarsi.** The principal cause is trachoma, **and the** full understanding **of the** pathological condition usually present will not be had until that subject has been discussed. At present I simply describe the operations which can be performed for its relief. The first is called canthoplasty, and was introduced by Von Ammon. I prefer to do it as follows: keeping the lids widely apart by a speculum, a narrow bistoury separates the skin from the mucous membrane at the outer canthus by an incision in the lid-border parallel with and around the angle. This incision will be 6 mm. long. From its middle a cut is then made straight out into the skin for a distance varying, according to circumstances, from six to ten millimetres. The two cuts form the letter ⊢⊣. In the space thus laid open, a pair of probe-pointed scissors dissects beneath the conjunctiva and cuts the fibres of the external canthal ligament above and below. As this is done, the skin is felt to give way, and the speculum can be readily lifted up. The conjunctival edge has been held by forceps while making the dissection, and now a cut is made into it toward the globe, only so deep as may be needful to bring the edge of the mucous membrane into proper coaptation with the skin, by three fine sutures. I adopt this mode of operating, because it is less apt to tear the conjunctiva, and requires less extensive incision into it

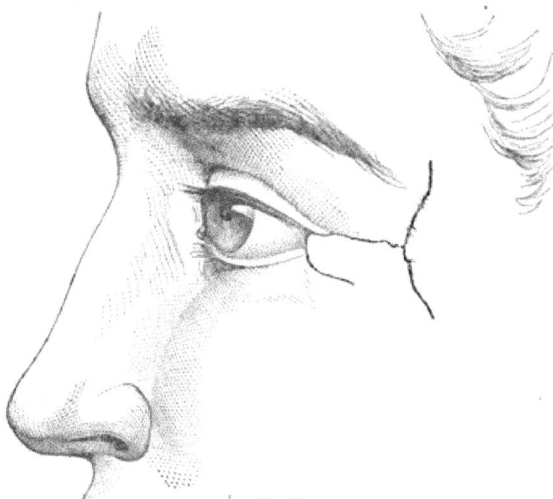

FIG. 70.

than must be made when all the tissues at the angle are cut at first with one clip of the scissors. The important parts of the proceeding are, first, the subcutaneous division of the canthal ligament, and secondly, the nice adaptation of the stitches. Usually but little reaction follows, while I have seen the contrary in cachectic subjects, and when decided conjunctival hyperæmia was present. It must be admitted that sometimes no permanent relief of the tightness of the lids is thus secured, and sometimes no benefit appears to result; but, in cases properly selected, I know that it is a valuable therapeutic measure.

Another operation to be mentioned is one which I have devised to meet the necessities of extremely difficult cases. They are those in whom atrophy of the conjunctiva and of the tarsus has deformed the latter and contracted the palpebral fissure to such a degree that the common operation just described can have no beneficial effect. The only way to separate the lids more widely and relieve the friction which their tightness exercises on the globe, is to introduce new material into the outer angle. I form a flap, as indicated in the figure (see Fig. 70) with the outlines as designated. The proper quantum of skin must be measured and the turn must not be too sharply made, otherwise an unpleasant wrinkling will be caused. The adjoining skin must be freely undermined with scissors to make it slide, and the whole raw surface must be covered. By observing these precautions, a useful alteration can be secured in the condition of the lids, and the patient's comfort greatly promoted.

ENCANTHUS

is a congenital malformation consisting in deficient development or absence of the levator palpebræ superioris muscle, in narrowness of the palpebral slits, with a tendency to slope upward at their outer extremities, in the presence of a crescentic fold of skin running from the inner end of the brow downward on the side of the nose, which more or less overhangs and conceals the inner canthus, and in great flattening of the ossa nasi. The deformity varies in degree, is apt to be hereditary, and to be associated with defects in the development of the eye, such as hypermetropia, or with albinism, or with nystagmus. I have seen three cases in one family. A striking case is figured by Von Ammon. The frontal portion of the occipito-frontalis muscle is constantly called into play as in cases of paralytic ptosis, and the head has to be thrown backward to look at objects which lie even a little below the horizontal meridian. In some instances nothing can be done, or requires to be done, while in other cases the wrinkle at the inner canthus is unpleasantly conspicuous, and can be remedied to some degree by excising from the root of the nose a piece of skin, and by undermining the adjacent folds very freely, so as to loosen them from their periosteal connections. The wound is closed by harelip or fine pin-sutures, and a vertical scar remains on the median line. Besides this, it is sometimes useful to perform the usual operation of canthoplasty at the outer angles. I have a photograph of a child who had a high degree of this deformity and aggravated converging squint. Nothing was ever done for her, because her drunken mother found the pitiful condition of the child a profitable means of mendicancy.

COLOBOMA OF THE LIDS

is a rare condition. I have personal knowledge of only two cases, one of which I operated on. There was a small notch in the middle of one upper lid, and through it, when the lids were closed, the cornea could be seen. It was congenital, and the child was nine months old when I operated.

INJURIES OF THE LIDS

are diversified in character, such as contusions, burns, wounds, and lacerations. It is needful to speak only of burns by gunpowder and of lacerations. For powder-burns, to which boys most frequently fall victims, the pain and inflammation are relieved by cold lotions, while the removal of the unburnt powder is to be effected by patiently picking out each little granule with a spud or cataract needle. The process is tedious and rather painful. Much of the powder will come out with the desquamating skin, while what lies below the epidermis must be picked out bit by bit, soon after the burn is produced. Blistering by cantharidal collodion has been suggested, but I should demur.

Lacerations of the lids are caused in many ways, viz.: by pieces of broken glass from bursting of soda-water bottles; by hooks caught under the lids; by the horn of a cow; by a thrust of a stick, etc. Besides the rules of ordinary surgery which apply to such cases, it is important to mention that pains must be taken to bring together the conjunctival and tarsal edges of the laceration, as well as the cutaneous borders. The neglect to unite the under portions of the lid results often in an irregular cicatrix, which by its contraction will cause ectropium or chronic swelling of the lid. The inner angle of the upper lid is most liable to suffer, and the line of rupture is along the fold over the upper edge of the tarsus. After cleansing the parts, the conjunctival wound should be carefully fitted together by means of fine silk, with small curved needles held with proper needle-forceps, and the stitches should include the tendon of the levator, or the tarsus, in case these parts are involved. Having completed this, the skin should next be united and the stitches should not go below it. It will often be difficult to procure a perfect adjustment of the inner angle, but the rule to be observed is to carry the lid inward to an extent somewhat in excess of what seems needful, because the contraction of the scar will drag it outward. One may even cut a notch into the skin on the side of the nose into which to lay the angle of the flap, and thereby avoid the objectionable condition of a large or irregular inner canthus. Should the lid thereby become a little shortened, relief can be had by canthoplasty at the outer angle.

I have been called on to relieve the deformity consequent on badly united lacerations, and have been obliged to excise the cicatrix and reproduce the original condition, and then, with the disadvantage of additional loss of substance, proceed in the way above described to remedy the mischief. By closely following this method, I have had a satisfactory result. It is not always easy to do this as neatly as one would wish, because the parts are slippery, they bleed freely, and are difficult of access.

BLEPHAROPLASTY.

The partial or complete restoration of the lid is to be attempted by methods which must be chosen to meet the peculiarities of each case. We have to consider two primary classes of deformities: first, those due to contraction of scars; and second, those caused by loss of substance.

The first class are often of little severity, and are commonly occasioned by caries of the bone with adhesion of the soft parts, or by the results of abscesses, or by chronic inflammation, or by burns. In this last category, some most frightful deformities occur.

The second class, in which loss of substance exists, may be caused by lupoidal, epithelial, or other disease, which, when removed, leaves a deficiency for which provision must be made.

The modes of repair consist of flaps, which are introduced by sliding adjacent skin, or of flaps which are twisted into position, or of portions of skin taken from a distant part and deposited upon the place to be filled up, and not having any vascular connection with surrounding tissues.

The choice of method depends upon the degree and kind of deformity, and on the condition of the surrounding skin. Of the method last mentioned there are two varieties: one which consists in using small grafts, which is known as Riverdin's; and the second, that of transferring a large patch of skin, is known as Wolfe's method—the largest yet used having been 3 inches by 1½ inch. This way of filling deficiencies is yet to a great degree experimental, and cannot claim the consideration which belongs to old and tried proceedings. It offers a solution of the most desperate cases in which no suitable integument can be secured from the vicinity, and it has done good service in eight out of ten published cases (see "Trans. Am. Oph. Soc., 1880," p. 43–52). It applies to cases of burns where all the neighboring skin is cicatricial, and this always makes bad material for transplantation. It will be described after discussing the operations in which the flaps have pedicles. Sometimes a sliding flap will survive which consists of cicatrix, but there is extra risk in using it, and a twisted flap of this kind is almost sure to slough. It is also bad to use a flap of healthy skin across which a cicatrix passes. In one such case, I had total loss of all the flap beyond the line of cicatrix. We are sometimes debarred from using sliding flaps, because the skin is tight and inelastic, and cannot yield to the requisite amount ; or, because, it must be taken from a region which will permit only a limited degree of displacement, like the bridge of the nose or the forehead. Twisted flaps are to be preferred in a great many cases, but it is desirable not to have to turn much more than through a right angle, both because their nutrition is imperilled by a sharper twist, and because the pedicle becomes protuberant. Care must be used that the pedicle be not too narrow or too thin, and that the natural distribution of the arteries be respected. While too large a flap makes a clumsy appearance, it is a fault far more serious to have the flap so small that it must be stretched tightly to cover the ground, because this endangers its vitality.

For the upper lid, a twisted flap is usually to be preferred, although one kind of sliding flap is available, viz., the triangular, first suggested by Wharton Jones (see page 141). For the lower lid, the sliding or twisted flap may be chosen, according to circumstances.

The following are some of the kinds of twisted flap resorted to : from the temple, with the base below and flap coming from the forehead, or with the base at the level of the lid and the flap in front of the ear. At the inner angle the flap may be taken from the middle of the forehead, with the base astride the nose, so as to draw nourishment from each side of the median line. To prevent a clumsy look by projection of the pedicle, it is only needful to dissect out skin enough from the bridge of the nose to make a bed in which to spread it flat. Its base may also get blood from the arteries coming from the upper and inner angle of the orbit, and its continuity lie along the side of the nose, with its apex near the ala nasi. This makes what I have called a naso-buccal flap, and I have used it on many occasions with success. It can be employed for the upper lid, in case of need, as the Fig. 72 shows, but is best suited to the lower lid. In planning a flap, about one-fourth more must be allowed in the length than the wound measures. It should not include much of the subcutaneous fat. In twisting a temporal flap, the outer limb of the including incision should extend beyond the level of the inner limb, so as to make the turn less abrupt. It will often be needful to cut out a gore or triangle from the skin at the extremity of the inner cut, to avoid a protuberant wrinkle. If the pedicle be very broad, this rejected piece of skin may be large. In one of my cases it was a triangle, measuring one inch on each side, and I transplanted it to the place from which the flap was taken, after the manner of Wolfe's operation, and it survived. The flap is best held in place by fine insect-pins, put in all parts which are likely to experience tension. The thread for interrupted sutures should be the finest black silk. What is known among the makers of sewing-machines as 00 or 000 is strong, smooth, and very fine. It is easily taken out on account of its color, and its fineness gives it all the immunity from irritation which is claimed for silver wire. The last I never use. For a long, straight seam, an overhand stitch with continuous thread and no knots, except at its ends, is a happy suggestion of my friend, Dr. Minor. A flap comes to its place more easily when the gap made by its being lifted is closed. To do this the adjacent skin should be extensively undermined, and sometimes relief cuts parallel to the sides of the gap have to be made. When, after lifting the flap, a large vacancy remained on the temple which undermining could not fill, I have sometimes constructed a second flap from the adjacent scalp, with which I covered the temporal space, and transferred the gap, which must be left to granulate, to a locality above or even behind the ear.

It is a rule which ought to be carefully observed, never to put a flap into position until all hemorrhage beneath it has ceased. Moreover, the clots of blood should be removed as cleanly as may be. Exposure to the air is the best way to stop bleeding from fine vessels, and larger ones may be twisted. A ligature will rarely be required to remain in the wound. Steady pressure and iced water, or extremely hot water, are the best hæmostatics. Sutures must be put in without stint, and the most accurate adaptation secured. No plastic operation should be done with haste. Nice adjustment and correct proportions are far more essential than rapidity. Mistakes in proportion are much more liable in twisted than in sliding flaps; but, when neatly fitted, they serve a most admirable purpose and give excellent cosmetic results.

The repair of loss at the inner half of the lower lid may be very neatly effected by a sliding operation which I have devised. If an epithelioma has been excised from this locality, even if it involve three-fifths or two-thirds of the lid, the deficiency may be well repaired by the following proceeding. It will be most successful with persons of loose skin. An incision is carried from the outer commissure horizontally toward the top of the ear, and may reach this point. Another incision is made from the wound down alongside of the nose to its ala, and perhaps to the angle of the mouth. The skin of the face is then dissected up and laid over, and when sufficiently loosened to be moved inward and upward to a degree sufficient to cover the gap, it is held in place by pin- and thread-sutures. Sometimes I have left the conjunctiva untouched, and taken the horizontal incision below and along the ciliary edge. At other times, when much of the lid had been sacrificed, I have cut through the inferior fornix and carried the remaining lid-border along with the flap inward. The operation is formidable in appearance, because the whole cheek may be turned over, and is liable to free hemorrhage, but it leaves no deformity, and the lines of scar are in inconspicuous places. Sloughing cannot occur, and no secondary deficit has to be made good. When a large loss of substance has occurred, the extensive movement of the cheek may diminish the palpebral slit unduly. This I have been obliged to correct by a subsequent canthoplasty.

Another sliding flap is described by Dr. Knapp as follows: after the diseased part of the lid has been removed, a cut is made from the outer canthus toward the top of the ear, and below it, at such distance as the case requires, another cut is made nearly parallel with it. The cuts, as they approach the ear, spread apart, to give the base of the flap more width than its free end. The flap is dissected up as extensively as needful, and dragged toward the nose. If it do not reach far enough, an additional quarter-inch may be gained by cutting the fibres of the internal commissural ligament, and loosening the skin overlying the lachrymal sac, which skin will stretch to meet the flap. In case this do not suffice, a flap must be made across the nose in form of a parallelogram, and long enough to unite with it. The nasal flap is apt to become unpleasantly prominent,

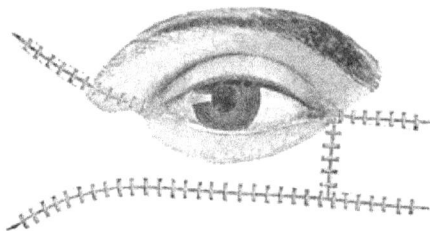

Fig. 71.

and is to be avoided if possible. The subcutaneous dissection at the inner canthus, which was my suggestion, is to be fully tested. This method is available when the whole lower lid is to be supplied, or when a large part — either the outer or the inner — is wanting (see Fig. 71).

When by such a method as the last mentioned, or by the one which I have suggested on page 151, a raw surface of skin is brought to lie against the ocular conjunctiva, it is most interesting to observe that a substitutional membrane appears upon the raw surface, which looks like and fulfils the function of mucous membrane. It has an epithelium, and after a time it is not to be distinguished by the naked eye from true mucous membrane.

Twisted flaps may be combined with sliding flaps, and sometimes a flap may be made useful both in the subtraction and addition of material—that is, when one lid needs to be lifted up by abstracting a piece of skin, and another needs the insertion of a piece. I did this for a young man who had great deformity of both lids, as a result of an abscess of the cheek caused by erysipelas. There is always occasion for new adaptations in plastic surgery, and much must be left to the ingenuity and mechanical skill of the operator. It has been my experience, that I could not determine precisely what seemed best to be done until all the wounds were made ; and when I had before me the extent and form of the space to be filled up, I could devise a plan of proceeding. One must consult authorities to

Fig. 72.

learn what devices have been used, and then work out his own way. Incisions should follow the natural lines of the face, and it must be remembered that contraction will go on in the scars for months, and especially is this true of wounds which heal by granulations.

A few words about the Wolfe operation of transferring skin from a distant locality. The skin must be dissected up free from subcutaneous fat; it should be laid in warm water, and applied to the raw surface without intervention of clots of blood. It need not be held by sutures, but it must be made to fit exactly into the space. It will shrink to two-thirds its linear dimensions when lifted from its place. It should be covered with gold-beaters' skin, the latter to be laid on dry and pressed down. I covered the gold-beaters' skin in my most extensive successful case with collodion—it is certainly well to hold down its edges in this way. Over all a mass of cotton and a bandage are to be placed, and the dressings are to be undisturbed for forty-eight hours, or several days. The greatest care is to be exercised not to permit the flap to be moved. If the flap "take," it will be seen to be simply white, and there will be no secretion of any amount. If it fail, copious and offensive fluid will appear and the flap turn black. It must be said, moreover, that when the new piece lives, it undergoes shrinkage for months afterward. In my case this amounted to a reduction of one-fourth the original size, and in a case of Zehender's it equalled one-thirteenth. This constitutes a serious drawback to the method. It is certainly well to remember that small, and sometimes large pieces of skin, can be utilized as grafts when more ordinary methods are unavailable.

The remark may be made about cicatricial deformities that much amelioration can be obtained in loosening and stretching the skin by the free use of sweet oil, and by traction on it practised day by day. This kind of massage has decided value in many cases which seem too formidable for operation. It is also possible to help some cases, where a scar is nodular and welted and adherent, by subcutaneous incisions, to be succeeded by systematic and long-continued traction.

Burns of the eye are not rare accidents. Most often it is by slaked lime, or fresh mortar, by melted metal, by nitric or other strong acids, or by ammonia—and red pepper may come under this head. When lime gets into the eye, it must be washed copiously with water, but the only effective means of removal is by mechanical means, viz., by forceps, or a spud or curette; wiping out the coarser masses with a bit of rag will serve, but the little pieces which remain eat their way into the tissue and become incorporated with it. At the inner canthus, and in the cul-de-sac, stuff will be found, and it must be dug out patiently and thoroughly. Often ether must be administered. After complete removal, syringe away all particles with warm water. The only relief for pain is in atropine and vaseline inside the eye, and cold water. Of course an anodyne may be given. The danger is of adhesion of the lids to the globe, and of deep opacity of the cornea. Nothing can prevent the formation of attachments, as a resulting ulcer slowly granulates and contracts. Shields of lead, and dressing with lint and sweet oil, and pulling the lid away, are all unavailing. The contraction will and must form, and at a future time it can be dealt with.

Burns by melted metal are often less severe than those due to lime, because as the metal solidifies it is taken out as a cup, and there is no continuously destructive chemical action. Nitric, sulphuric, and acetic acids, can do great mischief. They are to be washed out with cold water, freely applied, and the case treated for subsequent reaction. Ammonia causes a more superficial eschar, and is exquisitely painful. For red pepper, however dastardly its intentional use, it may be said that it does not cauterize. The ulcers caused by burns are damaging to sight in the ratio in which the cornea has been involved. But a better result can generally be expected than the first look of the case suggests. Several weeks or months will be needed to heal the ulcer, and meanwhile the eye is to be nursed and meddlesome medication kept away.

When the ultimate contraction has been reached, we have the adhesion of the lid to the globe, called *symblepharon*, and the question of repair is to be considered. The difficulties are in proportion to the extent of adhesion. For columnar bands, good results are possible; for total attachment of the whole lid-surface, the difficulties of an operation are great. No good at all is to be had by simply dissecting the tarsus and eye asunder—the lid is sure to go back to its old site. Naturally, the lower lid is the most frequent sufferer. Traction on the cicatrix, long kept up, will produce in it some stretching. The cure is, however, by means which shall not only separate the lid from the globe, but prevent readaptation. For columnar cases, the old proposal of inserting a lead wire through the mass at the fornix is of value. It has to be worn until a permanent hole is formed, which shall be lined by a kind of epithelium; then the adhesions may be cut, and the parts can be kept from growing together. But this operation is superseded by one of Arlt's. He dissects down the frenum, beginning on the globe, until he lifts it up to the fornix. It has a certain length, and through its free end a thread armed with two needles is passed, and these again pushed through the cul-de-sac to the surface of the skin, and the thread tied over a roll of plaster or bit of wood; by this device the outer surface of the column is laid against the globe. This raw surface is now covered over by bringing down flaps of conjunctiva from either side.

For more extensive cases of symblepharon, another operation, devised by Mr. Teale, is a happy contribution to surgery. The adhesions are dissected down to the fornix, then the vacancy on the globe is to be filled, while the lid is left to itself. Instead of sliding flaps of conjunctiva, a kind of sling is made in this way: from near one side of the vacancy an incision in the conjunctiva is carried around, just outside the cornea, to the opposite side of the vacancy. Then another incision, concentric to and outside this, is carried around, but its extremities must not come nearer to the vacancy than five to seven millimetres. It may even go up to the fornix, and with the first incision it encloses a band which may be from five to eight millimetres wide. Care is taken to make the ends of the band the widest part, by turning the extremities of the upper wound upward. The band is then loosened, except at its ends, and slipped down over the cornea to take its place in the gap made by removal of the adherent cicatrix. It is convenient to put threads into the edge next the cornea before making the outer incision they serve to hold and draw down the flap, and are used to fasten it in place. Some readjustment of intervening and adjacent conjunctiva is required, while the flap is carefully fastened by fine sutures in its bed. The spot from which it has been taken is left to itself. That in time, by granulation, becomes covered by a tissue which perfectly resembles normal conjunctiva. The result of the operation is most fortunate, and I have done it several times with great satisfaction. Operations by putting in separate flaps from either side, twisting them down, and uniting them by suture at their free ends, have been done by Knapp, Teale, and others. For cases which may be too bad for this proceeding, grafting of bits of conjunctiva from the rabbit is available, and would result in benefit. In most of these cases, the loosening of the adherent scar is to be done, for relief of pain and discomfort, without regard to sight. In many others additional operations, chiefly iridectomy, may be needed to gain better vision.

Anchyloblepharon is the adhesion together of the tarsal borders, and is also a result of burns or of **wounds**. It is easily remedied if there be any free spot or hole from which to **start in** separating the lids. If not, the attempt is useless. Sometimes the lids are both uniformly attached **over** the entire globe, and also adhere **to** each other **at** their edges. For **such** cases no interference is **proper.**

CHAPTER III.

THE CONJUNCTIVA.

Anatomy and physiology.—This membrane, so named because it joins the lids to the globe, presents for consideration its tarsal portion, the fornix **or sinus**, and the ocular portion. The tarsal **or** palpebral part is closely and smoothly adherent **to** the tarsi, and permits the Meibomian follicles to be seen through it. It has a faintly yellow **color** and a few vessels. The fornix, or sinus, **or** cul-de-sac, or fold of transmission, is very loosely attached to the parts beneath, and slips freely back **and** forth ; it has numerous folds, is of **a** turgid dark color, and has many glands. The depth of the fornix varies according to age and individual peculiarities. Sometimes **the whole of** the superior fornix **can** be exposed **to** view, and often **no effort will** display it. The **inferior** fornix can **always** be fully seen. **The** ocular conjunctiva lies smoothly upon the **globe**, but is loosely attached, and can be moved back and forth. It is **quite** transparent, and shows but few vessels. At the outer canthus the conjunctival sinus is deep, especially toward the lachrymal gland ; at **the** inner canthus, on the contrary, the sinus, **or** fornix, is shallow both **above and** below. At the inner end of the palpebral slit we have the **congeries** of glandular follicles, called the caruncle, and between it and the margin **of** the cornea is a slight fold of the membrane running nearly vertically, **yet** somewhat crescentic, called the plica semilunaris. It is bound rather firmly to the parts beneath, and is the analogue of the third eyelid of some animals. The conjunctiva consists of **a** basement structure of connective-tissue fibres, which in certain parts are elastic, and it has **an** epithelium which presents great variety in different portions, and has been a topic of controversy among histologists. There are also **in the** sinuses certain glandular bodies. On the tarsi, if it be closely inspected, the conjunctiva will be seen to be faintly velvety. This is due to the epithelium being raised in ridges and heaps, and also cleft by fissures **in** various directions. It is flat, cylindrical and spheroidal, and in many layers. This appearance was formerly described as identical with papillæ, but the name is not anatomically correct, although clinically useful. Above the tarsus in the superior fornix **the** layer of epithelium becomes convoluted in a deep and intricate manner, in some such way as we find the gray **matter** of the brain ; and **so** elaborately is it infolded upon itself that pits **and** tubules **are** formed, **which** have been erroneously described as glands. **There are in** the upper edge of the tarsus, and in the fornix, a number **of small** bodies called **the** acino-tubular glands of Krause. They **are** of **two** varieties. There is also a variable amount of lymphoid cell infiltration, or adenoid tissue in the upper tarsal portion and sinuses, and this is more abundant in children. The lymph-follicles, as they have been called, are considered to be a pathological product. Because of this elaborate glandular and epithelial structure, it is natural to expect a secretion proper to the conjunctiva, and this consists of tears, mucus, and a few epithelial cells. By this it fulfils its office of lubricant, and in this regard it has been classed among the serous membranes, and, like them, when the lids close it forms a shut sac. At the margin of the cornea it adheres most intimately to the sclero-corneal junction, and forms a structure called the limbus or ring, which has a thicker layer of epithelium than the neighboring tissue. The epithelium then passes over upon the cornea.

The nerves of the conjunctiva are branches of the fifth pair, and numerous. The blood-vessels come from the lids and from the globe. The lymph-vessels are very numerous and tortuous, and communicate with those of the cornea and sclera. The manœuvre of everting the upper lid requires some manual dexterity, and by some patients is greatly dreaded—even when done most smoothly. If managed properly, it is in most cases entirely painless, and need not excite any resistance. The patient is to look down to his lap; I place my left thumb on the brow, with its tip at the upper edge of the tarsus, while the other fingers of the hand rest on the forehead; the thumb shoves down the brow and lid, and gently feels the globe; with the fingers of the other hand I take the free tarsal border, and bend the lid over the thumb-nail as one would bend a card-board over a paper-cutter. I prefer the tip of my thumb as a fulcrum, to a probe or a pencil, because by using it I force down the lid to meet the upward movement made by the other hand. As the tarsal edge comes up, the tip of the left thumb catches it and holds it secure, and by pressing it backward can still evert the conjunctiva. If the lid be very short, or be rebellious, it may be pushed over the thumb in the following way: wrap around the thumb of the other hand a bit of muslin, and place its tip against the tarsal edge; push it up, at the same time bending it backward. The bit of muslin prevents the lid from slipping away, and when everted it is caught by the other thumb. Diseases of the conjunctiva contribute about one-half of the diseases of the eye, and its inflammations exhibit great diversity of features. We have, as general characteristics of conjunctivitis, hyperæmia, undue and abnormal secretion, swelling of the membrane and swelling of its individual structures, proliferations, and neoplasms, œdema beneath it, and frequently the neighboring tissues become involved. The prevalence of one or the other of the features thus sketched forms the basis of the classification of conjunctivitis into the following varieties.

PALPEBRAL CONJUNCTIVITIS

might, with more correctness, perhaps, be designated simply as *hyperæmia*, because the condition to be considered consists almost exclusively in congestion of the vessels. Yet this condition persists for a long time and causes annoyance. There is a little undue secretion, and at the inner and outer angles of the lids, where the tarsal conjunctiva is most hyperæmic, there may be some fine papillary outgrowths. The degree of congestion is variable. The membrane does not swell nor lose its transparency. Complaint is made of dryness of the lid, or of pricking or smarting sensations, or as if a mote were in the eye. The lids cannot be kept open to the full extent; there is discomfort in using the eyes, especially at night, and often there is some photophobia. Patients with good powers of language will add many others to these subjective symptoms, and it appears most frequently among the educated. **An important fact is that it may be indicative of some other trouble, or be merely idiopathic. If the latter, it will be controlled after a few days or weeks by local treatment alone ; if, however, it be symptomatic, the true cause must first be discovered. We find this cause, first, in whatever produces eye-strain ; second, in catarrhal conditions of the nasal mucous membrane. Under the head of eye-strain we have **the various errors of** refraction and of accommodation, viz., hyperopia, **astigmatism,** anisometropia (inequality of eyes), beginning myopia, **commencing** presbyopia **or spasm of** accommodation ; we have also fatigue **of the ocular** muscles, **and we have** irregularities or opacities **of the** cornea and incipient cataract. **All** these conditions, which make the use **of** the eyes fatiguing, are usually attended by hyperæmia of the palpebral conjunctiva, and this may be the condition which excites the patient's attention. Another accompaniment of palpebral conjunctivitis, and which is also causative, is nasal catarrh, and this must be dealt with and relieved **before the** eyelid trouble will yield. Analogous **to this is** the irritation **of the lids** which follows measles and other exanthemata, **and excessive** weeping.

Treatment consists in applications which are at the same time soothing and mildly stimulating. Among many popular remedies are weak salt water, a few drops of brandy in a cup of water, the addition of a few drops of fl. ext. of Hamamelis (Pond's extract) to water, the cold douche, bathing with cold water for a few minutes, water in spray by a double bulb apparatus. A few favorite formulæ are :

  ℞. Sodæ biboratis.............................. Ʒij.
   Aquæ camphoratæ........................... ℥ vj.
  M.

  ℞. Zinci sulphatis............................. gr. j.
   Aquæ...................................... ℥ iij.
  M.

  ℞. Acid boracic.............................. Ʒ j.
   Zinci sulph.............................. gr. iij.
   Aquæ.................................... ℥ vj.
  M.

  ℞. Fl. ext. opii.............................. Ʒ ij.
   Aquæ.................................... ℥ iv.
  M.

Of these the first is generally preferred, and the various ingredients may be combined as required. To find what suits the special case is always **the** indication. But more efficient than all these is the application of a weak nitrate of silver solution, viz., gr. j. vel iij. ad. ℥ j., two or three times weekly to the everted lids by a skilful hand. Clumsiness which inflicts pain in everting the lids will neutralize all the value of the treatment. To drop the caustic solution into the eye is not advisable. In touching the lids sometimes the small brush must be pushed quite up to the fornix to reach the troublesome spot. For several weeks the treatment must be continued, and, even in cases where the exciting cause is removed, the local congestion will require time and care. It may be added that sometimes the use of atropine, to control the eye-strain and aid in correcting the optical error, has a mitigating effect on the conjunctivitis. Other remedies are tannin, gr. x., glycerini, ℥ j. or stronger; a crystal of alum touched lightly on the surface; in some cases nitrate of silver is not so well borne as tannin. Persons present idiosyncrasies in regard to remedies, and one must not be astonished at anything in this respect.

Œdematous conjunctivitis is a form of acute inflammation in which there is not much discharge from the eye, nor intense hyperæmia, but a great deal of serum is effused beneath the ocular conjunctiva and into the lids. The chemosis or subconjunctival effusion is fluid, and sometimes of a tawny yellow color. There is not much pain, but the swelling is uncomfortable. The condition appears within a few hours or days, and occurs in strumous children or in adults whose tissues are watery. Mild treatment is to be adopted, viz., bathing with water lukewarm, or at a little lower temperature, and with such astringent solutions as alum, gr. v. ad. ℥ j., or plumbi acetatis, gr. ij.—vj. ad. ℥ j. The affection passes off in a few days, without any serious consequences.

CATARRHAL CONJUNCTIVITIS.

Considerable latitude must be allowed in the definition of this condition, but the characteristic symptom is secretion of a sticky, flocculent, mucous quality, and moderate in quantity. It consists of epithelial cells, serum, mucus, and a small proportion of pus. It gums the lashes together, and appears as strings and flakes in the palpebral sinuses. It is coagulable and semi-transparent. There is hyperæmia of the palpebral and ocular conjunctiva, more intense upon the former, moderate swelling of the lids, and œdema beneath the ocular conjunctiva. At the limbus corneæ there are often minute erosions of the epithelium. Early in the progress of the case, newly formed vessels may appear as a fringe of straight red lines for one or two millimetres upon the limbus corneæ. Pain may be slight or considerable, and in some cases there is photophobia. There is a feeling as if a foreign body were in the eye, and the lids cannot be opened to the full extent; the occasional lodgement of mucus on the cornea blurs the sight, which is restored after wiping the eye. A mild variety of catarrhal conjunctivitis takes place in consequence of the extreme heat of our summer; it concerns the lids more than the eyeball; has moderate hyperæmia and secretion, and may pass from one eye to the other, or affect both simultaneously. So, too, a coryza frequently involves the ocular as well as the nasal mucous membrane.

The term *ophthalmia*, formerly much more used than at the present day, is by consent now limited to affections of the conjunctiva, unless some qualifying word is used, such as ophthalmia interna, or sympathetic ophthalmia. It is a term which describes nothing accurately, and deservedly is being abandoned. All the forms of conjunctivitis which are accompanied by secretion are more or less contagious. The virulence of the contagion depends on the nature of the secretion, and in general, like will cause like. If the disease occur in endemic form in an asylum or public institution, it is communicated, not only by actual conveyance of secretion from one to another, by towels, handkerchiefs, etc., but also by infection of the atmosphere. At any rate, though the affected cases be vigorously isolated, new cases will appear among the apparently well. If isolation is not practised, the disease will attack most of the inmates. This is conspicuously true in overcrowded establishments where the ventilation is imperfect. Under such conditions the disease assumes a severe type, the secretion becomes co-

pious, it grows more purulent, the tissues become softened and infiltrated, ulceration of the cornea is likely to occur, and the conjunctiva acquires a thickened condition known as granular conjunctivitis, from which it will be a long time in recovering.

It is absolutely indispensable, when any form of conjunctivitis appears in a public institution, to put all the affected persons apart from the healthy; to give them rooms with the most abundant air-space, and enforce the most rigorous cleanliness in each individual, and permit no interchange of towels, dressings, etc. The apparently healthy should be frequently inspected, and all cases of slight palpebral irritation or latent forms of granular conjunctivitis be placed by themselves.

Duration, **complications,** *and sequelæ of catarrhal conjunctivitis.*—The attack **will** last a variable time according to its severity, viz., from a few days to two weeks or several months. The cases in which the sclerotic participates the **old English** writers called "Catarrho-rheumatic ophthalmia." In them **a** rheumatic or gouty element exists, and the disease is a mixture of **acute** scleritis and conjunctivitis. The secretion is more watery than glutinous, and pain and photophobia are much more pronounced than **in** simple catarrhal conjunctivitis. There can be no doubt that simple conjunctivitis of mild type may be due to the gouty diathesis, and it is a well established fact that sclerotitis is thus produced. Usually an attack of catarrhal conjunctivitis passes off without permanent lesion of the eye, but it may be accompanied by ulceration of the cornea, either minute or extensive, and even lead to perforation in weakly persons. A state of chronic palpebral hyperæmia frequently remains, and granular conjunctivitis not seldom ensues. The last-mentioned state is brought about in many instances, because the persons have had some of the latent forms of trachoma unrecognized, which have prepared the eye for worse conditions. In public institutions it may be expected that every case will result in granular conjunctivitis when the acute inflammatory stage subsides.

If catarrhal conjunctivitis assail old persons they seldom entirely recover from it; the conjunctiva becomes thickened, and if they are neglectful of themselves, they may get ectropium, lippitudo, chronic blepharitis marginalis, trichiasis, etc. Vascularity of the cornea, and various phases of chronic opacity, may ensue. The issue of this disease depends upon the care taken to insure scrupulous cleanliness and a healthy atmosphere, upon the vigor of the individual, and upon a rational treatment. While it may be only a trifling ailment, it may pass into conditions both pertinacious **and** deplorable. The line of division between catarrhal and purulent conjunctivitis is liable to be overstepped, and the prognosis thereby **become more serious.**

When the occasion of the attack has been measles, the child may have health enough to throw off the eye-trouble without harm, and promptly; but should this not be the case, ulceration of the cornea and grave possibilities may ensue; or, should nothing serious of this kind occur, the conjunctiva will long remain hyperæmic, and the eye weep.

Treatment.—As the quality of the disease is not severe, so the treatment should be mild. At the very onset it often happens that lukewarm fomentations **are** comforting, and relieve the eye by promoting serous effusion, but the mistake is sometimes committed, both by the ignorant and the educated, of keeping up this treatment too long and too constantly. For mild attacks of conjunctivitis, it is common for people to tie up the eye in a slippery-elm bark poultice, or to bind on a napkin wetted in cold water, and keep these on all night. Often the trouble vanishes in the morning.

But a prolonged attack leads to a continuance of such methods, which may be put in the form of scrapings of raw potatoes, a raw oyster, a bread-and-milk poultice, a rotten apple, and substances too disgusting to be mentioned. Under such fomentation the œdema of the lids and of the conjunctiva is greatly increased, the hyperæmia becomes intensified by relaxation, and ulceration of the cornea is promoted. The softened tissues become water-logged, and less capable of recovery than if the inflammation had been left to itself. It is not a rare thing in a public institution to see eyes which have been poulticed to destruction.

It is a safe rule, long ago stated by English surgeons, to put on an inflamed eye either warm or cold applications, according to the choice of the patient. It will be found that the vigor of the tissues and the severity of the attack will safely determine the temperature. The point to be guarded is not to continue either warm or cold applications for too long a time. The suitable way is to let them remain for a few minutes, or for half an hour, and then take them off for half an hour or an hour, to provide intervals of abstention, and to regulate the application and the interval by their effect and by the gravity of the process. The advantage gained by this intermitting proceeding is that the eye thereby is well cleansed, whereas the constant poulticing retains the secretions and secures all the hurtful effects of their accumulation and decomposition. The suitable manner is to fold a piece of muslin into a compress of a few thicknesses, and, soaking it in the water, change it frequently for a fresh one, and keep up a succession of them for the time thought advisable, and during the interval to tie the eye up lightly with a handkerchief, or cover it with a shade.

For medication, the indication is to use the mild astringents. A distinction is to be made between the medication applied by the surgeon and that which is entrusted to the patient or to the nurse. Nitrate of silver is a remedy which is so potent for good and for evil, that the surgeon should reserve it for his own hands. When judiciously dosed, both in intensity and frequency, and skilfully applied, it often supersedes the need of any other remedy. For cases such as we now have in view, it may be taken at from one to five grains to the ounce, and put upon the everted lids, both upper and lower, with a small camel's-hair brush. It is assumed that the surgeon can turn over the upper lid without inflicting the pain of a surgical operation, but the writer has often been told that the attempt to do this was a worse infliction than the disease, and he has sometimes been obliged to desist. He is not too modest to assert that in performing the manœuvre he employs about as much skill as experts claim. If the strength of the solution seems too feeble, it must be remembered that a weak solution will cause greater effects in proportion to the quantity applied and to the time during which it is kept upon the surface. It ought not to cause sharp pain, and for a few minutes no effect may be felt ; soon its cauterizing influence comes on, and is followed by free discharge of serum. During its flow, and when this ceases, the eye feels better. While the serous flow continues, keep cold compresses on the eye. Such an application will not be needed more than once daily. For the patient's use, prescriptions may be made of solutions of alum, gr. v. ad. ℥ j. ; of borax, gr. v. vel. x. ad. ℥ j. ; boracic acid, gr. xv. ad. ℥ j., or four per cent. solution. Boracic acid is purely an antiseptic, and as such has marked value. It may be used in place of water, because it has no irritating effect, and its antiseptic properties give it an extremely valuable influence. It has grown largely in favor within two years. It may also be combined with zinc-sulphate, viz. : ℞ Zinci sulphatis, gr. iij.; acid. boracic. ℨ j.; aquæ, ℥ vi. The old

11

formula of sugar-of-lead **and opium** is best made as follows: R. Liq. plumbi subacetat., ʒ ij.; fl. ext. opii. deodorat., ʒ j.; aquæ, ʒ vj. M. **Some** decomposition takes place in the mixture, and it is one *not* to be commended, especially if any abrasion exist on the cornea; the precipitation of lead on such abrasion is a most serious mishap. But old usage gives it currency, and it may be given sometimes without harm. Sulphate of **zinc and** sulphate of copper are too irritating to be frequently employed. **The lead** lotion should be an outward application, while the alum, borax, and especially **the boracic** acid, may come upon the conjunctiva as well as upon the **skin.** For these outward applications, the patient need not be strictly governed by injunctions; he may be left a great deal to the guidance of his **own** sensations. One more item in local treatment is the use of simple cerate or vaseline upon the edges of the lids to prevent their being stuck together when wet applications are intermitted for some hours. This will often be the case at night, when there may be no need to continue the lotions. For most cases of catarrhal conjunctivitis there will be no occasion for purgatives, but they may be used at discretion.

The details above given apply to the great majority of cases; a few **words** may be given to some special cases. It is not to be understood that nitrate of silver is to be universally used; many cases get well with only the common collyria, or with water. To cases of œdematous inflammation, especially in old people, the best application is tannin, gr. v., aquæ, ʒ j., both to the conjunctiva and to the skin. For a severe attack of "catarrho-**rheumatic** ophthalmia," it may be needful to apply two to four leeches to **the** temple to allay pain, and sol. sulph. atrop., gr. ij. ad. ʒ j., may likewise **be** required for the same reason. No astringents should be used until mucoidal secretion begins, and the local application until then should rather be of an agreeable temperature. A saline purgative will be useful, and sometimes small doses of Rochelle salts, frequently given for diuretic effect, will help **to eliminate the** materies morbi. To such sequelæ as lesions of the cornea, or granular conjunctivitis, etc., consideration will be given under these heads. In the greater number of cases the disease disappears without leaving any tissue-changes either in the conjunctiva **or in** other parts of the eye. It hardly need be said that, during the continuance of an attack, the patient shall not attempt to read, or write, or smoke, even though the trouble be slight, and will remain in a dimly lighted, well-ventilated room.

Chronic catarrhal conjunctivitis is a condition in which the surroundings or habits or constitution of the person prevent a perfect restoration of the inflamed conjunctiva. It remains red, becomes thickened, and produces undue secretion. This happens to such as work amid dust or heat, **to old** persons, to habitual drinkers, etc. The common ultimate result of **the disease** is to cause redness and hypertrophy of the palpebral conjunctiva, **and** especially of the lower lid. It is apt to become everted, the surface dry and glazed, the punctum lachrymale displaced, and epiphora follows. The condition will persist indefinitely, and causes much discomfort by itching, watering, and irritation. On the other hand, many stolidly accept it as of no consequence. The treatment is, mild astringent lotions, and in **case they fail, to** touch the lids with sulphate of copper crystal, or tannin, Ɗj. ad. **glycerine,** ʒ j., once in two or three days. In extreme cases, where decided ectropium exists, an operation will be necessary, which may include the skin (see page 141), or require nothing more than excision of a strip of thickened conjunctiva along the whole length of the lid.

PURULENT CONJUNCTIVITIS.

Under this head we consider first the eye-troubles of new-born babies, known as *ophthalmia neonatorum*. The subject is conveniently dealt with as if it were a distinct disease, whereas we meet with every variety of conjunctival inflammation in the newly born. The serious cases are the purulent variety, and need careful attention, but the great proportion are not serious and not purulent in any important degree. A red and sticky condition of the eyes is common to many infants when they begin their extra-uterine adventures. The nurse washes them out with warm milk-and-water, and soon the eyes are clear. A little attention may be required for several days, the principal point being the careful removal of all secretion as fast as it appears, and soon all is well. Should there be a little swelling of the lids, and secretion be rather troublesome, a compress wetted with a solution of borax or of alum may be laid on the eyes for fifteen minutes, once in two or four hours, and the edges of the lids smeared with simple cerate or vaseline. The great proportion of the cases of this disease will not require more serious attention, and will not cause the physician any anxiety. A physician in large obstetric practice told me that he had had only one serious case in an experience of twenty years. In hospital practice the situation is otherwise. Of late, obstetricians have adopted the practice of treating the mother by astringent and antiseptic vaginal injections for some weeks before delivery and during the time of confinement, so as to get rid of the obnoxious secretions which often cause the infantile malady. But there are cases in private practice which need active attention. Of these I have seen two varieties. The less frequent are some which have more the character of granular than of purulent conjunctivitis. There is very little swelling of the lids—secretion is almost wholly watery. There is little or no hyperæmia of the ocular conjunctiva, while the palpebral conjunctiva is reddened, and at the fornix is considerably thickened and swollen. This hypertrophy is the essence of the trouble. While this continues the eye keeps watering and a little sticky, and the condition goes on for weeks. The cornea does not get hazy, and there is no special danger. For the milder cases, where an anxious mother insists on something being done for the baby, I deprecate anything more than washing with a solution of boracic acid several times a day. But if a child is a month old, and the discharge continues and the fornix exhibits decided swelling, I have been obliged to use solutions of tannin and glycerine as strong as Ͻij. ad. ℥j. before the condition would yield. I had tried nitrate of silver in mild solution, and, unwilling to make it more caustic, had taken a solution of tannin, gr. x. ad. glycerine, ℥j.; but this had only a temporary good effect, and the disease was not subdued until the strong solution was applied. It was done every second day to the everted lid, and was of course quite painful.

Much more important is the purulent conjunctivitis which is generally known as *ophthalmia neonatorum*. It presents itself under various degrees of intensity, with swelling of the lids, with a yellow, thick secretion issuing from the eye, or, if dried upon the tarsal edges, it glues the lids to each other, and the conjunctiva, both ocular and palpebral, is reddened and swollen; there may be chemosis, and the palpebral conjunctiva be thickened and intensely red and spongy, with ridges and prominences, and be cleft by fissures between the enlarged papillæ. The tumidity of the palpebral conjunctiva, which increases up to the fornix, is a notable feature.

The cornea may remain clear, but its integrity is the point of anxiety. Its invasion may show as its first sign, a diffused haziness, or a local spot of purulent infiltration, or an ulceration, while in weakly infants, of whom premature births and foundlings are often extreme examples, the cornea may suddenly break down with a general infiltration, and become a mass of yellow putrilage. The point of ulceration in the cornea may be anywhere; if upon the periphery, and if it perforate, the iris falls into the opening and becomes permanently adherent, and from this a partial staphyloma may ensue; or, in case of less extensive destruction, nothing more than a distortion or concealment of the pupil. Should the ulcer be central, and not too large, so that if it perforate and the sphincter iridis cannot be drawn into the opening, the lens will then come forward and rest in contact with the posterior surface of the cornea. Sometimes a fistula will give rise to prolonged contact of the lens-capsule with the cornea. As a consequence, there will be an opacity in the centre of the lens, the middle of which will be a small white dot, and around it will be a fainter zone, and the whole will be apparently two or three millimetres in diameter. Sometimes, long after the eye is well, a fine thread is seen running from the lens to the middle of the cornea, which is the attenuated vestige of the material deposited while the surfaces were in contact. The opacity belongs to the capsule chiefly, and is both upon it and overlaid by it, sometimes having a distinct pyramidal form. It will be again referred to in speaking of cataract. In cases so marked as the above the secretion will be thick, creamy, and copious. The patients do not seem to suffer much pain, and usually nurse well. The skin of the cheek is apt to be excoriated by the discharge.

Treatment.—The first requirement is thorough cleansing by bits of rag, or a small brush to aid the moist applications. These will be at the outset simple water, which may be made cold by ice if the swelling be severe, but usually a moderate temperature suffices. Nothing is better as an antiseptic than boracic acid solution, four per cent., or alum two per cent. solution. But when secretion is considerable and begins to be puriform, the remedy will be nitrate of silver. The lids must be everted, first the upper and then the lower, and each in succession receives the application by a brush; its strength will be from gr. iij. to gr. x. ad. ℥ j., seldom so strong as gr. x. It will be used once daily. When a case comes during the stage of papilliform swelling, the mitigated caustic stick (nitrate of silver one part, nitrate of potassa two parts) is well suited to touch the outgrowths accurately and with proper vigor. The cornea is not touched by the caustic until the lids are replaced. The above measures will usually control a case in a healthy subject, if seen early enough. The measures will be of a similar kind to those to be described under the next heading, but less severe. Incision at the outer canthus is not practised, because loss of blood is serious to infants, and the tissues are so soft that there is less tendency to strangulation in spite of swelling. Unremitting care in keeping the conjunctival sac free from pus is the great reliance; and the astringent effect of nitrate of silver is the trustworthy remedy in bad cases.

2d. *Purulent conjunctivitis in adults* may be described in almost the same terms which have been used when speaking of the disease in infants. It may be spontaneous, or be caused by contagion from a similar case, or from gonorrheal virus, whether from the male or the female, and even after the secretion has passed into the stage of chronic gleet. In its early stage the secretion may be thin and like a puriform mucus rather than like pus. There may be little papilliform swelling of the palpebral conjunctiva; or

there may be on the palpebral, and to a less degree on the ocular conjunctiva, a plastic exudation in a thin layer, which, with more or less difficulty,
may be wiped off. This plastic, or sometimes called croupous material,
may come off in rolls, or may stick so closely that, if forcibly removed, the
subjacent surface bleeds. Chemosis will usually occur and may be extreme, sometimes overhanging the border of the cornea for several millimetres. The surface of the conjunctiva will not only be very red, but it
may be speckled with minute hemorrhages in its substance. The tumid
lid will be of a dusky red. The symptoms appear rapidly, and at the beginning there will be much pain. This is not only superficial, as denoted
by burning and smarting, etc., but deeper and more severe, and referred to
the interior of the eye and to the surrounding parts of the face and head.
The secretion is copious and apt to be retained in the folds of the conjunctiva and beneath the lids.

The gravest symptom is mischief to the cornea. This will commonly
not appear until signs of abatement can be seen in the activity of the disease. Just when we may felicitate ourselves that the climax of the attack
has been passed, the surface of the cornea may in some spot become
clouded. There may be a variety of ways in its manifestation. Either the
whole may have a uniform opacity of varying density, or this may appear at
the centre or periphery. Ulceration may occur in company with it or soon
follow it, or may be the first invasion. A furrow at the margin of the
cornea, beneath the overhanging chemosis, may completely or partially
encircle it, and it may eat through the cornea in a very few days. Ulceration at the centre, or at any point, is equally prone to occur. The degree
of corneal mischief can never be predicted. It is the product of three
local factors, namely, the pressure of the swollen tissues, the corrosive
action of the secretion, and direct continuity of inflammation to the substance of the cornea. It is also favored by unhealthy constitutional conditions, and by bad hygienic surroundings. It is impossible to assign to
each factor its exact measure of influence, or to assert that a peculiar lesion
of the cornea must be attributed to predominance of one or the other of
them ; in most cases, all concur in the disastrous effect.

The culmination of the active symptoms commonly arrives within ten
days, and the duration of the disease is from four to twelve weeks. The
final result upon vision, in case the whole cornea is not destroyed or the
eye escapes suppuration, is apt to be better than would seem possible during the severity of the symptoms. So much of the cornea as may remain
will become far more transparent and serviceable than might seem at all
likely.

Treatment should be vigorous from the outset. In robust subjects, or
with intense initial swelling or pain, four to six leeches may be put upon
the temple. Bits of muslin taken from iced water, or from a block of ice,
should be kept upon the eye constantly, and changed every minute or two
as they grow warm. In some cases a bit of ice, wrapped in muslin, may be
held upon the eye if its weight can be endured. Continuous cold, to the
degree which can be tolerated, is the rule. Absolute cleanliness of the inside of the lids is equally imperative. To effect it, the lids must be drawn
apart and the secretion removed by a soft camel's-hair pencil inserted between them, or by dripping an antiseptic solution upon the eye from a small
sponge, or by injecting such a solution with a dropping-tube. Some one
of these methods may be chosen, according to convenience. This process
will be repeated every five or fifteen minutes during the days and nights of
active secretion, and at longer intervals as the discharge lessens. The

proper fluid for cleansing, is a solution of boracic acid, four per cent. ; **of** chlorinated water, ten per cent. ; of chloride of sodium, one per cent. ; **or** of chlorate of potassa, three per cent. It is sometimes well to use **an** elevator in separating the lids, but by sensitive persons this will not **be** allowed, and extreme delicacy is needful in all manipulation, especially when ulceration has appeared upon the cornea. In certain cases in which the discharge is not thick and creamy, but ichorous, gruel-like and thin, nothing but this constant cleansing, by antiseptic lotions and the iced water, are needful ; no caustic need be applied ; and the chief concern will be the cornea, and for that we may use atropia, and possibly paracentesis. I have just had a remarkably good recovery of a case like this, in a young man who acquired the disease from gleet. It is a severe ordeal for a patient to undergo with the incessant cleansing by night and by day, and sometimes morphia must be exhibited in quantities sufficient to benumb undue sensibility. The fidelity and thoroughness with which this is done have the greatest influence in the strife for preservation of the eye. Should the cornea become invaded, a solution of sulphate of atropia, gr. ij. ad. $\text{\openbullet}$ j., should be instilled every three to six hours. It will be seen that, to perform this laborious nursing, not less than two attendants, and they both strong and faithful, are indispensable.

Another indication is to be borne in mind, **viz.** : the relief of pressure by incising the swollen tissues and unloading the vessels. Mr. Tyrrell laid stress on scarification of the chemotic conjunctiva, and it has benefit ; but the infiltration does not flow away through the cuts, however deep, because it is of a plastic and coagulable quality ; but the depletion is of use, and the incisions may be repeated on more than one occasion. Furthermore, not a little relief is afforded by incising the lids at the outer angle down to the bone, and for a half inch or more in length. Canthotomy freely done unloads the vessels and lessens the pressure of the lids by weakening the orbicularis and giving room for the infiltration to spread. I consider this proceeding indispensable when great swelling of the lids exists, and do it with little swelling if the cornea be threatened. It may be needful to again snip the tissues with scissors in the site of the cut, after a week or more, because healing quickly takes place, and conditions of partial strangulation by the lids may be reproduced. In resorting to such proceedings, a surgeon will require to act with discretion as well as boldness, appreciating the significance of the symptoms, the danger to the eye, and the general health of the patient. As against the propriety of incisions, the partial interruption of the process of cleansing is to be considered ; but this need not be wholly suspended ; its method may be so modified that **the** lid need not be greatly disturbed for the next twenty-four hours.

Finally, we have to consider the utility of nitrate of silver, for there is no other remedy to be thought of in local treatment. It will have no place during the period of invasion, and of thin, gruel-like, puriform secretion. It may not be admissible at any stage of the disease. But if the secretion become creamy and distinctly purulent, and the conjunctiva be velvety, a solution gr. v. vel x. ad. $\text{\openbullet}$ j., may be tried on the everted lids. The longer the brush is held in contact with the surface, the greater is its effect, and this may not be small. Pain will ensue for an hour, or longer ; for a time the secretion is suspended, or becomes watery, and if after twenty-four hours the swelling is less, and the secretion not so copious, the remedy has been well chosen and may be repeated once in twenty-four hours. It must be remembered that the first application is tentative, and to be carefully watched in its effects, especially on the cornea. When good results from its use, the gain is very decided. This solution, however, may not be strong enough, and twenty grains may be cautiously tried.

The issue of an attack may be in perfect restoration of the conjunctiva to its normal texture, in case it were healthy before, but it not seldom passes through a period of papillary hypertrophy (granulations) which may continue for weeks. The case may be under treatment for this condition long after the ocular conjunctiva has recovered its normal state, and at this stage, sulphate of copper crystal, tannin and glycerine, ⊙j. ad. ℥j. or stronger, will be the most useful applications. In the event of the cornea being damaged, all eventualities are possible, between a slight degree of opacity, almost amounting to transparency on the one hand, and staphyloma cornea, or atrophy of the globe, on the other. A partial staphyloma, or anterior synechia, for which an artificial pupil may be practicable, is exceedingly common. It is of course possible that both eyes may be lost, but the right is the one most often affected; and it is common, when a second eye is attacked, for it to be less severely inflamed than the first. The reason is, because the secretions are less noxious during the decline of the inflammation. For this reason, the pus of a florid urethritis will cause much more severe eye inflammation than that taken during the stage of gleet. The fatality to sight of this disease is great, and is set down by Klein as follows: Out of 40 eyes, 16 became totally blind, 9 retained some vision, and 5 had useful vision, or could obtain it by operation; in 30 the cornea was involved, while in only 10 did it escape. That is, corneal trouble occurred in three-fourths of the eyes.

The protection of the fellow eye, in case it be intact, is of the utmost importance, and my experience has been that the endeavor to seal it hermetically by a bandage or plasters, is unavailable. It cannot be carried out effectively, because of the wretchedness which it inflicts in making the patient for the time wholly helpless, and because of the excoriation of the skin of the cheeks and lid, and of the impossibility of preventing the patient from slipping off the covering during the night, and unwittingly exposing himself to contagion by his soiled fingers or accidental communication. In fact, the need of handling the eye to change the dressings, which should be done twice daily, involves more exposure to risk than to warn the patient and attendants of the need of utmost caution, and to insist that the patient sleep, lying on his back, or on the side of the affected eye. A device, whose authorship I do not know, has answered well in some cases, viz., to cover the sound eye with a watch-glass, attached by means of rubber plaster. It need not be removed oftener than once in several days, as secretion may demand. The mystery which sometimes attaches to the transmission of the inflammation to a healthy eye is less surprising, when it is known that the pus when diluted, 1 to 1,000, still retains decided contagious properties. It ought also to be stated that nitrate of silver, by its power of coagulating albumen, antidotes in high degree the noxious character of the pus, and it has been shown that a one-fourth per cent. solution, or gr. ij. ad. ℥j., renders it innocuous. This fact explains why so few cases of contagion occur in public clinics, where the same brush may be used for different patients, when it has passed through a solution of nitrate of silver. I instance this fact, not to encourage economy in the use of brushes, but as a point worth knowing. Every patient needing an application to his lids, should have his own brush; and purulent cases should be most vigorously quarantined from communication in any manner with other patients.

Another matter to be noted is that when the cornea is deeply infiltrated or ulcerated, it may be needful to puncture it as if there were no conjunctival implication. The rules on this point will be stated under the head of keratitis.

Croupous, or diphtheritic conjunctivitis.—Under the description of purulent conjunctivitis, I have mentioned that sometimes the secretion is to a degree plastic, either forming shreds or rolls, or adhering to the mucous surface. No special significance is to be attached to this fact, but cases sometimes arise where the plastic quality of the secretion is a distinctive peculiarity, predominating over all other features of the case, and, therefore, deserves special mention. Various grades of this quality appear, and if we find that by a little effort the layer of false membrane can be rubbed off from the conjunctiva, there may be no need of regarding the case as differing essentially from a blenorrhœa, because it is subject to the same laws of development, and will soon exhibit the same features by transformation after some days of the plastic substance into the common purulent secretion. Such cases may be spoken of as croupous, they may originate per se or by contagion, and no rules are to be laid down for their treatment other than those given under the previous head. When plastic exudation is not confined to the surface of the conjunctiva, but infiltrates its **substance** and renders it thick, stiff, pale, grayish, or buff; when the lids **are** difficult to be everted, because they are stiff as well as swollen, and the eyeball instead of a bright red, has a dull lead-color; under these condi-**tions** we have a disease whose virulence may be unbounded. The infiltration may be on the palpebral conjunctiva alone or also on the ocular. There will be the usual signs of severe inflammation in swelling of the lids and pain, but there will be only a thin turbid discharge from the eye, and the vascularity of the conjunctiva moderate; the temperature of the lids will be increased, and they are often not only swollen, but stiff and hard.

The disease may be constitutional as well as local, being attended by fever and prostration, and there may or may not be diphtheritic exudation on other mucous membranes. I have seen a few cases of this character, but happily they are rare in this country, or, at least, in this part of it, and are nowhere so often seen as in Northern Germany.

The stages of the disease are those of plastic infiltration, succeeded by a period of resolution in which the material is dissolved and cast off and may leave ulceration of the membrane, and this may eventuate in a gradual return to health, or to an opposite result, viz., cicatrization and shrinking of tissue. The cornea is almost certain to be attacked, and in cases of extensive ocular infiltration, it will slough. Hence, the prognosis is serious.

Treatment is, for severe cases, powerless. The discomfort may be as-**suaged.** Sometimes ice-cold applications are best, and at other times warm water must be used, the comfort of the patient being the guide. Incisions into the ocular conjunctiva, however swollen it may be, do no good, because no fluid can escape, scarcely even blood. The substance cuts like cheese or bacon. Division of the outer canthus gives some relief, but the wound becomes infiltrated with exudation and a fresh source of propagation. Leeches to the temple may be cautiously tried if pain be great, but the disease is asthenic and the patient does not tolerate reducing remedies. The leech-bites are apt to become infiltrated; nitrate of silver cannot be used, because it simply adds to the destruction of tissue. The only remedies are such as have antiseptic quality, and the best will be boracic acid, four per cent., or chlorine water, washing out the eye frequently and giving the patient such internal remedies as the symptoms indicate, particularly quinine in full doses; chlorate of potassa has been mentioned, but I cannot give an opinion upon it, the doses being from **gr.** ii. ad. gr. v., every three hours. In Germany, mercurials are insisted on, but I should demur. A full, nourishing diet, and administering to the patient's comfort, is all that can avail in this sad condition. Very early the cornea **becomes** attainted, and will either ulcerate at some spot or turn to a yellow mass of infiltration. The most dreadful case of this disease which I have **seen was** a full picture of all the worst features above portrayed. Children **under five or ten years are** most frequently affected. At the stage when discharge appears, **and the** plastic exudation begins to give way, the nitrate of silver may be used as indicated in cases of blenorrhœa, and that is the type into which the disease passes ordinarily. But there are cases in which the **stage of** resolution does not occur, but the plastic material seems to dry up **and** never to be cast off; it grows thinner, is absorbed, and atrophy of the con**junctiva** takes place with destruction of the cornea. When the attack is spent, the conjunctiva frequently undergoes shrinking and atrophy, especially at the retro-tarsal folds, where bands of attachment, as if from burns, will be formed. So, too, the tarsal conjunctiva suffers and may give rise to trichiasis or entropium.

GRANULAR CONJUNCTIVITIS, OR TRACHOMA.

Trachoma.—Under this name, an essentially chronic **disease** is designated, which exhibits itself under a great variety of phases and regarding whose essential pathology there has **been** much debate. Its clinical picture is easy to be recognized, but unless **we** are able to associate the clinical features of a disease with its morbid anatomy, we have no correct basis of classification, nor have we that perception of the principles **of** treatment which will enable us to pursue an intelligent plan or to form **a** rational prognosis. **Within** twelve years, much has been done to clear up the obscurity which has hung about the pathology of granular conjunctivitis. A special morbid product, known as the trachomatous granule or lymph follicle, and when in diffused masses, known as trachomatous infiltration, has been signalized **as the** characteristic of the disease (Arlt, Stelwag). It has been likened to tubercle, although not claimed to be iden- tical with it, and it has been asserted that its progenitor is found in certain lymph follicles, or in the adenoid tissue, normal to the healthy conjunctiva. The weight of authority denies the existence of these lymph follicles in the tarsal portion of the conjunctiva, and admits them only in the retro-tarsal **fold.** The trachoma follicle is a small rounded **mass, the size** of a rape**seed,** consisting of lymphoid cells, in which connective **tissue cells are** also **to be found,** and surrounded by a capsule of fibrous **tissue ;** it is traversed **throughout by a** reticulum of minute fibres **and** blood-vessels. The lym**phoid cells are** the essential feature. **That** they appear in globular **masses is** best explained by proliferation of cells from a central point and **the formation** of a fibrous capsule, which is usually imperfect, by the condensation of neighboring connective tissue. Such lymph-cells are scattered promiscuously and in great numbers through the conjunctiva, forming adenoid tissue. But, besides the lymphoid tissue, whether globular or diffused, there **is in** granular conjunctivitis a large hyperplasia of epithe**lium.** The peculiar arrangement of the tarsal epithelial surface must be remembered, that it is cleft in all directions by fissures, that these branch **and** subdivide in compound and intricate ways, forming diverticuli and pockets, and such formations may, with a certain justice, be called glandules as authors have styled them, because the prolongations are often tubular and are lined by epithelium. In granular conjunctivitis, the epithelium becomes hypertrophied, the pocculi and clefts become crowded together. Their communication with the free surface is in many cases cut off, their cavities become filled with epithelial cells, and in certain cases, severe effusion **can occur in** these occluded spaces, which will cause a cystoid degeneration.

There is, moreover, hypertrophy of the papillæ and basement membrane or connective substance of the conjunctiva, and a general formation of products of inflammation, including blood-vessels. In brief, we find accumulation of lymphoid cells in various forms, hypertrophy of epithelium with exaggeration **of** its anatomical peculiarities, increase of connective tissue, and the usual products of inflammation, all of which constitute the essential elements **of** granular conjunctivitis.

When we subdivide the disease according to its clinical characters, we have the following forms :

1st. Papillary trachoma, in which enlargement of papillæ, hypertrophy of epithelium, and inflammatory products without notable lymphoid hyperplasia are the characteristics.

2d. Miliary trachoma, in **which lymphoid** aggregations in minute **masses**, with or without moderate inflammatory products, occur.

3d. Diffused trachoma with lymphoid infiltration, and inflammatory products and epithelial accumulations, and papillary hypertrophy.

Papillary trachoma is the sequel of acute purulent conjunctivitis, or blenorrhœa, and is capable of entire disappearance without structural injury to the conjunctiva. Such is its usual course, and for this reason it is commonly set apart from ordinary trachoma.

Miliary trachoma, also called follicular conjunctivitis (Saemisch) and swelling of the conjunctival follicles (Schweigger), is a condition of the lids in which the only deviation from a healthy state is the presence of a number of seed-like bodies which lie just under the surface, and of whose presence the bearer may be wholly unconscious. They are semi-transparent, can be picked out with a needle, and the surrounding conjunctiva may be to appearance entirely normal.

General trachoma includes all the pathological lesions in varying proportions, and becomes more or less distressing, according to the degree of inflammatory complications, according to the stage of the disease, and according to the alterations it has impressed upon the structure of the conjunctiva and lids, and according to the effect it has had upon the cornea. It is capable of disappearance, but the conjunctiva is left in a condition of **more** or less atrophy, because there is always notable development of connective-tissue which inevitably undergoes contraction in the course of its history. This propensity is the chief cause of the mischief entailed by granular conjunctivitis, so far as the lids are concerned, while the lesion directly or indirectly inflicted on the cornea, often causes deplorable results on vision.

Miliary and general trachoma generally require a long period for their evolution ; the greater the amount of lymphoid infiltration, the longer the time needed for absorption. The general health of the person and his hygienic surroundings have a marked influence on the duration of the disease. The continuance is to be reckoned by years, and while some cases arrive at a condition of comfort in less time, it is not rare for ten, or even twenty years to elapse before the final stage is attained. Under such protracted continuance, it is inevitable that important mischief should be done to the structure and function of the parts concerned.

Let us discuss the several forms in detail. *Papillary trachoma*, whether or not succeeding blenorrhœa, presents a condition of general redness over all the palpebral conjunctiva, with thickening of its tissue, points of greater prominence which look like the elevations on a wart; these are sometimes single and acuminated, sometimes flattened or cleft. There is moderate muco-purulent secretion, and the lid cannot be fully lifted, except by the finger. At the retro-tarsal fold the infiltration increases, and wrinkles parallel to the palpebral border are found. The surface of the cornea may be hazy, by proliferation and maceration of its epithelium and inflammation of its superficial layers; it may be faceted extensively by minute erosions. Often it is quite intact and clear. Rarely does it become vascular, except with a fringe of short, straight vessels at the limbus, not often more than from two to four millimetres in length. The bulbar-conjunctiva is hyperæmic, but not thickened or papillary.

Treatment consists in daily applications of mild solutions of nitrate of silver, gr. v. ad. ʒ j. After a time, as muco-purulent secretion abates, using tannin, gr. x. vel xxx., ad. glycerine, ʒ j., and as a substitute to be less frequently used or reserved to a later stage, when the membrane has become torpid to other stimulation, the smooth crystal of sulphate of copper. Within two to six months, in healthy subjects, the conjunctiva will recover its proper condition; but in old persons, or those who neglect their eyes, or are uncleanly, the condition may be indefinitely prolonged.

Miliary trachoma is more frequent upon the lower lid, and the retro-tarsal fold is apt to be swollen and wrinkled. There is usually no secretion, and there may be little irritation or discomfort. The granules do not materially increase in size, but multiply, and thus come into contact. They may continue latent for a long period, but a mild conjunctival hyperæmia without thickening is liable to occur, and this brings the patient for advice. It is attended by smarting and worry, with aggravation at night, amounting, perhaps, to inability to use the eyes, and attended by a little photophobia and lachrymation.

Treatment.—The removal of attendant irritation is to be sought by such mild measures as would be used for conjunctival hyperæmia. The second indication is to procure absorption of the miliary granules. This requires prolonged use of such stimulants as will not injure the tissues but will quicken the nutritive activity. Both indications can frequently be combined by application to the lids of tannin, gr. x., glycerine, ʒ j., daily, and when the conjunctival irritation abates with less frequency. A smooth crystal of alum is a useful daily remedy, and nitrate of silver need not be stronger than gr. ij. ad. ʒ j., but is to be commended for only a brief period, because of its tendency to cause staining of the tissues. The disappearance of the granules will require months or years, and may be finally achieved without any advance of the disease to the more serious condition of general trachoma. If neglected, it is highly probable that such an issue will develop, and such is the fact in many cases.

General or diffused trachoma.—The conjunctival surface presents very great variety of aspects, according to the severity and the stage of the disease. It may be covered by small nodules, projecting above a thickened membrane, and these rounded and hard or flattened; there may be numerous papillæ, with yellowish or jelly-like lymphoid infiltration interspersed, and becoming more abundant at the fornix. The lid may be swollen and dense, with a smooth, glazed, red surface, or it may have patches of red and buff, intermingled with lines of cicatrix, running through it. No brief description can record the multiform phases of this disease. The appearances vary, principally according to the preponderance of one or the other of the elements which constitute the morbid product, whether of inflammatory infiltration, of hyperplasia, of epithelial prominences and papillæ, or of lymphoid infiltration. The last may be in small masses, which have been likened to frog-spawn, or be diffused. Furthermore, it is the natural evolution that connective tissue is formed and appears as whitish or gray lines traversing the tissue, and because it always shrinks it gives rise to degeneration of the membrane and deformity. In an old case, the connective tissue, under some cicatricial form, becomes the conspicuous factor. The mucous membrane will have lost its proper texture, and as the tarsus has been infiltrated by inflammatory products, it shrinks as these undergo metamorphosis and becomes shortened both in its length and breadth; it first becomes incurvated, and at a later time, takes the shape of a solid roll. The contracting tendency causes the retro-tarsal fold to be reduced in depth, and in extreme cases it is thrown into vertical ridges or fræna and becomes pale and glazed. As a last stage of degeneration, the conjunctiva loses all appearance of a mucous membrane, becomes dry and cuticular, a condition denoted as *xerosis*.

The cornea undergoes superficial inflammation with notable development of vessels in most cases. This may be confined to such portion of it as is habitually covered by the drooping lid, viz., about the upper half, or it may, and does in time, extend over the entire surface. By reason of this complication, the patient often suffers greatly from pain, photophobia, and lachrymation, and grave mischief occurs to sight. It is characteristic of granular conjunctivitis, that attacks of acute inflammation are extremely liable to come with slight provocation. It is the cornea which bears the brunt of the trouble, and while the patient's sufferings are not to be lightly esteemed, the additional loss of vision is the most serious harm. It is not usual to have deep ulceration of the cornea, but such may occur, and lead to the usual results of prolapsed iris and partial or total staphyloma.

It is not very uncommon for iritis to be added to the lesions already recited, and an adherent pupil renders the ultimate state of the eye considerably worse.

The description above given applies to the disease as it is found under a *chronic* form, and which is by far the most frequent ; but there is another phase of it which has been known as *acute trachoma*. An accurate pathological statement of its character may not be possible, but it exhibits a sudden and severe beginning with great pain, lachrymation, photophobia, thickening of the palpebral conjunctiva, and injection of the ocular conjunctiva. The palpebral mucous membrane shows, at an early stage, marked infiltration in its substance, with little papillary elevation, and a notable absence of catarrhal secretion. The attack would be supposed to be one of ordinary acute conjunctivitis, but for the rapid and firm infiltration and the absence of mucoidal or purulent secretion. It is, in fact, a sudden and considerable inflammatory hypertrophy of the deep parts of the palpebral conjunctiva. There is not always evidence of lymphoid infiltration, but such attacks are likely to befall those who are in hospitals, or in places where the air is impure, and who may have unconsciously contracted lymphoid infiltration before the acute attack. One of the most remarkable cases I remember, was in the person of one of the internes of Bellevue Hospital, in whom the disease lasted between one and two years, and finally left the membrane, to all appearances, in a normal condition. This fact argues against the probability of considerable lymphoid deposit. Attacks not unlike this are liable to occur to those who have chronic genuine trachoma, and they greatly aggravate the existing malady, besides causing prolonged and keen distress.

Another phase of trachoma might be considered in connection with miliary granulations. It consists of deposits scattered over the tarsal conjunctiva of gelatinous, soft, isolated or confluent masses, which are softer than common miliary lymphoid bodies, and aggregate at the retro-tarsal folds into broad wrinkles and ridges, or a diffused infiltration. The looseness and softness and pale pink of this material constitutes a conspicuous clinical picture. There is considerable mucoidal secretion, and moderate conjunctival irritation. The lids are a little swollen, and cannot be freely opened. This condition does not greatly impair the cornea, and does not inflict severe discomfort, except in inability to use the eyes. The disease is obstinate, and special suggestions are to be made as to its treatment.

Treatment.—The rationale consists in promoting absorption of lymphoid and inflammatory infiltrations, in removing hyperæmia, in restraining the development of the connective-tissue, and promoting the clearness of the cornea and the maintenance of its proper curvature. To attain these ends, special regard must be paid to the stage of the disease, to the general health of the patient, and to the condition of the cornea. There is also an idiosyncrasy in many individuals which will compel the employment of means especially suited to their case.

A broad distinction must be made as to whether there may be acute inflammation of the cornea. If this be not present the lids alone will call for attention, and the local stimulants will be steadily pursued. But if there be acute keratitis, with severe pain, lachrymation, and photophobia, stimulation of the lids must in most cases be omitted, and attention confined to measures which shall allay the corneal trouble. In case there be no active keratitis, the treatment of the conjunctiva will be guided by its appearance and the quantity of secretion. If soft, and pulpy, and red, with some mucoidal secretion, the best remedy is nitrate of silver, gr. v. ad. ℨ j. as an average solution. This may be used once daily to the everted lids, and its effect will depend on the quantity used, and the time it is kept in contact with the surface. Usually there is no need to wash away excess with water, nor to neutralize with solution of salt. The brush should be of fine hair, and not too large. The upper lid should first be touched, and then the lower. The pain, which comes in a few minutes, is to be assuaged by spraying with cold water, while the most effective mode is by a fountain douche. Tears flow copiously, and the irritation will last from one to four hours, sometimes more. In place of solutions, the crayon of nitrate of silver and nitrate of potash, one to two, is sometimes preferred, because it can be limited with exactness to certain spots, or to prominent folds near the cul-de-sac. It can be made stronger, or milder, by touching heavily or lightly, and in experienced hands it works well. It is usually best to neutralize excess of caustic thus applied by a saline solution. Scarification of the tissue often aids the result, by the depletion of the vessels. It is proper to guard against the temptation to make with a so-called mitigated stick a more serious cauterization than is suitable, or than may have been intended. In the hypertrophied and succulent state of trachoma, no remedy will take the place of nitrate of silver. It is to be persevered in month after month, daily, or every second or third day, according to effects. A repetition should not be made until the eye has recovered from all reaction caused by a previous touching. It is better to let the patient enjoy a day of comparative comfort before giving another dose. It is always bad practice to use caustics severely, and the mischief consists in the increased production of connective tissue, which active cauterization promotes. **The** aim is to stimulate absorption, and not to destroy redundant tissue.

Often it is advisable to use sulphate of atropia, gr. ij. ad. ℥ j., three **times daily, when there is moderate** irritation and haziness of the cornea, **the condition being such as not to** preclude the stimulation treatment **while calling for attention to the** cornea.

Sometimes the conjunctiva is extremely **sensitive and will tolerate** only the mildest remedies, **viz., argent.** nitrat., gr. j. ad. ℥ j. (one-fourth per cent.), or alum, gr. v. ad. ℥ j., **one per cent.,** or warm **water fomentations** or cold water. When, **however, the more** passive **condition appears, during** which the lymphoid **substance is** predominant, **sulphate of copper** becomes very suitable. It is most commonly used in smooth crayon, **and** must **be** lightly or harshly used according to circumstances. It is always **a** painful remedy. Other things must sometimes be substituted, and the favorite of the milder kind are tannin, **gr. x.** vel xxx., **ad.** glycerine, ℥ j., two **per cent. to six** per **cent., or a smooth** crystal of alum. In fact, in the **treatment of** trachoma, **all the** remedies ever proper to the conjunctiva must **be invoked,** according to the variations which will be presented, **and much tact has to be** employed in adapting them to suitable exigencies.

When we deal with *acute trachoma,* all stimulating methods **are** to be avoided. Bathing with cold or warm water continuously for intermitting periods as the patient finds agreeable, exclusion of light, rest in bed, anodynes as required ; if there be notable **secretion, the** trial of solution of nitrate of silver, gr. ij. **to** v. ad. ℥ j., or sol. **acetat. plumbi.,** gr v. ad. ℥ j., provided the cornea has no abrasion. **This will be** the programme until the **acute stage passes.** Scarifications may **be** cautiously tried.

For certain cases of miliary or follicular trachoma, **when the outgrowth** is exuberant, **either in** voluminous soft folds at the sinuses, **or sometimes** in detached hard **masses, it is** proper to remove them with the **scissors.** Care must **be used not to take** away more tissue than can well be spared, **and one must keep in mind** the tendency to shrinkage which **persists** through the whole **course of the disease.**

Entropium **or trichiasis will be treated as has been already** indicated ; and if **only a few cilia** are **faulty, it is better to pull** them out repeatedly than to **do** any operation. **In advanced stages of** atrophy, the palpebral slit becomes shortened and **the tarsi incurved. For** this the operation of cantholysis (canthoplasty) may be done, **and often with** marked advantage. **The** chief element in its benefit **is the division of the** fibres of **the** external **canthal** ligament (see page 146). **In** atrophic conditions yet further advanced, where it may be impracticable to **do** the simple operation, **a true** plastic proceeding may be employed, which emphatically deserves **the name** of canthoplasty, and which is described **on** page 147.

The state of the cornea keeps pace commonly with the progress of the conjunctiva, and grows more clear as the latter improves ; it is, in fact, the best **index** of the success of treatment. But **in** very old cases, the cornea may be densely opaque **with pannus,** and the conjunctiva may have passed into atrophy. There is little use in stimulating the lids, and the following suggestions may be made : **A solution of tannin and** glycerine, 3 j. ad. ℥ j., may be dropped into **the** eye, **or a solution of** sulphate of copper, gr. x. to xxx. ad. ℥ j., **directed against** the cornea by a spray apparatus for a few minutes. The operation **of** *peritomy* has been practised to induce disappearance of the corneal vessels. It consists in dissecting up a zone of conjunctiva from four to six millimetres wide all around the margin of the cornea, using fine forceps **and** sharp-pointed scissors, and taking pains to pick up every shred of connective tissue and vessel which can be lifted. The wound is left to granulate, and the eye is treated by cold compresses. The benefit of this proceeding does not appear for some weeks and must not be too highly rated.

Another proceeding, which must be employed with great caution, and only in cases of complete pannus, the subjects having some vigor, is to inoculate the eye with blenorrhœa and leave the inflammation to take care of itself and run its course. Necessarily there is danger of thereby destroying the eye by an excess of reaction, and this limits the proceeding to cases where both eyes are totally covered with pannus. I have done this in a few cases, and, observing the above cautions, have had satisfactory recovery of clearness of the cornea.

It happens that patients cannot always remain under the hands of the surgeon for the completion of the treatment of this tedious disease, and it becomes advisable to instruct a friend in the manipulation of turning the lids and applying either sol. tannin and glycerine, alum crystal, or the sulphate of copper crystal. In case the patient must depend wholly on himself, he may be supplied with an ointment of sulph. cupri., gr. v—x., ad. vaselinum, ʒj. The vaseline may in warm weather be stiffened with powdered gum arabic or starch.

The necessity of remitting active treatment when relapses of inflammation occur, as they may frequently, must not be forgotten, and then warm water and atropine will be the best relief. The use of the latter is often kept up during the continuance of the local stimulants. Sometimes with prominent eyes the lids in the atrophic stage of trachoma are left so shortened as to press disagreeably on a rather prominent eye. Canthoplasty may have done all that it can, and the skin operation above described may not be desirable. In such cases, continual pulling at the skin of the temple, forehead and cheek, a kind of massage, will, in time, loosen the subcutaneous connective tissue and secure some relief.

Xeroma of the conjunctiva, which is the stage of consummate atrophy of the membrane, only admits of palliation by using emollients, such as vaseline or olive oil several times daily.

Sequelæ of granular conjunctivitis.—Among the results which follow a long continuance of the disease are various deformities of the eyelids, viz.:

Distichiasis, trichiasis, which mean inversion of a few or of many eyelashes. The incurvation of the tarsus, which involves both the inversion of the lashes and shortening of the palpebral slit, and usually atrophy of the conjunctiva. Blepharo-phimosis, which is partial closure of the palpebral slit, and is due to atrophy.

A condition of ectropium of the upper lid occasionally arises from great hypertrophy of the conjunctiva after injuries, or as a result of purulent conjunctivitis. Moreover, the orbicularis keeps up the eversion by bending the lids at the upper edge of the tarsus. This may be in a measure relieved by applying india-rubber plaster But, if this be ineffectual, a cure is effected by making the two lids grow to each other for a space of eight to ten millimetres at the middle of their borders. The posterior edge of each is shaved off by a narrow knife at opposite points, and two sutures put through the lids and left for several days.

Amyloid degeneration of the conjunctiva is a disease of extremely rare occurrence, and somewhat resembles trachoma. It affects chiefly the upper cul-de-sac and semilunar fold, and may cause such hypertrophy of the tissue that it shall project between the lids. The conjunctiva is converted into a thick, soft, gelatinous substance of a yellowish color, and containing some semi-transparent granules larger than ordinary granulations. Other parts of the membrane may be quite intact.

The only treatment of this condition is excision of the morbid tissue. I have seen but one case.

PHLYCTENULAR CONJUNCTIVITIS—HERPES CONJUNCTIVÆ—PUSTULAR CONJUNCTIVITIS.

Under these names are included several varieties of inflammation, whose type consists in an isolated elevation like a vesicle upon the ocular conjunctiva and at the apex of a triangular area of blood-vessels, which point toward the cornea, and whose base may extend to the cul-de-sac. The vesicle may be semi-transparent or yellowish, or be semi-solid and gelatinous. It may be near or remote from the cornea. There may be one or many. The following varieties may be noticed. Sometimes nothing appears but a limited area of vascularity, which remains for a few days and disappears without any eruptive exudation. Again, a series of minute transparent vesicles crop out on the limbus corneæ, and are attended by an area of vascularity in adjacent parts; they rupture and form minute ulcers, and after a few days disappear.

Again, a typical phlyctenula may form on the conjunctiva, be attended by vascularity of limited extent, and soon disappear. But this same eruption, beginning at a distance from the cornea, may gradually advance to the limbus, be attended by severe irritation, and invade the cornea as an ulceration. It may continue for weeks, and there may be a number of others, either upon the corneal margin or on the ocular conjunctiva, each of them attended by its own vascularity. Again, a most serious form begins at the limbus corneæ, with marked infiltration and suppuration, and gives rise to ulceration and perforation of the cornea. This phase of phlyctenula belongs as much to the cornea as to the conjunctiva, and indicates that the disease is one which finds its nidus in either tissue, as may happen. Again, phlyctenular conjunctivitis may remain upon the ocular conjunctiva and penetrate into adjacent structures, and develop episcleritis. In many cases an attack of phlyctenula is attended by little annoyance, and soon passes away. At other times it is accompanied by more or less catarrhal symptoms in secretion, swelling, lachrymation, pain, and photophobia. It also appears in scrofulous children, who may have eczema of the face, severe nasal catarrh, with fissures and ulcers and scabs at the nostrils, pouting lips, and cachectic aspect. There will be aggravated photophobia, and lachrymal discharge. The lids will be tightly shut, or the patient will furtively cast a sidelong look away from the light. They will have fickle appetite, fretfulness, constant worry, etc. When the lids are forcibly opened the whole ocular conjunctiva may be injected, and the palpebral will likewise be involved. With these conditions the cornea is very likely to become affected.

The cause of the affection is obscure, but is probably to be sought in irritation of fifth-nerve twigs, as has been shown by Iwanoff. It occurs seldom in adults, is almost peculiar to children; it affects those in apparently good health, but is most frequent among the delicate and ill-nourished.

Treatment.—In mild cases very little attention is needed. Warm water, solutions of borax or of alum, and abstinence from use will suffice. In more active forms the favorite treatment, when severer symptoms have been allayed by lotions of lukewarm water, is to dust the eye with finely levigated calomel, from a dry brush, once daily, and in cases more obstinate or torpid, to put between the lids a small bit of Pagenstecher's ointment, viz.

R. Hydrarg. oxid. flav........................ gr. j.–iij.
Amylo-glycerini vel vaselini .... ℥ j.
M.

Of this the quantity used may be no larger than the head of a pin, and is to be rubbed over the eye by pushing the lid about with the finger. Care must be taken to begin with a minute quantity, because the irritation of larger amounts may become excessive. The ointment is by some used stronger than I have indicated, but my own experience has been against it. When strong, or used in considerable quantity, it may cause an eschar and ulceration of the conjunctiva in the cul-de-sac, and severe reaction. The cases whose general health is poor, the so-called scrofulous, require much patience and skilful nursing, as well as medication. Their treatment becomes identical with that of phlyctenular keratitis, and will be considered under that head. For such as are accompanied by catarrhal symptoms, with mild, muco-purulent secretion, the treatment will be weak solutions of nitrate of silver, gr. j.—iij. ad. ℥ j., or simple astringents, and simple cerate to the edges of the lids at night. In all these cases, strict attention to hygiene, in good air, in plain and nutritious food, bathing and proper clothing, are of the utmost necessity. Often a change of air, especially from the city to the country, is valuable.

There are other eruptive diseases of the conjunctiva, such as pemphigus, chancre, and certain forms of tertiary, syphilitic ulceration, and lupus, all of which are rare, and need not be described.

Sub-conjunctival ecchymosis is a common and unimportant occurrence. It presents a dark-red spot, usually upon the inter-palpebral portion of the globe, sharply defined, and not attended by any hyperæmia or by any pain. The spot may be small, or cover the whole globe. Usually the subject is unconscious of its occurrence. It happens from a great variety of causes, such as the straining of whooping-cough, and it appears in elderly persons, in whom it indicates atheroma or brittleness of the vessels. Sometimes there is a little dimness of sight, because some of the hæmatine finds its way into the aqueous humor, and it may even give it a visibly yellow tinge.

Absorption of the blood will occupy from one to three weeks, and no treatment is of value or necessary.

PTERYGIUM.

This condition appears most frequently at the inner side of the globe, and consists of a thickening of the conjunctiva, which may touch the border of the cornea, or extend over its surface and be attached to its substance. It is triangular in form, its apex at the cornea, its base toward the equator of the globe. It may be thin and almost transparent, narrow, and encroach but little on the cornea. It may also be very vascular, thick, broad, and reach to the centre of the cornea, or even beyond. It

12

occurs in adults, and especially after middle life, and more among those who are habitually exposed to the weather or to an irritating atmosphere, such as sailors, farmers, etc. Among such persons chronic conjunctivitis is common, and if an ulceration occur at the border of the cornea, and ex-

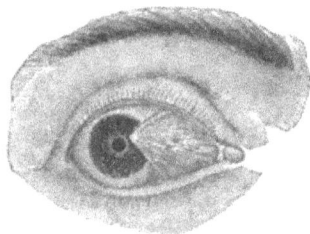

Fig. 73.

tend to the adjacent conjunctiva, the ulcer in healing will drag over a little of the relaxed mucous membrane to the cornea. With renewed attacks of inflammation, this process advances and the apex of the pterygium is preceded by a slight ulceration. The membrane in the meantime grows thicker and more vascular, and with a greater laxity of the conjunctiva, more of it becomes dragged into the advancing triangle. The special looseness of the conjunctiva at the inner side of the globe, where we have the plica semi-lunaris, and the predominance of the act of convergence above other movements of the eye, accounts for the prevalence of pterygium in this region. It occurs next in frequency on the outer side and may occur in any region. It may be multiple, and often is symmetrical in both eyes. There is, however, another view taken of the development of the lesion. In many cases there is no evidence of an ulceration in the cornea. On the contrary microscopic examination shows that the corneal part of the growth has epithelium, not only on its anterior surface, but also on its posterior or applied surface (Schweigger). Also on the sides of its pointed part there is an infolding of the tissue, under which a fine probe can be thrust a little way. This indicates a process of adhesion between the overlapping conjunctiva and the cornea, without destruction of epithelium. Alt points out that the apex of the pterygium buries itself a little beneath the superficial layers of the cornea. In whatever way produced, whether by irritation of a marginal ulcer or as an outgrowth of the conjunctiva, the advance of the structure is very slow. It remains thin and semi-transparent for years, and when progress occurs it is with intermittent attacks of slight inflammation, which may finally assume a persistent chronic form.

Treatment.—The only effectual proceeding is removal by excision. Strangulation by ligature is figured in many books, but is painful and unsatisfactory. There is no occasion to interfere with the disease in its incipient and quiescent state, because no harm comes to the eye, and no actual blemish is produced. If large enough to attract notice, the growth may be removed, while if it have extended over one-fourth or one-third of the diameter of the cornea, it ought to be removed. The operation is done as follows, either with or without an anæsthetic. Seize the structure near its tip with fine-tooth forceps, shave the apex from the cornea by a cataract knife. Take care not to lift a needless amount of corneal epithelium. Then, with pointed scissors dissect back toward the base as much as the thickness and extent require, and cut off the superfluous material. A raw surface will be left on the sclera, which is to be covered by loosening the adjacent conjunctiva, by incisions above and below, parallel to the periphery of the cornea, and sliding the flaps toward each other. They will be held together by sutures of fine black silk. This proceeding is simple in plan and easy to execute, and when the amount of material removed is duly proportioned, the result is satisfactory. The denuded surface on the cornea heals slowly and becomes vascular. Hence an opacity is left, which continues for months. It may ultimately disappear, or, if the pterygium has been large and thick a secondary one may seem to take its place. This, however, is nothing more than the development of blood-vessels as a necessary part of the healing process in the cornea, and the new growth is always smaller than the original, and may not show any disposition to extend. If it should, there would be no objection to a second operation. In regard to the recovery of sight, it must be remembered that haziness of the cornea extends an appreciable distance beyond the pterygium, and the reaction due to the operation will render this area temporarily larger, hence there is not at first that gain in vision which will be finally secured. This circumstance is to be borne in mind in treating cases where the growth has reached the border of the pupil. After the operation the eye should be bandaged for twenty-four hours, and afterward treated by cold water dressings. The sutures may remain three to five days.

It may happen that pterygium has been allowed to grow to the centre of the cornea, and seriously damaged sight.

It is conceivable that the most suitable way of dealing with a case might be to perform iridectomy downward and outward, and the removal of the growth would be optional.

TUMORS OF THE CONJUNCTIVA.

Pinguecula frequently appears in the membrane in the form of a small yellowish elevation between the semi-lunar fold and the corneal edge. It consists of connective tissue, elastic fibres, and epithelium, and contrary to the import of its name, it does not contain fat. It is more conspicuous in later life, is not always to be distinctly recognized, does not cause any mischief, and need not be meddled with.

Dermoid growths are sometimes found in the conjunctiva, and may contain hairs. They are congenital, pale yellowish, smooth, and may slowly increase. They are not always to be distinguished from epithelial growths, especially when seated at the limbus corneæ.

Another congenital malformation is a dense white structure of cartilaginous feel, which may occupy the outer canthus and project in a semi-lunar form half way toward the cornea. It remains stationary and need not be disturbed. Dermoid tumors should be removed.

Cysts of the conjunctiva are not very uncommon, and are easily recognized. A peculiar form is one which is excessively thin and transparent, of small size, and seems to consist, as Hirschberg suggests, of distentions of the lymph-vessels.

Lipomata sometimes appear beneath the conjunctiva, toward the equator of the globe, as outgrowths of the orbital fat.

Pigment-patches sometimes occur, and they may be stationary and innocuous, or may be associated with malignant growths.

Polypoid, or granulation growths appear on the conjunctiva, and more especially upon the caruncle.

Epithelioma may attack the conjunctiva, especially at the corneal margin. It may be white or pigmented. It is usually adherent only at the limbus, and admits of removal, which should be done early.

Sarcoma also finds a habitat in this tissue, and is of more serious import than the above. It may, or may not, be pigmented, and will probably destroy the eye. For a discussion of this subject, **see** *Archiv. of Oph. and Otol.*, July, 1879, article by the writer.

Malignant ulceration likewise attacks the conjunctiva, and penetrates **to the deeper tissues.** The only treatment will be enucleation of the eye.

Foreign bodies on the conjunctiva.—Usually they are easily gotten away, **but** if they are caught in the tissues, as bits of iron may be, it is often needful to excise a little of the membrane together with them. Sometimes a small stick **or** a beard of grain may get into the upper cul-de-sac, and remain **a** long time before being discovered.

Lacerations of the membrane seldom need **treatment, while sutures** might be appropriate.

Burns by melted metals, **acids, caustics,** etc., are **not rare.** Constant dressing with cold water is the only treatment, and subsequent proceedings according to the amount of harm inflicted, whether by adhesions of the globe (see Symblepharon, page 154), or by opacity of the cornea.

Powder burns are frequent, and when only superficial, **inflict** harm by the deposit **of** the powder in the tissues. It must be picked out with a **spud,** or clipped out with scissors, including a bit of the conjunctiva. If **much** has been lodged, it is hopeless to attempt its entire removal.

CHAPTER IV.

THE CORNEA.

Anatomy.—We have a transparent structure shaped like a watch-glass, but in reality an ellipsoid with a basal diameter of 12 mm. in the horizontal, and less in the vertical meridian, and curved upon a radius rather less than 8 mm. Its thickness at the edge is 1.2 mm., and in adults its middle is about 1 mm. We have five layers for consideration, viz.: an anterior epithelium, and a posterior endothelium ; an anterior elastic lamina (Bowman's membrane) and a posterior elastic lamina (Descemet's membrane) ; and in the middle, the proper corneal substance. The anterior epithelium is in several layers—the lowermost cuboidal and the exterior flat. This structure is continuous with the conjunctival epithelium. The anterior elastic lamina is a dense, transparent membrane, closely adherent to the corneal substance, formerly described as structureless, but now said to consist of the most minute fibres. The corneal substance has been much studied, and presents great difficulties. The latest account of it is as follows : it consists of minute fibrillæ, held together by a matrix which is equally transparent and has the same index of refraction. The fibrillæ are a modified form of connective-tissue ; they are united into bundles, and these again into lamellæ ; their general direction is parallel to the surface, but they are intersected by others crossing them at various angles ; they are continuous with the fibres of the sclera. In the matrix is a system of minute channels, or canals, which expand at certain points into lacunæ of a lens-like form. The canals anastomose freely, and the lacunæ or spaces send out fine processes, which pass to and communicate with neighboring spaces. In the spaces are to be found the stationary corneal corpuscles, or cells, which have a large nucleus, and possess fine processes which fit into the branching tubules of the lacunæ. They do not fully occupy the corneal spaces. There are also small cells, not different from white blood-cells or the lymph-cells which make their way through the corneal canals, and are called the wandering cells. Thus, we have in the corneal tissue fibrillæ composing laminæ, a basement-substance, a system of canals with spaces and anastomoses, and we have stationary cells (corneal corpuscles) and wandering cells.

The posterior elastic lamina is structureless, very elastic, and separable from the corneal tissue. Upon it lies a single layer of flat cells, the posterior epithelium (endothelium).

The posterior surface of the cornea reaches farther back than does the anterior, that is, it underlies the edge of the sclera. At the extreme periphery of the cornea, and nearer its posterior surface, is the canal of Schlemm, or circular venous sinus. It is plexiform in character, and in parts is a circular channel of ovoid section, and in other parts it consists of several channels. It is regarded as a large lymph-vessel by Bowman and Leber. From the iris certain fibres pass at the angle of the anterior

182 DISEASES OF THE EYE.

chamber to the posterior elastic lamina, and form a mesh called the pillars of the iris, or the pectiniform ligament. In lower animals, like the ox, etc., there is also found in this situation a channel called the canal of Fontana.

The distance from the summit of the exterior of the cornea to the front of the lens is about four millimetres. The attachment of the iris is about two millimetres behind the exterior or apparent border of the cornea, but not the same around all its circumference.

The limbus corneæ is a structure differing a little from the rest of the tissues, because the epithelium is thicker, and the fibres merge into those of the conjunctiva. In the cornea are no blood-vessels, except such as run for a distance of two to four millimetres into the anterior third of its thickness. In the fœtal eye the whole anterior surface is covered with vessels. The nerves are numerous, and more so in the anterior third. They run forward to the epithelium, and reach to its very surface. Their precise terminations are unknown , they are branches of the fifth pair. The anterior chamber contains a watery fluid, which is secreted from the iris and ciliary processes, and is regarded as an enormous lymph-space, and the fluid finds exit by way of the canal of Schlemm.

Pathology.—Inflammation declares itself by opacity, by loss of substance, by development of blood-vessels, by formation of pus. The epithelium undergoes proliferation and destruction in the most various degrees. In the corneal substance we have proliferation of the corneal corpuscles and immigration of lymphoid cells, both by way of the natural canals, and they also force themselves through the substance of the cornea. We have degeneration of the fibrillæ and imbibition of serum. The membrane of Bowman is liable to partial or total destruction. The membrane of Descemet is much more capable of resistance. The endothelium may proliferate and may undergo fatty degenerations with or without proliferation. The whole cornea may soften and degenerate into a necrotic mass. The result of a partial destruction is an ulcer, and if this perforate, the iris comes forward to fill the hole, and remains attached, constituting anterior synechia. In central perforation the lens-capsule may become adherent, and either so continue, which is rare, or subsequently be pushed away, and leave a mark on the cornea and also on its own surface. In large ulcers the lens may for some time be exposed, or may be expelled. During the healing of large ulcers the regular curve of the cornea is seldom retained, and, if the loss of substance be great, the cicatrix bulges, and we have a partial or total staphyloma. In the beginning this may be nearly transparent, but it subsequently becomes opaque and increases in thickness.

Opacity of the cornea is due to irregularity of structure in the reparative process, by which the homogeneousness of the tissue is disturbed. It is often a true scar, and indelible. Opacities have the utmost varieties in degree and kind. The names nubicula, macula, and leucoma have little value, and need not be observed.

To detect faint opacities of the cornea, we find the greatest aid in the method of oblique or partial illumination (see p. 21). So too the ophthalmoscopic mirror with *dim* light will detect both slight opacities and irregularities of the curve. If opacity be dense a glance usually discovers it, while looking sidewise at it from different points will decide its location.

KERATITIS.

Symptoms of corneal inflammation.—Besides the changes above mentioned, of which opacity is the chief, there will be pain in the eye—sometimes radiating to parts around—lachrymation, and photophobia. The last symptom is especially typical ; in children it is intense, and if—as may happen—it be wanting, attention must be given to possible anæsthesia of the corneal surface. In all cases we have hyperæmia of the neighboring vessels, both conjunctival and scleral (ciliary injection), as well as of the palpebral conjunctiva. We may have conjunctival secretion. The pupil is small, but contractile, and the iris is hyperæmic. Both iritis and conjunctivitis are frequent complications. The whole globe may become involved. For convenience, keratitis may be considered under the following heads : 1st, superficial keratitis, including phlyctenular, vascular, and traumatic ; 2d, parenchymatous keratitis, including diffused and syphilitic ; and, 3d, suppurative keratitis, the last both as abscess and ulcus corneæ serpens ; mention is also to be made of the condition called, 4th, Descemitis, which is really a more extensive disease, involving also the iris and choroid, and will be again referred to under that head ; and we have also to consider, 5th, neuropathic keratitis.

PHLYCTENULAR KERATITIS.

A point of opacity, perhaps not bigger than the head of a pin, appears either on the edge or on the surface of the cornea, and a bundle of vessels may run to it. It soon turns to an ulcer by loss of epithelium, and is very liable to be accompanied by one or more similar exudations. An areola of faint haziness will surround it. There will be a copious flow of tears, the lids be tightly shut, there will be extreme photophobia and pain in the eye. There may be conjunctival symptoms in swelling of the edges of the lids, and muco-purulent secretion. The subject is usually a child, and though sometimes seeming to be quite healthy, is most likely to be otherwise. In fact, the delicate, the ill-fed, or overfed, and the scrofulous, are the most common victims. The conditions of health portrayed under phlyctenular conjunctivitis repeat themselves in cases such as we now consider, and the features need not be rehearsed. If the case continue long, blood-vessels will appear in the cornea continuous with the conjunctiva. The process may, in bad subjects, go on to purulent infiltration, or to penetration of the cornea, and even to perforation. It is characteristic of unhealthy subjects that one phlyctenula may follow another, and the disease be kept up for months. This is readily comprehended when we take into account Iwanoff's investigations, which show that a phlyctenula makes its appearance upon a terminal nerve-twig, which is imbedded in a mass of cells and other inflammatory material. After a time the acute symptoms abate, and we find upon the cornea a leash of blood-vessels running up to a gray ulcer. There are mild cases which come to an early termination and leave no mischief behind, while in other instances of longer duration, or of attack of the middle region of the cornea, serious injury to sight will remain. These cases were formerly described as scrofulous ophthalmia. This form of keratitis is similar to that which ensues sometimes after measles, chicken-pox, and scarlet fever ; in those cases the debilitating effect of the constitutional disease is apt to render the local process extensive and destructive.

Treatment.—The preponderance of symptoms of nervous irritation must constantly be remembered. For this reason depressing remedies are to be avoided. Cold water in the outbreak is not well borne, the application should be of warm water, or, what is sometimes better, of dry warmth, that is, by a folded napkin heated by the fire or upon a hot iron. Drop into the eye a solution of sulphate of atropia, gr. ij. ad. ℥ j., three times daily, and use warm compresses for half an hour, three to eight times daily. Carefully wipe away secretion and open the lids to let out the tears; guard against excoriation of the skin and fissures at the angles by simple cerate, and in some cases, by an application of a pointed crayon of caustic. If there be conjunctival secretion, use a three-grain solution of nitrate of silver to the everted lids, once in two or three days. For protracted cases, where the child has acquired a habit of refusing all light, and the local conditions are not severe, dropping iced water between the lids has been shown recently by Dr. Oppenheimer to be useful. Dusting fine calomel into the eye is sometimes needful when the acute stage is over, and then, too, the yellow oxide of mercury ointment, gr. ij. ad. ℥ j., may be carefully tried. Blisters, scarification, and leeches, are not to be advised. Tincture of iodine on the forehead is less objectionable, but its value is not great. For severe cases, cantholysis has been practised, but it should be reserved for such only. Hygienic and constitutional measures are of great importance. The bowels must be kept in order, small doses of rhubarb and soda, or of hydrarg. c. cretæ, are, in the beginning, often useful. At a later time quinine will be often needed, and cod-liver oil. But proper food takes precedence of all medicine, so soon as vitiated secretions are discharged from the bowels. Establish a simple and nutritious diet in such form as to be easily digested—the prohibition of alimentary gimcracks of every form, insistance on milk, beef, bread, and plain boiled potatoes or rice, daily bathing, and friction of the skin, taking the child into the fresh air, or change of air, and not permitting excessive avoidance of light. Some children will keep up this habit long after the real need has passed. A careful trial of conium maculatum tincture, as mentioned in speaking of blepharospasm, may be tried in very bad cases; but the drops of ice-water, and plunging the face into a bowl of water, are first to be given a fair trial.

The treatment of keratitis following exanthemata will be guided by the character of the lesions, but will be essentially like what has been sketched, viz., atropine, lukewarm compresses, mild astringents, if there be muco-purulent secretion, and vigorous tonic measures, because debility plays an important part. Sometimes stimulants may be needed, besides nutritious food.

VASCULAR KERATITIS.

As the result of repeated attacks of phlyctenulæ, and sometimes without being able to trace the mode of development, we find the whole cornea covered by haziness, the epithelium rough and faceted, and a thick network of vessels overspreading it. We find circumcorneal injection, moderate photophobia, and perhaps no severe complaint of pain. The case has been long in coming to this state, and may be either in a youth or an adult. Under other circumstances we have the same condition as the result of granular conjunctivitis, and by far the greatest number owe their origin to this cause. A condition of extreme vascularity is called pannus. It may be more or less dense—it may cover the half or the whole of the cornea. It may be in a state of acute irritation or otherwise. Its essential pathological element is the development of connective tissue beneath Bowman's membrane, and the great abundance of vessels. This condition is the essential mischief of granular lids, and it has been referred to in that connection. When it reaches its fullest development, it forms an almost cuticular surface upon the cornea, and the tissue is not only dense and vascular, but almost devoid of sensibility.

Treatment.—Mild cases will abate under sulphate of atropia and moderate use of warm water, and for granular lids nothing need be said. The utmost prudence must be observed not to venture on severe stimulation while acute symptoms exist. On the other hand, in the torpid stage, only moderate benefit to sight can be expected, and one must not be too bold lest suppurative processes be excited. It has been shown by experience that what promotes nutritive activity of the eye favors the disappearance of pannus, and hence, intermitting poultices have been useful, say for an hour, three or four times daily. Schweigger recommends spray of sol. sulph. cupri, gr. v. ad. ℥ j., or stronger, directed on the cornea. I have spoken of inoculation and how it is to be guarded. I may also mention peritomy as less risky, and capable of doing good. Touching the pannous cornea with pure nitrate of silver is not to be commended. The yellow oxide of mercury ointment is often suitable, or tannin and glycerine in some cases. Operations to relieve tension of the lids sometimes do good service, as in trachoma. Under any treatment imperfect sight will remain.

KERATITIS DIFFUSA—PARENCHYMATOUS KERATITIS.

The majority of these cases exhibit no roughness of the epithelium, although it may appear as if rubbed with sand-paper in certain cases of dense opacity. For the most part, we have a smoky haze, which may begin at the middle or at the margin, and which soon involves the whole structure of the cornea. It may be more or less dense, and can be a deep bluish white. There may be a fringe of vessels in the marginal part of the cornea, but they do not reach inward. There will be circumcorneal injection and moderate subjective symptoms. Oftentimes photophobia will scarcely exist. The loss of sight, even with mild opacity, will be great. A few days may suffice for the whole cornea to become hazy, but months may elapse before its disappearance.

This disease is in almost all cases constitutional. Mr. Hutchinson showed that hereditary syphilis is its cause in a large number of children, and he indicated the frequency with which notching of the two middle upper incisors accompanies it. But imperfect development of the teeth exhibits a variety of appearances in these cases; they may be peg-topped and stunted; sometimes the enamel has crumbled at a sharp line all around the tooth, leaving the distal portion abruptly smaller. I have sometimes called them telescoped teeth. The bad condition of the teeth is a strong reason for believing that hereditary syphilis exists in a case of diffused keratitis, but it is not conclusive. Other evidence should be sought for in the previous history of the child, also in the glands, the skin, the nasal passages and the bones, and the general nutrition. It is doubtless true that much which is called scrofula (a vague and unsatisfactory term) is hereditary syphilis more or less remote.

The corneal disease may come among the secondary symptoms of acquired syphilis. It then appears in adults, and is certainly not frequent. I have also seen cases in which the same lesion was due to gout, either as a clear diathesis or in the more obscure form of rheumatic gout. Finally, we cannot account for the cause in some cases.

Treatment.—For these, as for all other forms of keratitis, when symptoms of local irritation are acute, we use atropia sulph.; but if there be little pain or photophobia it is only useful to prevent possible pupillary adhesions, and needs but occasional use. The most efficient treatment is strenuous application of hot fomentations. The water should be as warm as can be comfortably borne, and the pads of muslin should be removed every two minutes, or often enough to keep up the heat, and the stuping of the eye continued from four to six hours daily. Of course this is wearisome, but it affords occupation when nothing else can be done, and its efficacy is undoubted. Not that all cases will thus be cured, nor that any will get well quickly, but I am convinced of the value of the method. Should unpleasant conjunctival irritation be caused, a solution of borax, 3 j., aquae camphorat., ℥ viij., is useful. I have ordered this solution to be applied hot, as water would be, and to advantage. In the same way hot fomentations of infusion of chamomile flowers is a favorite German remedy. So much for local treatment, but the constitutional requires not a little judgment. If hereditary symptoms exist, mercurial inunction must be practised for weeks or months with moderation, so that no ptyalism shall be caused, but the influence be realized. In place of the ointment, the oleate of mercury, five to twenty per cent., may be substituted, as more cleanly and suited to a delicate skin. Usually iron and bitter tonics will have to be steadily kept up to combat the meagreness of the blood and to aid digestion. The most diligent attention must be given to the food, that it be of nutritious quality and that a suitable quantity be digested. Often the meals must be small and frequent. All the helps that can promote digestion will have to be held in readiness. Of the same quality are to be ranked cod-liver oil and extract of malt. As a case improves, the duration of the fomentations will be shortened. I have seen the disease, when syphilitic, affect the same person several times between childhood and twenty-five years of age. The debility caused by nursing has seemed to be the occasion of an outbreak, and the prolonged influence of malarial poisoning has also had weight, as it seemed to me. In the case of one patient, a girl of twelve years, marked improvement followed an iridectomy. The opacity had disappeared about the edge of the cornea after several weeks' fomentation, and what remained on the centre, with much greater rapidity cleared up after the operation. Her second eye was attacked a year or two later; the opacity was dense, and covered the whole cornea. She had, on this occasion, full trial of fomentations and mercurials, but to little purpose, and an iridectomy was not attended with benefit.

When a gouty diathesis lies behind the malady, it is more apt to occasion pain and photophobia, and adults are more liable to be attacked. In such cases the aqueous humor is apt to be hazy, betokening iritis. Of course atropia must be employed. Sometimes the mydriatic has had a bad effect, because, very soon after putting it in, the eye has grown painful and has flushed up. Under such conditions a hypodermic injection of muriate of pilocarpine, ℈j. ad. ℥ j. (four per cent. solution), about twelve to twenty drops in the arm may be tried. Its effects of salivation and perspiration may cause a favorable change. If it do, it may be repeated once daily. But a remedy of immediate efficacy when there is pain and slight plus tension, is paracentesis of the anterior chamber. Some skill is required to do it neatly—a steady patient and a good instrument. Not more than three drops of fluid should be taken, and this *can* be done without giving more than the slightest pain. I use a knife with a very thin blade and sharp point, which enters so gradually as not to cause a sense of resistance, and when about three millimetres inside the anterior chamber withdraw it a little, and with gentlest pressure permit a little fluid to trickle out. There is a so-called paracentesis-needle, which is apt to penetrate with difficulty, because the

Fig. 74.

blade is too thick. It causes pain, and ether may be required. It is, however, a safe instrument, and for this reason to be commended (see Fig. 74). I do not think any good is gained by paracentesis, except relief of pain. For constitutional remedies, I am apt to begin with sal Rochelle, ℥ j. in water every two hours. If this does not succeed, give full doses of lithia-water, or take wine of colchicum-seeds. I have resorted to daily Turkish baths, with apparent good. Iodide of potassium has seemed to be inefficacious, or even harmful.

SUPPURATIVE KERATITIS—ULCUS CORNEÆ SERPENS.

The titles just given are not meant to be understood as identical, but both classes of cases may be considered together. Pus may appear in the substance of the cornea without ulceration of the surface, and certain cases of ulcer are attended with suppurative infiltrations, and are very liable to develop pus in the anterior chamber. The cases may be the result of injury or be spontaneous. Persons of all ages are possible subjects. The presence of pus means a large influx of wandering (lymphoid) cells, with proliferation of the corneal corpuscles and destruction of the fibrillæ and soakage with effusion. How the pus should appear in the anterior chamber is accounted for in three ways. The wandering cells force their way through the corneal tissue to its margin, usually going downward, as a white streak to the lower edge of the cornea often indicates, and emerge through the meshes of the pectiniform ligament into the aqueous humor. This explanation covers the majority of cases, especially the slighter forms of keratitis with hypopyon. Another mode is by perforation of an abscess through the membrane of Descemet. This is the least frequent. Another way is by proliferation of the endothelium of the posterior surface, accompanied by a low grade of iritis, which also yields lymphoid cells. In whichever way we have hypopyon, we find that, clinically speaking, there are two cases: one in which there is extreme pain and sharp inflammatory action, the other of an opposite type. The latter class will be either in old or feeble persons, or be dependent on paralysis of the fifth pair.

To consider first the sthenic cases. The attack comes suddenly, with or without injury by a foreign body; there will be acute pain, sometimes agonizing; moderate ciliary injection, photophobia; pupil narrow; on the cornea, near its middle, a small excavation with grayish bottom, and surrounded by a hazy areola; at the bottom of the anterior chamber a yellowish line, which has been likened to the lunula of the nail, but its upper border is horizontal (see Fig. 75). This may rise high enough to be easily discerned, and may be so small as to be visible only by looking down upon it from above. From day to day the ulcer increases, more of the cornea is infiltrated, and the exudation in the chamber rises higher. The case may present itself after having gone for a week or more, with a large extent of the cornea denuded of its epithelium, its anterior layers sloughing; what remains, yellow with pus, and only the margin retaining a degree of translucency; the hypopyon may occupy one-half the anterior chamber. By natural progress the corneal tissue gradually bulges, then bursts; the pus finds vent or projects in a thick, yellow plug. As the sloughy material becomes eliminated the iris shows itself as a black bead or nodule, or larger mass. Sometimes it presents in several portions, and, as healing occurs, a staphyloma is established.

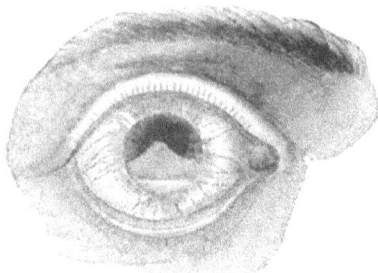

FIG. 75.

Treatment.—At the outset great relief is afforded by the frequent use of sol. atropiæ sulph., and the moderate use of warm water; an anodyne may be required if pain be intense, and the eye will naturally be protected. Commonly no other treatment will be needed, if, as often happens, the process takes a favorable turn promptly. The ulcer slowly heals, and the hypopyon quickly disappears. It is often a question whether one must let off the exudation by paracentesis. If the symptoms be not urgent, if the ulcer be not large and not progressing, one need not actively interfere. I have seen pus mounting to one-third the height of the cornea become absorbed entirely. But if there be a doubt, it is better to let off the exudation. The opening must be made with a Beers cataract-knife, or by a broad, thin blade capable of making an incision about three millimetres long at the margin of the cornea, with the patient lying down and possibly under the influence of ether. The exudation will not flow readily, because of its plastic constituents, and therefore requires a rather roomy aperture. The lips of the wound may be held open by the partly withdrawn knife, or, if the material do not escape, an iris-forceps will withdraw it. Generally a second paracentesis will not be needed. For cases where the surface of the cornea has been rather extensively eroded, say over a surface 6 mm. wide, and the remaining cornea is infiltrated and hypopyon exist, a different proceeding must be adopted, for which we are indebted to Saemisch. The whole of the ulcer and adjacent cornea is to be slit by a narrow Graefe's knife. Introduce the point at the outer edge, where the cornea is not yet destroyed, and let it emerge at a corresponding spot diametrically opposite; divide all intervening substance, avoiding the iris, and the contents of the chamber will at once follow. If a portion stick in the wound, withdraw it with iris-forceps. Apply borated lint and a bandage for a few hours, and then assiduously use hot fomentations of boracic acid four per cent. It is not advisable to meddle with the wound by probing it every twenty-four hours, as has been recommended. The cornea collapses, the chamber is annulled, and the iris remains against the cornea. Iritic adhesions may not always form, and the secretions commonly flow off spontaneously. Generally the necrotic tendency in the cornea is arrested. The destroyed tissue is cast off; a clean ulcer, possibly somewhat enlarged, ensues and reparation begins. From the time of section pain is substantially ended. For complete healing several weeks will be required. A staphyloma of moderate height may form, for which repeated punctures may be needed; a more or less extensive and central opacity will ensue. If it be dense and large, an iridectomy may confer vision, perhaps qualitative, possibly only quantitative, but something vastly better than any non-operative treatment would have afforded. It cannot be pretended that Saemisch's section will save all cases of serpiginous ulcer with hypopyon, for in some the suppurative action will advance and attack the deeper structures, and set up panophthalmitis. For this deplorable state one of several proceedings may be selected: first, more extensive incision, dividing the ciliary region with intentional partial escape of vitreous; or secondly, the wiping out of the contents of the sclera after making the incision (evisceratio bulbi); or thirdly, enucleation of the eyeball. Unless one of these three methods should be employed, the process of suppuration will go on for many weeks; scleral staphylomata with spontaneous openings for escape of pus, great swelling, febrile reaction, disgusting discharge, etc., will be features of an unhappy and protracted experience. The choice of one of these methods is to be determined by circumstances. I have found the evisceration an admirable resort, because it yields a better stump for an artificial eye, and more quickly terminates the process than simple division. Enucleation is sometimes rather a severe operation, is troublesome to do because of hemorrhage and orbital infiltration, and leaves a less favorable nidus for an artificial eye. It does, however, bring the sad drama to a conclusion more quickly than any other proceeding, while a few cases of fatal issue after such an operation have been recorded. None have come within my personal knowledge. In hospital practice enucleation is done with more frequency than evisceration, because the subjects are more speedily enabled to resume their work. Yet the other is often practised, especially for women.

There is another kind of corneal ulcer, which has but once come to my notice in a pronounced form. Its characteristics were that it was confined to the surface, not going deeper than just below Bowman's membrane; that there was little opacity at its bottom, no suppurative infiltration, but at one time there was slight hypopyon; it lasted six months. It went over the lower half of the cornea, and did not yield until at length the ulcer was split by Saemisch's section. Mooren has described the disease as rodent ulcer of the cornea.

Another and most dangerous type of ulceration attacks the old and feeble, in the form of a furrow about two millimetres inside the margin of the cornea, which rapidly travels around the entire circumference, leaving the undisturbed centre environed by a trench, and with equal pace penetrates the depth of the tissue. It may not be attended with grayish infiltration, but not the less destructively does it ravage this structure, and usually spoil the eye. As it progresses, the middle of the cornea becomes cut off from nutritive supplies and perishes. Such patients often have atheromatous vessels; they are asthenic from various causes, and the only thing left to be done for them is vigorous nutrition and stimulants, with occlusive bandage, alternated with hot water.

Again, the surgeon sometimes witnesses passive, painless, total suppurative infiltration of the cornea. It occurs in the old, the starved, the absolutely decrepit. I have twice seen it in the very obese and old. It may confront the operator who has removed a cataract, and, because the patient has been free from uneasiness, he may have left the dressings undisturbed for several **days.** The opened lids disclose a yellowish, moist, insensible, and opaque cornea. The process goes on to total suppuration of the eye. The cause lies in the incompetent nutritive power of the patient, either from diseased vessels or from general causes. Happily this unmitigated disaster seldom takes place. A similar kind of suppuration sometimes follows severe cases of scarlet fever, and may occur in scurvy, but it is rare.

NEUROPARALYTIC KERATITIS

is very apt to **become** suppurative ; yet there are not **a** few cases that do not advance to this stage, and a somewhat careful observation has shown me that moderate degrees **are** more frequent than is usually recognized. The **frank** and ordinary attacks of keratitis are attended with increased sensibility, but a percentage by no means inconsiderable will be found to permit the cornea to be brushed with a thread or with a few fibres of cotton without exciting any shrinking. Of course this implies reduced sensibility of the fibres of the ophthalmic division of the fifth nerve. The cases of partial insensibility are often not recognized, while those of total anæsthesia thrust themselves upon one's notice by their absolute torpidity **and** tendency to suppurative ulceration. What should arrest attention when-**ever** the cornea is inflamed is a sluggish kind of hyperæmia and œdema of the conjunctiva, and, above all, the absence of photophobia and complaint of sharp pains. There is an air of mildness about the attack which does not belong to ordinary superficial keratitis. Such conduct should instantly suggest testing the sensibility of the cornea. There is another type of keratitis, with blunted sensibility of the surface, which depends on malaria, and usually exhibits neuralgia of the supra-orbital nerve. My colleague, Dr. Minor, has published a number of my cases in detail (see *American Journal of Medical Science*, 1881). It is needless to say anything about the features of the **corneal** lesion ; they are simply ulcerative, and may **be** suppurative. The important facts are : first, the diagnosis, and secondly, **the** treatment.

Treatment.—The prime necessity is protection by closure **of the lids,** or by covering the cornea with something to exclude the air and foreign particles. A bandage is ordinarily better than plasters, and Theobald's bandage, which somewhat resembles Liebreich's, is far better than a roller. It is **easily** changed, and is not cumbersome. The **lids** must be opened and washed out as often as secretion collects, and borated gauze with borated cotton are to be preferred for padding. For cases not severe, or for some of exceptional type, protecting the cornea by smearing it with vaseline, and in hot weather with cosmoline, is satisfactory. One instance in my own experience was that of a man who had paralysis of many cranial nerves. In one eye total atrophy of the optic, in the other eye total paralysis of the facial, and consequent perpetual exposure of the cornea, because the lids could not be shut ; in the same eye total anæsthesia of the cornea, followed by suppurative ulceration of its lower two-thirds. Occlusion of the globe and iridectomy upward made sight contingently possible, and vaseline enabled him to get the benefit of it. He put a small bit between the lids two or three times daily, and wore a bandage at night. The main resource in such cases is protection, and the usual application of warm water and atropine will be in order. In one case I seemed to realize striking benefit from the daily use of a constant galvanic current—about eight cells of a Grove's battery ; but I have not been oftentimes induced to use it, because the above-mentioned suggestions have proved sufficient. It is proper to add that in herpes zoster ophthalmicus it is not a rare experience to find some anaesthesia of the cornea, notwithstanding that the great feature of the disease is the intense and painful irritation of fibres of the fifth nerve. This fact indicates not only the propriety of an occlusive bandage, but the value of galvanic treatment—one pole over the closed eyelid, the other at the back of the neck, the current to be kept up for ten or fifteen minutes daily, and of appropriate strength.

Much space might be taken in attempting to picture a great variety of lesions which occur in the cornea, for instance.

First.—On its surface we have calcific deposits, especially in old staphylomata ; they usually consist mostly of phosphate of lime.

Second.—We have lead deposits when solutions of lead salts have been incautiously used as collyria, in eyes whose corneal epithelium has been ulcerated.

Third.—On the limbus the epithelial layer sometimes becomes greatly thickened, and forms a bluish or grayish skin, which concentrically encroaches upon the cornea. Its centre remains clear and the process is of very slow growth. It may reach a width of 4 mm., and of course reduces by so much the apparent size of the cornea. It is rare, unyielding, and more apt to affect the old.

Fourth.—A form of opacity which begins on the horizontal meridian, at the outer and inner margin of the cornea, within one millimetre of the limbus, and advances very slowly toward the middle, occupying months or years in its progress. It ends in a form of glaucoma, or of a condition consecutive to glaucoma or in some other internal chronic disease. It was described and pictured by Graefe under the name of belt-like (*band-förmige*) keratitis. It will be found, on close inspection, to be made up of minute specks deep in the cornea.

Fifth.—There is a subacute form of keratitis which begins as a faint cloud in the corneal substance, and, on examination by a lens and with oblique illumination, is resolved into a cluster of dark molecules which slowly spread over more or less tissue ; it goes through its evolution in weeks or months, without lymphoid infiltration or ulceration. There is little pain or photophobia, or hyperæmia ; it is little controlled by treatment, and finally comes to a stand with more or less opacity. Protection of the eye and suggestions derived from inquiry into the general condition of the patient are all that I know about treatment. It resembles to some degree Descemitis or serous iritis.

Sixth.—There is a process of sclerosis which affects the corneal substance at its margin and sometimes has a slight yellowish tinge, and forms a zone whose breadth is considerably wider at some portions than at others.

Seventh.—Isolated gray dots occur in the surface of the cornea in certain chronic cases of irido-choroiditis with sub-tension of the globe.

Eighth.—Arcus senilis is a fatty degeneration of the marginal part of the cornea within one or two millimetres of its edge, which appears most often at its upper part, next in frequency at the lower part, and sometimes completely encircles it. It appears after middle life, and has a gray color. It does not contraindicate operations in this region.

Descemitis.—Minute pigment-dots on the posterior surface of the cor nea, most thickly strewn on its lower third, and attended with haziness of the aqueous, with discoloration of the iris, and possibly with posterior synechia, with faint circumcorneal injection, are the principal external signs of this disease. The epithelium of the posterior elastic lamina proliferates and causes the minute pigmentary and opaque specks. But the disease has recently been shown not to be proper to the cornea, but to be a choroido-iritis with involvement of the corneal endothelium, and extending back also to the endothelial lining of the sheath of the optic nerve. It will be referred to on a future page.

STAPHYLOMA CORNEÆ.

This deformity is a usual outcome of extensive ulceration, and is explained by the pressure of the intra-ocular contents acting upon the fresh reparative material before it has gained strength for proper resistance. For a period lasting sometimes for weeks, while the neoplastic structures are young, they have a certain translucency which deludes the patient into supposing that a useful measure of sight will remain, and the inexperienced physician may share the same belief; but gradually, as the tissues gain thickness, discernment of objects becomes more vague, and finally nothing but quantitative perception remains. The new membrane,

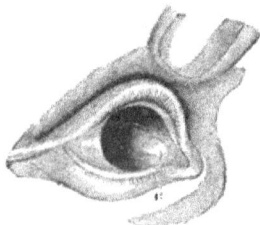 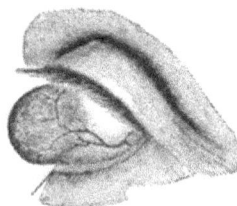

FIG. 76.   FIG. 77.

which takes the place of the destroyed cornea, gradually forms, and as it bulges forward it acquires more resisting power, and finally an irregular conical or rounded prominence of more or less opaque hue remains, which constitutes a corneal staphyloma. In some cases, where the whole cornea melts away, the transparent lens with undamaged capsule presents itself, and persuades the patient that the sad prognosis which had been given cannot be true; but a veil of opaque tissue slowly forms which shuts off sight. It is sometimes better, under these conditions of exposure of the lens when tension of the globe is increased, to open the capsules and let the lens escape. But when staphyloma is fully formed, we are called upon for a decision as to its treatment. If small and not advancing, and the subject young, it should be let alone. If it be partial and some comparatively clear cornea exist, an artificial pupil can be made (see Fig. 76). An iridectomy, with broad excision of iris, will reduce intra-ocular tension and confer a measure of sight.

For total opaque staphyloma (see Fig. 77) the following hints are offered. While it is in process of formation, puncture with a needle or a cataract-knife once in two or three days will lessen its projection and final development. When it has fully established itself, if confined to the cornea, the proper treatment will depend on its form and size. I have already spoken of iridectomy where that offers the least chance of improved vision ; it should never be lost sight of, notwithstanding the mechanical difficulties of its performance may be very great. A very narrow Graefe's knife must be carried along the outermost rim of the cornea, and by counterpuncture make a suitable wound. If a counterpuncture should be impossible, into the wound of entrance, a scissor-blade can be insinuated, and with this a suitable wound be made ; then the atrophied iris may be pulled out, sometimes only in shreds, either by a curved iris-forceps or by a hook, and an opening for light secured. The iris-forceps may be of a kind which I have had made, with projecting teeth, whose prehensile power is superior to the common type. If no clear opening is made, the filtration cicatrix will reduce the growth of the staphyloma, and a future iridectomy or other operation may gain the desired artificial pupil. In no part of the body is conservative surgery so important as in the eye. In certain cases one may shave or slice off the apex of a conical and thick staphyloma, and leave it to heal gradually. The aqueous may or may not escape. Something may be gained in the clearness of the summit. In other cases, where better sight is not to be hoped for, but only reduction of deformity, I would cut through one side of the cone by transfixion with a Graefe's knife, and with scissors complete the excision of a piece like a segment of an orange ; then, if needful, bring the edges together with very fine sutures. These should remain until they drop out. It need make no difference whether in these cases the iris should or should not be adherent to the cornea. If the staphyloma be spherical or bulbous, a more extensive amount of substance must be removed. In some cases the projecting mass has been excised and no measures taken to unite the edges. The lens has been evacuated by opening the capsule (ether being used), a pressure-bandage applied, and gradually cicatrization has covered up the vacancy and made a satisfactory stump. But sometimes serious intra-ocular hemorrhage and grave inflammation of the interior of the eye follow this simple proceeding. In view of this not infrequent liability, operations have been devised for closing up the wound, which should also serve to prevent the loss of the vitreous. Mr. Critchett carries needles armed with sutures through the ciliary region, and leaves them in place until next he abscises the staphyloma, and if needful lets out the lens, then he draws through and knots the sutures. Dr. Knapp makes use of the conjunctiva with the same view, dissecting back a strip of it from the base of the staphyloma, and through this passes the proper number of threads, excises the staphyloma and draws the conjunctiva over the opening. The danger attending all operations for staphyloma is in proportion to the distention of the globe, and especially of the ciliary region. In Mr. Critchett's method the needles perforate this dangerous ciliary zone and may cause serious reaction, including the possibility of sympathetic trouble of the other eye. By Dr. Knapp's method, which is less objectionable, we do not, as I have found, obtain certain immunity from the deep inflammation. If the whole pre-equatorial part of the eye enlarge, the case will be very likely to cause trouble ; especially will this be true in delicate persons. Therefore, with large staphylomata whose walls are thin, I advise against both Critchett's and Knapp's operations. I prefer either enucleation, which I might with great reluctance decide to do, or I would cut off the anterior part of the globe and wipe out its contents. A stump would be gained for an artificial eye. Sometimes, in young subjects, the

13

splitting of a large staphyloma (so-called ciliary staphyloma) and evacuating the lens, will reduce the eye and meet the indications. For young subjects, every palliative proceeding is to be tried in preference to enucleation, especially in females.

A case of megalophthalmos originating in intra-uterine iritis, which I saw when the boy was eight years old and the globe double the size of its fellow, was treated with brilliant success by total avulsion of the iris. Its full narration will belong to the chapter on iritis. For some of these cases, sclerotomy has been practised successfully.

CONICAL CORNEA—STAPHYLOMA PELLUCIDUM.

This condition is essentially one of atrophy of the middle of the cornea, which therefore yields to the intra-ocular pressure and gradually protrudes. There are no inflammatory antecedents, and usually but little pain. The symptoms are dulness of sight for distance, and need of holding objects close, with inability, pain, and fatigue in reading. No glasses fully correct or even considerably improve the vision, either distant or near. To this, in advanced cases, may be added the visible abnormity in the curve of the cornea, the uncommon brilliancy with which light is reflected from it, the distorted and elongated form of the reflected corneal image. By the ophthalmoscopic mirror we get a peculiar play of shadow around the apex of the curve when a feeble light is thrown upon the eye, holding the mirror at twelve or more inches—this being a phenomenon of total reflection. No ordinary concave glass will correct the vision satisfactorily, although some improvement is often gained, but V never $= \frac{2}{3}\frac{2}{3}$. It is to be noted that convex cylindrics, combined with concave sphericals, making mixed astigmatism, often procure the best optical correction. Sometimes convex cylinders, by correcting the error on the side of the cone, serve the best purpose. Dr. Thomson, of Philadelphia, has reported such cases (see "Trans. Am. Oph. Soc."). It is of the utmost value to do all that may be practicable by glasses, and many interviews will be required to exhaust all their possibilities; but if no satisfactory success results, and vision remain less than the common needs of life require, an operation may be tried. Paracentesis, however often repeated, has little value. Two proceedings are available: 1st, trephining the cornea as Bowman devised; or 2d, Graefe's operation. Trephining is to be done by an instrument like a delicate punch, sharpened at one end, and with a milled head for a handle at the other (see Fig. 78). This is to be rotated over the apex of the curve until it has gone through at some point, and the removal of the disc completed with scissors and forceps. The eye is to be bandaged and left to heal spontaneously. There may be prolapse of the iris, and subsequently an iridectomy may be required. I have done this with success. Graefe proposed to shave off the apex of the cornea with a cataract-knife until half of its thickness should be removed, to touch the raw surface the next day with a point of caustic, and repeat every three or four days until suppurative reaction was excited, then to treat the ulcer like any other, by warm water and atropine; when the ulcer healed, contraction would flatten the cornea, and an iridectomy some months later would complete the proceeding. Iridodesis—an operation advocated by Mr. Critchett—consists in causing prolapse of the iris at the periphery of the cornea and engaging in the wound, not

FIG. 78.

the edge of the pupil, but if possible, a more exterior part of the iris. This gives a small and lateral pupil, which is optically beneficial, but liable to provoke cyclitis, and it has therefore not been generally accepted. Iridectomy makes the case, as regards sight, worse.

Keratitis vesiculosa, and *K. bullosa*.—These two forms of disease are excessively uncommon, and their names indicating their nature, their mention alone is needed.

OPACITIES OF THE CORNEA.

Treatment of them is demanded by patients with great urgency, both because they impair sight and constitute a blemish. Sometimes they give rise to divergent strabismus, or convergent strabismus. If slight, recent, and in young subjects, they will disappear spontaneously in weeks or months. So long as blood-vessels remain in the tissue the reparative process is relatively more active, and hope of improvement is brighter; but, even without vascularity, a gain may be reasonably anticipated. So, too, with ulcers which are attended with vessels, their repair is facilitated. When non-vascular, especially when we have the clear, flat, so-called absorption ulcer, their filling up is vexatiously slow. Under these conditions, stimulating treatment is in order. Dense opacities, with semi-transparent edges, will greatly clear up under favorable conditions. But if a dense, white, cicatricial spot persist after trial of stimulating applications for a year or more, some other mode of treatment must be tried, or the case given up.

The stimulants for corneal opacities are legion, viz.: calomel powder; Pagenstecher's ointment; the whole list of astringents, except nitrate of silver; hot fomentations; the tr. opii, 1 to 10; tannin, ℥ss. ad. glycerinum, ℥j.; sol. sodii chloridi, 1 to 5, or 20; sol. iodid. potass., 1 to 3, or 1 to 2; mixtures in vaseline or cosmoline, molasses, etc. The object is stimulation which shall not cause reaction for more than five or twenty minutes, according to the irritability of the eye and the patient. Popular remedies are without number. None have specific or definite efficacy, and one must choose and vary, and play upon them as the character of the case and of the patient obliges one to select. The use of stenopaic glasses, with a slit or hole for central opacity, was strongly commended by Donders. It may have limited application for near work, but I have ceased to place dependence upon it. The performance of iridectomy for corneal opacity is not to be done until all prospect of gain of transparency is absolutely extinguished, and this applies to all the fainter kinds. The effect of the operation, as I have repeatedly verified, is to quicken the process of recovery, and after a brief increase of opacity, due to the wound, absorption sets in with new intensity, and may carry off the original spot. If this should be the case, the enlarged pupil will not only be unsightly, but detrimental to sight. Iridectomy is therefore suited to old cases and to dense opacities. It should always be very narrow, and a Tyrrel's blunt hook is better than iris-forceps. The size of the wound determines the size of the pupil. If the case is such that incision of the sphincter iridis is possible by Wecker's scissors, or in any other way, this makes the best pupil. A very late suggestion of Donders and Snellen for moderately dense opacities, is to prick them with a needle at intervals of some days, making one to four punctures into the corneal tissue, as the eye will tolerate. They claim decided benefit by this proceeding. I have not yet attempted it.

For very dense and total cicatrices, it has been proposed to insert a bit of glass formed like a minute shirt-stud, and let it remain. Another idea has been the transplantation of a piece of cornea from an animal, in the hope that it would remain. Neither of these methods has given much encouragement. For very dense cicatrices which involve the whole cornea, the trephine may sometimes be serviceable, because often the newly formed tissue is less dense than the old, and better quantitative perception is secured, and possibly a degree of qualitative. The blemish caused by opacities may be much abated by tattooing with India ink, as Wecker proposed. A thick emulsion is made in water, a drop placed on the cornea, and numerous pricks of a bundle of needles, or of a grooved, wide needle, made to drive the coloring matter beneath the corneal epithelium (see Fig. 79). Often ether is not demanded, and if the patient be courageous, an extensive opacity can be blackened in two or three sittings. Usually but slight reaction follows. The staining lasts for six to twelve months, and is apt to disappear by gradual absorption of the molecules of carbon. They have been found in remote parts of the cornea, whither they had been carried by the wandering cells. There is no objection to the repetition of the operation when the pigment has disappeared. The discoloration is a benefit to sight, because it stops out so much light which otherwise enters in a diffused way and blurs the retinal image. It is also done in cases where iridectomy has been made, and is frequently the second step in their operative treatment. Sometimes transparent cornea is thus treated, as in some cases of unduly large artificial pupils, the pigment acting as the means of reducing the aperture. It is then applied at the margin.

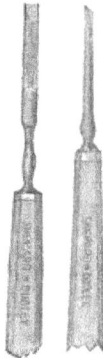

FIG. 79.

CHAPTER V.

THE SCLERA.

Anatomy.—The sclerotica (sclera), constituting the greater part of the outer tunic of the globe, has as its principal qualities toughness, resistance, and a little elasticity. Its structure is very like that of the cornea, while its fibrillæ are coarser and less regularly arranged. They are gathered into bundles, and cross each other in various ways, and are united by a homogeneous cement. There are fixed corpuscles, and also wandering cells, and a little pigment which in the African race becomes considerable. Its greatest thickness behind is 1 mm., while at the equator it is re-enforced by the tendons of the muscles, and just behind their insertions we find its thinnest portion, viz.: 0.4 mm. In front the tissue is covered by the conjunctiva, and beneath this is a loose episcleral connective tissue. Behind it, fifteen degrees to the inner side of the macula lutea, and a little (three degrees) above the horizon, it is perforated by the bundles of fibres of the optic nerve. The mode of entrance constitutes a sievelike perforation called the lamina cribrosa. The sheath of the optic joins the sclera. The inner surface of the sclera is lined by an endothelium which has an imperfect layer of large polygonal cells and pigment. On its outer surface the layer of connective tissue constitutes part of the capsule of Tenon (oculo-orbital fascia). The posterior part of the sclera, for a space about ten to twelve millimetres in diameter, of which the optic nerve is nearly at the centre, is pierced by blood-vessels and nerves known as the posterior or short ciliary. Immediately around the entrance of the nerve a few vessels anastomose, and compose a circle which forms the only medium of connection between the blood-supply of the retina and that of the choroid. The ciliary vessels go to the choroid, ciliary body, and iris (uvea). At the front the sclera is penetrated by the terminals of the muscular twigs known as the anterior ciliary vessels. The nerves penetrate the sclera behind, in the same region with the vessels. They are divisions consisting of twigs from the ciliary ganglion, whose roots of origin are the oculo-motor (3d), the ophthalmic (5th), and the sympathetic ; it lies nearly at the orbital end of the optic nerve to its outer side. There are the short ciliary and the long ciliary nerves. The junction of the sclera with the cornea is by continuity of fibres, which have no distinct line of demarcation.

SCLERITIS.

There is a form of acute inflammation which presents a pale pink injection of the anterior part of the globe, without signs of iritis or keratitis, and with no signs of conjunctival secretion. It is attended with dull, heavy pain, it fades gradually, or in some instances it invades the interior of the eye. Under the first alternative it lasts a few weeks and goes away without doing any harm. Under the second alternative it causes haziness of the aqueous and vitreous humors, increased tension of the globe, and the development of acute glaucoma in florid type. Such attacks as either of these are rare, and they are most probably of rheumatic or gouty origin. In one instance, an autopsy clearly demonstrated gouty kidney in a patient whom I treated having the above type of glaucoma (see "Trans. Am. Oph. Soc.," 1873).

Episcleritis.—This is a limited form of scleral injection, more frequent than the above described, yet not very common. There may be some thickening of the suprascleral tissue and a few large vessels may appear, but the greater number are fine and of dull color. The hyperæmia attacks a limited spot, and the rest of the sclera may be normal. I have seen such a condition completely encircle the cornea. There is no secretion, and the lesion takes a sluggish course, with moderate pain and little or no impairment of sight. It is, however, liable to aggravations by sudden extension of the injected area, and sometimes this will be attended by severe pain. Persons who have once enjoyed an attack are prone to have others. It is a disease most prevalent after middle life. While gout and rheumatism are its most frequent causes, we cannot always discover its etiology.

Sclero-keratitis is an inflammatory attack located in the limbus, which is liable to excite both iritis and opacity of the cornea. It is as obstinate as all other forms of sclerotic inflammation, and has capabilities for mischief to sight, both by opacity of the cornea and by penetrations into the ciliary region. It may attack either the feeble or the robust. In the worst cases it may lead to ciliary staphyloma.

Treatment.—For all these cases the first suggestion is atropia and warm water. But an exception must be made for the glaucomatous type, which would properly be treated by eserine at the beginning, and, as soon as the symptoms have reached great tension and pain, would be submitted to iridectomy. For general scleritis and for episcleritis free and vigorous doses of sal Rochelle will sometimes give help and sometimes be utterly inert. The next remedy to be tried is iodide of potassium, gr. v. vel x. ter in die. But recently some auspicious results have been reported from the hypodermic injection of muriate of pilocarpine, gr. $\frac{1}{4}$ vel $\frac{1}{12}$ at a dose. The acute diaphoresis and salivation have seemed to exercise immediate control, and bring to a close within two or three weeks an attack which, under previous experience in the same person, had lasted six and eight months. The use of pilocarpine in the eye seemed valueless. Should malarial poisoning exist, the sufficient use of quinine is self-suggested, and for similar cases arsenic is sometimes suitable. In some cases with obstinate pain and infiltration, I have done iridectomy with great relief, making the incision at the seat of the vascularity. For less intense, yet obstinate cases, deep scarifications into the sclera have been recommended. Great care in protecting the eye from irritations and exposure are essential, and oftentimes good nursing is more efficient than any other means.

CORNEA GLOBOSA—STAPHYLOMA SCLERÆ—HYDROPHTHALMUS.

The diseases known under this head may not begin as maladies of the cornea or sclera, but of the internal tissues of the eye. But, as their conspicuous features are in the anomalies of the form of the cornea and sclera, they may here be briefly mentioned. We have cornea globosa, in which the enlargement is principally of the cornea and **extends in** some degree upon the ciliary region. The cornea will **be** thin **and more** or less hazy, the ciliary region bluish **and** traversed by large and **tortuous** vessels. The anterior chamber will **be** very **deep, the** pupil large, the lens either **clear** or opaque, possibly dislocated. It may or may not be possible **to examine** the eye with the ophthalmoscope, but no good view can **be obtained.** Vision is generally wanting, **even** as to perception of **light. There are other** cases in which a similar deformity depends on iritis, **and deeper inflammations with** total occlusion of the pupil and distention **of the** posterior **chamber.** The iris may be pressed against the cornea at its middle, **while** its periphery is drawn back and pushed out with the yielding scleral **ring.**

Another variety of scleral staphyloma gives the eye a resemblance **to** the shape of the seed-capsule **of** a poppy. A ring of bluish prominences appears upon the ciliary part **of the** distended globe, and the cornea is enlarged and hazy.

We also have the eye uniformly and enormously stretched, converted into a bluish sphere, over which it may be impossible for the lids to be drawn, notwithstanding they have in the progress of the malady themselves undergone great elongation.

There are other cases in which the **sclera has given way at a** single spot, **and on this a** tumor or staphyloma has **developed** in **a large,** dark hemisphere. **The most** common point is the **ciliary region and** beneath the upper lid. **Sometimes** the tumor is lobulated, **and** there may be considerable sight, or none at all. Most of these cases are due to cyclitis, and will be alluded to again.

Posterior staphyloma scleræ is by far the most frequent deformity **of the globe,** and is the essence of cases of myopia. It occupies a small region **at the** posterior pole. It need not, like the above-named deformities, **be** supposed to be of inflammatory origin. **It** has already been sufficiently considered under the head of Myopia.

Treatment of ocular deformities will vary with their degree and quality, and has been spoken of under corneal staphyloma. Many will be let alone ; for many only enucleation will serve ; for some sclerotomy will be of value. For one case, where the iris was greatly distended from behind and the pupil attached to the lens, making a funnel like a convolvulus flower (morning-glory) the total removal of the iris (iridorhexis) cured the deformity (see "Trans. Am. Oph. Soc.," 1880, case mentioned on page 194). Partial removal of the staphyloma, with or without sutures, risks suppuration and its attendant protracted suffering. On the other hand, suppuration has been purposely induced by running a silk seton through the globe and letting the process take its own course. The object is to gain a better stump than enucleation permits. It is no small calamity to be obliged to enucleate a greatly enlarged globe from young persons and from females. There has been so much atrophy of orbital tissues, and sometimes absorptive expansion of the orbital walls, that upon such a nidus an artificial eye sinks down and never serves a satisfactory cosmetic purpose. It is therefore important to attempt **to save a** portion of the globe, and in doing this the greatest attention should **be** given to the patient's general health, that he may possess such reparative power as to rise above the tendency to profuse suppuration. The same rules and methods must ofttimes be observed which would be practised if a patient were preparing for a capital operation in general surgery.

CHAPTER VI.

THE IRIS.

Anatomy and physiology.—We have to do with a highly organized structure, consisting of muscular fibres, pigment, epithelium, connective tissue, blood-vessels, lymphatics, and nerves of every type, and ganglia. It is a curtain adherent to the sclera, beyond the transparent border of the cornea; its periphery is attached to the endothelium of the cornea by the fibres of the ligamentum pectinatum (pillars of the iris), and behind is continuous with the ciliary body. The membrane is perforated by a round opening which appears to be in the centre, but is really a little to the nasal side. It rests upon the anterior capsule of the crystalline, over a large area. Between it and the margin of the latter a space is formed by the convexity of the lens which necessarily has the shape of a ring, and is called the posterior chamber. In section it has the general form of a triangle, into whose base the ciliary processes project. In front is the iris, behind is the lens-capsule and the suspensory ligament or zonula of Zinn. The anterior and posterior chambers compose the aqueous chamber, and in its fluid the iris floats, giving the most perfect chance for its muscular fibres to exert their force. They are flat, arranged in bundles, and are of the unstriped variety. Certain fibres are arranged in curves about the pupil, constituting the sphincter, which is rather nearer the back than the front surface, and can be readily recognized, while other fibres run in radii and are more deeply situated, and constitute the dilator pupillæ, although this statement has been much debated. They join each other near the pupil in curves or arcades which are often conspicuous. The sphincter is under the control of the third cerebral nerve, the motor oculi, and the dilator responds to the sympathetic. The stroma of the iris consists of layers of connective tissue, in which are found spindle and wandering cells and fibres. Its cells do not contain pigment, and have been confounded with the muscular fibres. On the front of the iris is a layer of endothelium, whose cells overlap on their edges, and on the back of the iris is a much thicker layer of endothelium deeply charged with pigment, and called in a restricted sense the uvea. This word should, however, be reserved as describing the whole internal pigmentary structure of the globe, viz., iris, ciliary body, and choroid. The front of the iris is checked by numerous threads and pits, and is therefore quite rough. The brown pigment is scattered upon it irregularly and the differing hues of its surface are to be explained by referring them to interference phenomena. The pigment is of the same quality, no matter what may be the effect of the whole in causing the irides to seem to be blue, brown, hazel, pied, etc.

Most of the prominent lines upon the front of the iris are blood-vessels. They are extremely plenty, making the tissue highly vascular, and their walls are especially thick. The sensibility of the iris is acute, and I have seen that the sensibility of the cornea may be totally lost through paralysis of fibres of the fifth, while the **iris nerves were** unaffected, as I saw demonstrated by iridectomy in a case of trigeminal paralysis. The pupil varies in size by reflex influence of light upon the retina. It also contracts when the accommodation is in action for near objects, and enlarges when the accommodation relaxes. **Both** irides **act** consensually and the centres of the pupils **of the two eyes are on an** average fifty-eight **millimetres** apart, but the range **of variety in** interpupillary distance is large. Irritation of **the** cornea **causes contraction of the pupil.** The object of the iris is **to act as a** diaphragm **whose** aperture **shall** vary with the quantity of light, **and it cuts** off the peripheral rays which would be least correctly focussed.

IRITIS.

If attention be fixed upon the two cardinal signs of inflammation **in the** eye already insisted on (see page 18), it will be easy to recognize **iritis.** The disease is by no means rare, and to overlook it is to damage the patient's sight. The signs to be noted are alteration of tissue and impairment of function. Under the first head we have loss of the brilliant hue **which is** habitually reflected from the iris; its **tracery** and pattern seem **blurred, and hence its color is changed,** being made darker **as** well as **muddy; close inspection will show that this depends in part on** turbidity **of the aqueous fluid,** which changes **the clear black of the** pupil to a **smoky tinge;** moreover, the pupil will be **small, and** on its edge the pig**ment will seem** to have increased, and possibly spots of attachment to the **lens will be** noted. Such attachments will be sure to appear **in** the great majority **of** cases, and, if not discernible at first, will be apparent when sol. sulph. atrop. is dropped in the eye. Posterior synechiae, as these are called, are conclusive as to iritis.

I have said that **the** pupil will **be** small; this is the natural effect **of** swelling of the tissue, which makes it encroach on the free space of the pupil, and this swelling is composed both of effusion and of enlarged vessels. The pupil will not move on alternations of light and darkness, **nor** in common with its fellow, nor will it partake in the changes which occur in the act of accommodation. Necessarily, sight is reduced, and, perhaps, to an extreme degree. In this fact, and in the non-performance of the normal activities of the pupil, are found the fulfilment of the symptoms grouped under the head of disturbance of function.

An accessory symptom is hyperaemia of the ciliary vessels. In mild cases, or during the retirement of the disease, this corona is limited to the near vicinity of the cornea, forming a zone about six millimetres wide, of fine, pink, and nearly straight radiating vessels. They run out farthest toward the **recti** muscles. Let the disease be severe, and the whole front **of the** sclera and also the conjunctiva will be of an intense red; the palpebral conjunctiva will be but little engorged, and no mucoidal or purulent secretion will form, but tears will flow and the light be distressing. A prominent symptom will be pain, situated at first in the globe, but soon radiating along the branches of the fifth nerve, chiefly up the twig which goes through the supra-orbital notch and spreading over the forehead and up the vertex; sometimes the twigs to the inner side of the orbit or on the side of the nose, and frequently over the malar bone, manifest the irritation of these branches of the ophthalmic division of the fifth. Pain of this quality, and which is often extreme, will always make one look to the iris as likely to be its source. The globe will not bear to be pressed with the finger, and its tension is liable to be somewhat augmented. The pain is usually less severe in the early part of the day, but toward evening and during the night, it wakes to full activity. In most severe cases, chemosis may occur and the lids swell moderately. Both eyes may be affected together. Generally one only is concerned, and one may follow the other. Repetitions of attacks are frequent. The effect of atropine, when it succeeds in stimulating dilation of the pupil, is to give it various irregular forms, according to the number and extent of the synechiæ (see Fig. 80).

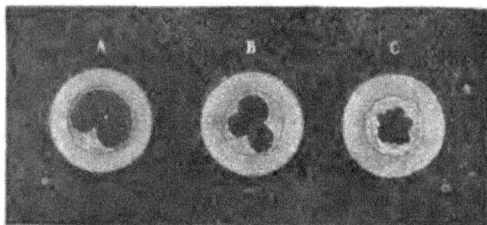

FIG. 80.

In cases where atropia has been used early, and has brought the pupil to a maximum, adhesions to the lens are still liable to occur, and do usually take place, which proves that in this dilated state the iris either continues to rest on the lens, or is in very close proximity to it. But while plastic exudation so frequently forms, both at the pupil and on the posterior surface of the iris, there occurs in certain cases a more copious, yellowish exudation which infiltrates any portion of the tissue, and comes out in globular masses on its surface. If in very large amount, it will fall to the bottom of the anterior chamber like hypopyon. Naturally, the pupil is in such cases liable to be filled up ; and what more frequently occurs is the formation across it of a false membrane. Should the attack of iritis be severe, not only will the aqueous be muddy, but, by extension to the ciliary body and choroid, the vitreous will become hazy by molecular infiltration or by formation of membranes. The reason for the prompt appearance of pupillary adhesions in the beginning of iritis is found in the fact that the aqueous humor is saturated with plastic material coming from the whole iris, and this, by precipitation, naturally glues together surfaces which already touch each other (Arlt).

An attack of iritis may be so slight as to pass in a few days, or, if neglected, may continue for months, and its end be atrophy of the globe and hopeless loss of sight.

The varieties of iritis are to be distinguished by their pathological features, while it is also customary to talk of varieties which depend simply on differences in causation. It is, however, utterly impossible to allege with more than probability that a given case of iritis owes its origin to one morbid cause rather than to another. There is no specific sign in any kind of iritis which points to its paternity. Holding, therefore, to the pathological varieties, we have serous, plastic of various degrees, spongy, and suppurative iritis.

One type of *serous iritis* was mentioned under the head of inflammation of the posterior layer of the cornea, viz., *Descemitis* (see Fig. 81). The disease formerly described by that name has been shown, by an autopsy re-

ported by Knies in 1879, to have been an affection of the whole uveal
tract, and of the inner sheath and stem of the optic nerve up to the chiasm.
It produces upon the cornea minute pigmentary deposits on its endothelium, chiefly at its lower half, and which to a great degree destroy the
cells; the aqueous is cloudy, especially its lower portion; the iris is dull,
and the pupil reacts slug-
gishly; vision is reduced,
and the eye feels a little uncomfortable without being
really painful. The hyperæmia about the corneal border is insignificant, and sometimes entirely overlooked.*
The disease lasts for weeks,
and is exceedingly apt to be
disregarded by patients, and
overlooked by physicians.

Fig. 81.

Serous iritis may also present itself in more emphatic
form, and be more truly a
local disease. We have the aqueous humor increased in quantity, pushing the iris back, sometimes enlarging the pupil and giving it a muddy
hue. The pupil may also be somewhat contracted, and adhesions occur.
The iris will be changed in texture and color, vision be impaired, there
will be pain and hyperæmia. The great depth of the anterior chamber
and the fewness of the pupillary adhesions make up the chief points in the
picture.

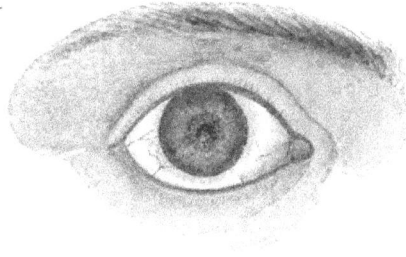

Plastic iritis, as it usually appears, has been the type of the description
given in the beginning of the chapter, and this need not be repeated.
Adhesions of the pupillary edge occur at a very early stage, and the pupil
becomes extremely small. In other cases the plastic exudation comes out
upon the surface as a yellow substance, and may be in one mass or in
many. It often is spoken of as gummy exudation, which it really is. It
may be traversed by blood-vessels, visible to the observer, and may be attended by hemorrhage into the anterior chamber. Sometimes this is in
quantity, and then is called hyphæma; at other times it is scattered in
minute streaks and specks over the front of the iris. When thick masses
of exudation disappear they leave the iris tissue not only densely adherent
to the lens, but in a state of atrophy. At a subsequent period, when acute
symptoms have all faded away, their sites will often be recognized by the
gray look and frayed fibres of the iris. Sometimes the tissue becomes so
thinned that to the ophthalmoscope light will shine through. This atro-

---

* In Knies' case, besides deposits on the lower side of the cornea, the entire iris
was infiltrated with round cells; masses of wandering cells were heaped in clusters in
its substance; the ciliary body was similarly affected, but its pigment was intact, and
so was the posterior pigment-layer of the iris; the whole capillary layer of the choroid
was infiltrated, but the pigment was unaffected; the retina was not involved, the vitreous at its posterior part was liquefied and detached, and its anterior part was permeated by numerous granular cells and membranes; the optic nerve was inflamed and
infiltrated with cells which did not pass beyond the papilla into the retina but for a
very short space, yet were present along its orbital part up to the chiasm, and the pia
mater of the sheath was notably filled with cell-infiltrations. Both eyes were alike.

It is evident that the disease, of which the above is the only autopsy on record, is
a more serious affair than is commonly thought, and is a true uveitis and more.

phy takes various phases, and in **all cases the iris will be seen to have lost** its proper complexion and markings.

In cases where pupillary adhesions become complete, sealing up the aperture, **a** peculiar result may in time arrive. At certain spots where exudation has been great and atrophy considerable, a projection will appear and the rest of the iris seem retracted. In other cases fluid will accumulate behind the iris in greater quantity than can **be** conveyed away, and the iris be pushed forward **in a** convex shape, **with** its pupillary edge deeply drawn in and its periphery pushed against its corneal border (see Fig. 82.) Sometimes the anterior chamber is in this manner almost abolished,

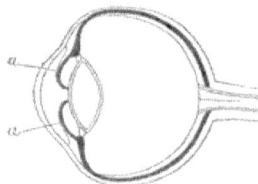

FIG. 82.

because so much of the iris touches the cornea. In other cases the whole posterior surface of the iris is glued fast to the lens, the periphery is deeply retracted, showing a circular furrow, which indicates that the tissue has acquired adhesions to the ciliary processes, **and** that the posterior chamber is annulled.

Spongy iritis is a rare disease. **It** shows a very muddy aqueous and great obscuration of the iris. The material is a low form of plastic exudation of a smoky hue, and when it disappears it begins to absorb around the edge. It has been mistaken for **a** transparent lens dislocated into the anterior chamber. The iris is pushed far back, and should any doubt arise about the nature of the case, **puncture** will let off the substance and display a clear iris.*

Suppurative iritis.—It **most frequently has a** traumatic origin, either from operation or from injury.

It also belongs to the make-up of a case of panophthalmitis, and may be part of a metastatic choroiditis. It has no special features to be depicted which will not be suggested by its name and by its associations. It is not to be confounded with gummy (plastic) iritis, above described. Naturally it is likely to have a disastrous termination as regards sight.

Complications of iritis.—The iris may be said to be the middle member of many ocular inflammations, for which its position and its vascular connections specially adapt it. It often becomes consecutively involved when the cornea and sclera are inflamed, and both with and without perforation of these tissues. On the contrary, the reverse order of sequence **may** take place, although perhaps with less frequency. Upon the other **side** the communication between the iris and the ciliary body, and between **the** latter and the choroid, is so intimate that irido-cyclitis and irido-choroiditis, in which latter the ciliary body will be included, are extremely

---

* This form of exudation is especially described by H. Schmidt, **in** Zehender's Klin. Monats-Blät., IX., p. 94. He speaks of some cases in which it seems to come out of the iris like a cyst, and in other cases gives essentially the same description as that in the text, which I have personally seen. Arlt speaks of it (Bericht der Oph. Congress, p. 68, Heidelberg, 1879) as a cloud floating over the pupil, an exudation which gradually shrinks and makes a thin membrane on the capsule, and finally disappears without synechia. Alt (l. c., p. 90) describes it at a sero-fibrinous and hemorrhagic iritis, in which the fluid part of the blood separates from the cell-elements, and the latter remain in the iris. Of the fluid portions, two layers are formed in the anterior chamber: one composed of a network of very fine fibres, like the exudation of croupous pneumonia, and the latter is a formless, gelatinous mass. Alt has had the opportunity of microscopically studying the substance, and gives a figure of its appearance. The **name** "spongy iritis" was first used by Dr. Knapp.

common. It is doubtless true that with every iritis there is hyperæmia of the ciliary body and choroid. But inflammation of the ciliary body concurs with iritis, in a form so pronounced as to command special attention, and this will be spoken of later. Likewise irido-choroiditis is to be specially individualized. The effect of these complications is to cause exudation and proliferation in these structures, with secondary atrophic results, and also to cause opacity in the vitreous. Of course the retina will likewise sustain tissue-changes and damage to function. But it has been pointed out by Schnabel that there are in many cases complications of iritis which do not appear through the medium of the ciliary body and choroid. He has shown that hyalitis (inflammation of the vitreous body) can occur per se, and cause serious opacity of its structure ; while we have been accustomed to say, whenever this takes place, that cyclitis or choroiditis must have been present. He has also called attention to the co-existence of retinitis with iritis ; that the region near the optic nerve is affected ; that it shows a grayish or yellowish infiltration, with hyperæmia of the papilla and of its own vessels ; that it may last much longer than the iritis ; that it explains many cases of impaired sight whose cause has been obscure ; and, on the other hand, that it may be recognized in unmistakable features by the ophthalmoscope. It must be said that the hazy aqueous renders inspection of the fundus difficult, and our conclusions uncertain. It will be necessary in the future to give attention to these assertions, and try by the ophthalmoscope to decide upon their value. It is not an easy question to settle.* The optic nerve will always be found red in iritis, and just as the sun looks when viewed in a fog.

The common sequelæ of iritis are : 1st. Adhesions to the capsule of the lens, and, if these have been torn by treatment, pigmentary spots remain. 2d. A membrane may form in the pupil by organization of exudation, and this is sometimes vascularized. Of course it impairs sight, and, if very dense, it will be indistinguishable from cataract, and sometimes is called spurious cataract. 3d. It may happen that the capsule of the lens undergoes thickening by proliferation of its epithelium, and cataract is a common result of chronic iridocyclitis. The cataract thus formed, and styled sometimes inflammatory cataract, is unlike the usual form both in color and in shape. The capsule is generally thickened, the hue yellowish or densely white, or it is chalky ; cretaceous degeneration sometimes occurs. Usually it may be assumed that there has been simultaneous deeper inflammation. 4th. Complications of the choroid and vitreous have been mentioned. If but two or three adhesions remain after iritis, and these not broad, no material injury may follow. But when many and broad adhesions are left, it has been a general belief, originating in an aphorism of Graefe, that a tendency to recurrent attacks is thereby occasioned. For this reason, operations have been made to guard against future attacks, and their necessity has been called in question. Schnabel makes a strong argument against the value of this belief. He thinks that the deeper-

---

* Since the above was written, I have seen a boy who had syphilitic gummy iritis, and whose pupil was fairly dilated. The vitreous was clear enough for a distinct inspection of the fundus. About a month after the inception of the attack, there was seen a well-defined neuritis optica, with some swelling, hyperæmic vessels, etc The disc at first was red, and later became white, as from atrophy. In the second eye iritis also occurred, and this was succeeded by a less pronounced, but undoubted neuritis optica. Of course an explanation can be given that the attack upon various tissues may only depend on new and independent localizations of the constitutional poison, which in most cases is, and in this one was, syphilis.

seated concomitant lesions which he has pointed out cause the injury to sight which has been ascribed to the subtle depreciating influence of the synechiæ. It must be admitted that iridectomy or iridodesis have not always been able to restore sight—not even always to protect against repetition of inflammation. The question will come up again under the head of **treatment.**

The *causes of iritis* **are** numerous, being local and constitutional. As local, **we have injuries with or** without penetration of foreign bodies, extension from **adjacent structures,** viz., the cornea, choroid, a swollen lens, detachment **of retina, etc.;** and **we** have also the communication from the other eye, known **as** sympathetic ophthalmia, of **which more hereafter.** As constitutional causes, we have syphilis secondary, tertiary, and hereditary, sometimes intra-uterine ; also rheumatism, gout, scrofula, gonorrhœa, variola, febris recurrens, malaria, and conditions which are unknown. The local appearances give no trustworthy clue to the constitutional causation, except that gummy iritis may be affirmed to be almost always syphilitic ; and the most frequent cause in general is syphilis, amounting to nearly fifty **per cent.** Rheumatic and gouty are next in order, and they are apt to be **of the** serous type, and to be both obstinate and recurrent. Gonorrhœa, which may or may not be associated with gonorrheal rheumatism, I have recognized to be a cause by cases which have fully removed my own original scepticism on this point.

Treatment.—The essential and master remedy in iritis is sol. atropiæ **sulphatis** ; it is the beginning, middle, and **end.** The prevailing fault is to **use it with** too much caution. Its potency when the iris is inflamed is far less than when the eye is normal ; the reasons are as follows. The power of endosmosis through **the** cornea is impaired because its tissue is surcharged with fluid, and the tension of the globe is increased. The swollen condition of the iris, the necessary inaptitude of its muscular fibres to contract, the hyperæmia and the adhesions combine to oppose its effect even when the solution has entered the aqueous chamber. For these reasons, a solution, gr. iv. ad. ℥ j., must be used in such frequency as will effect the purpose. This will vary in different cases. It will be dropped in once an hour or once in two hours ; or it may be put in six times an hour three times daily, or four times an hour three times daily. For iritis after extraction of cataract, I use a solution, sixteen grains to the ounce—the condition not permitting frequent instillations, and for this reason their strength is increased. So long as certain pernicious effects presently to be described do not occur, the effort to dilate the pupil is to be perseveringly **pushed** until it is actually accomplished. But there are certain possibilities of harm in atropia not to be overlooked. They are chiefly its poisonous constitutional effects. Patients quickly complain of dryness of the throat, and it will be seen to be red and the saliva scanty—this is not to **be** heeded as dangerous ; but **when** a flushed face, a quick and feeble pulse, nausea, prostration and fainting appear, and when, as sometimes occurs, delirium, at first talkative, afterward with delusions and violence, shows itself, **the** situation is sufficiently alarming. Some persons are specially susceptible, and are disturbed by small quantities. When such signs appear the atropia must be stopped, alcoholic stimulants given, and, if violent delirium exist, hypodermic injections of sulphate of morphia, gr. ¼ to ½, repeated as needful. To prevent poisonous symptoms, care should be taken to drop the solution into the outer rather than into the inner angle, and hold the head so that the fluid does not readily flow toward the puncta ; pressing with the finger over the puncta and sac is of service to hinder passage of the solution into the throat. Another, but less frequent and less serious effect of prolonged use of atropia, is that it causes conjunctivitis. This may be measurably counteracted by dropping between the lids a solution of alum or of boracic acid. Mixing the atropia with vaseline, gr. iv. ad. ℥ j., will to a degree correct the unpleasant conjunctival effect. A very small bit is placed between the lids, and allowed to melt—just as the watery solution would be used. If atropia must be abandoned, we have a substitute in duboisia to be given in the same strength of solution, or hyoscyamine ; but both these are in a degree liable to cause like constitutional effects. Usually all mydriatics must be renounced until the toxic symptoms subside, and then resumed in such degree as may be tolerated. One may not expect the full effect on the pupil at the beginning, and if there be great hyperæmia, the use of leeches to the temple will promote its absorption ; if the anterior chamber be deep and the eye tense, paracentesis will greatly aid its effect. It is a common experience that as soon as the pupil enlarges to a considerable degree, say to about six or eight millimetres, the symptoms speedily give way and recovery sets in. This will take place even though some adhesions remain. Yet in rheumatic iritis this happy sequence does not always appear. The aqueous remains turbid and in large quantity, and pain may continue. It will be well to apply two leeches to the temple, and paracentesis may be admissible. It is also efficient to give a hypodermic injection of morphia. Another remedy is the hypodermic injection of muriate of pilocarpine, gr. ⅛ or ¹⁄₁₆. My own experience is small, but I am prepared to credit the favorable assertions of Schweigger and others about it. Iritis may occur in persons of gouty diathesis as the first token of their constitutional tendency. If it appear when there is great depreciation of health, the disease may not be violent in intensity, but is likely to be most pertinacious in duration, and aggravating in its ups and downs. Nothing but general hygienic measures will in some cases be of any value—except, always, atropine.

An additional application is warm water, and the temperature to be as the patient prefers. Sometimes for suppurative iritis, especially traumatic, water as hot as can be borne is to be kept continually applied for two or three hours, and this repeated three or four times daily. To less severe cases dry warmth is grateful—a heated napkin, or a bat of cotton and bandage. Cold lotions often disagree with iritis, except the traumatic kind, and in vigorous persons. For a large proportion of cases the local proceedings will control the attack. There will be nocturnal pain, for which hot fomentations are to be used, and morphia or some kind of opiate administered. Rubbing into the forehead the oleate of morphia is a resource of special efficiency when the pain is not very severe. Friction over the forehead and temple, with extracts and ointments, are uncleanly, and if they do good, it is simply by benumbing the cutaneous nerves. The friction is often to be commended, but the substance should be cleanly and anodyne. Chloroform liniment, cautiously applied to avoid its getting into the eye, may be comforting. To severe cases, especially with tendencies to relapse, confinement in the room is indispensable, until the intensity markedly abates. To debilitated subjects this rule also applies, but may need modification in the use of a thick bandage, and permitting them to walk on a piazza or where no wind blows, yet fresh air can be enjoyed.

Constitutional treatment will, in some cases, be indispensable, while even in the syphilitic cases, providing their type is mild, mercurials are often not given until the local symptoms subside. They are not the great weapon of success, while atropia and the above proceedings are, and only need skilful use. But if the attack be severe, if the pupil remains closely adherent, if there be gummy exudation, if the syphilitic poisoning be profound or there be hereditary syphilis, mercury must be given in its most effective way. Inunction is, in many cases, the best method—rubbing into the arms, sides of thorax, or inside of thighs, about half a drachm of blue ointment. If for any reason this is to be avoided, the protiodide of mercury, gr. j. ter in die, may be given, while a rapid salivation is to be had by gr. $\frac{1}{16}$ of mild chloride every hour. Oleate of mercury, twenty per cent., is sometimes with delicate skins to be preferred to the blue ointment. In all these cases salivation is to be carefully avoided. At its first sign, a gargle of tannin, or of chlorate of potash, and diligent use of a soft tooth-brush, are to be insisted on. Mercurial vapor-baths, viz., ℥ j. of black oxide of mercury upon a red-hot iron, added to the usual arrangements for a vapor-bath, is an admirable way of getting in the mercury without disturbing the stomach, giving it once daily. These proceedings are directed against a case of severe inflammation. But it is very common to give patients the so-called mixed treatment of biniodide of mercury, iod. potas., and keep it up long after the eye has recovered, to counteract the constitutional poison. For special details as to the treatment of syphilis, I must refer to other treatises. See Keyes on syphilis, in this series. Syphilitic iritis usually comes with a papillary or roseolar eruption, about four to six weeks after chancre, while it may be delayed to the second year or to the tenth year, and it may come as a tertiary symptom. For children the mercurial ointment is the best treatment, combined with careful attention to nutrition, cod-liver oil, healthy and clean skin, etc.

A case of rheumatic or gouty iritis will call for alkalies in full doses—Rochelle salts, liquor potassæ, lithia water—and especially useful is the Turkish bath. Sometimes colchicum will prove its superiority, and, in general, regard must be paid to the phases of the constitutional diathesis, and to the remedies which have proved useful for other symptoms. Salicylate of sodium has been vaunted as of special value in rheumatic iritis; but its claim, while worthy of attention, is not yet fully established. Turpentine oil in doses of five drops in capsules, three times daily, has been recommended. I must also insist upon the proper appreciation of a patient's general condition, whether sthenic or asthenic. In the former, purgatives and sweating and diuretics may be exhibited with freedom ; in the latter they must be most cautiously given, and often quinine will be the potent remedy, and stimulants be needful to build up strength. A feeble patient will always have a protracted attack, and care must be taken to supply the means to carry him through the long misery.

Gonorrhœal iritis will be found to keep step with the success of the treatment of the urethra. I have known iritis to follow the introduction of a sound in treating urethral stricture, the same patient having had iritis with gonorrhœal attacks. Iritis is often cured with perfectly normal pupil, or synechiæ may remain. As has been said above, this fact has been held to account for the tendency to relapses which many cases exhibit. That this is measurably true must be believed. That this is so frequently the case as has been claimed, is not true. Many cases of extensive posterior synechiæ are to be remembered, in the experience of every large practitioner in eye disease, which have not had relapses. The operation of corelysis was devised by Streatfield and modified by Passavant, to detach such adhesions. The former used a notched spatula, which was inserted between the iris and the lens, to pull away the attachment, as with a blunt hook. The latter used a pair of fine forceps to grasp and pull off the iris at the adherent spot. In both cases the wound should be in the cornea and oblique, so that while admitting the instrument the loss of aqueous should be a minimum. Care should be taken not to permit the iris to be caught in the wound, and not to injure the capsule. The iris tolerates such an interference well, and the operation may be as many times repeated as the number of the attachments may require; but it has not seldom happened that the synechiæ were re-established. It is wise to wait to learn whether, in a given case, there be need to interfere, and then to choose corelysis or iridectomy, as the condition may dictate. For close and numerous and broad adhesions, iridectomy at the place of attachment should be chosen; for a single or two broad synechiæ, with an irritable eye, or with tendencies toward neuralgia, corelysis with fine forceps may be tried, viz., Passavant's method.

The iris is sometimes glued to the lens over a large extent of surface, the pupil occluded, and the tissue atrophied. To obtain sight, an iridectomy will be made, and when the iris has been excised, a good artificial pupil with jet black color may result; but, when healing is complete, no better sight is gained. On examining by the ophthalmoscope no reflex is gotten, and by oblique illumination it is discovered that, by splitting of the iris, its uveal lining remains upon the lens, and looks like black velvet. For such a condition there is no remedy short of removing the lens, and at a later time perforating membranous obstruction to secure a final clear opening.

It happens in the course of iritis, while the process seems to be going on well, the media not being very turbid, the pupil fairly dilated, and the hyperæmia not great, and while the ophthalmoscope gives a good reflex, that the vision is not in such ratio as would be expected. It is not to be supposed that the fundus can be easily seen and the details made out in the acute stage of iritis, nor is it prudent to subject to a strong light an eye in such condition; but if one has a good reflex, and other symptoms are satisfactory, a certain degree of sight may be expected. If the patient cannot count fingers at from three to eight feet in moderate light, there is reason to fear deeper complications, such as choroiditis, hyalitis, retinitis. Then it may be that optic neuritis, as above described, will be found. The patient should be kept under restraint and treatment for some time after the iritic symptoms have ceased. He will remain indoors, and be subjected to all the rules of prudential restraint as respects light, exertion, quality of food, sleep, mental activity, etc., and very probably will need persistence in the antisyphilitic treatment which has been already instituted.

Sometimes we have to deal with chronic iritis with extensive synechiæ, and we find the local and constitutional means almost ineffective. The eye remains hyperæmic; the globe is perhaps tense, perhaps yielding. If tense, we may conquer the trouble by iridectomy; if soft, the operation may be attended by excessive hemorrhage and followed by phthisis bulbi. Yet this is not always the issue; at any rate, if left to itself, the eye will be lost, and the operation gives the best chance. Such cases really involve the ciliary body and choroid, and are always of grave significance.

14

CHAPTER VII.

THE CILIARY BODY.

THIS is a zone five to seven millimetres in width, situated between the periphery of the iris and the ora serrata. It adheres near the sclero-corneal junction firmly at one spot, and is at other parts more loosely attached to the sclera by connective tissue. It is rather narrower on the nasal side than on the temporal. Its forward portion is thrown into projections from seventy to eighty in number, which are called its processes. These become lower behind and run into a flat surface. The zone is thus divided into a plicated and non-plicated portion. The posterior limit of the latter is the place of beginning of the choroid, and at this point, moreover, is the ora serrata of the retina. On the interior face of the ciliary body lies the zonula of Zinn, a transparent membrane which is continuous behind with the envelope of the vitreous, and also consolidated with the prolongation of the retina, and in front sends forward a continuation which attaches itself to the rim of the crystalline; there it is called the suspensory ligament of the lens. Between the ciliary processes and the sclerotic is to be found a grayish substance, previously called a ligament, but known now to be muscular. It really consists of fibres, some of which run in meridional lines to join and be lost in the choroid, others beneath them pass in oblique directions, and another and innermost bundle goes around the eye parallel to its equator. The last is called Müller's muscle, as the meridional and oblique parts are known by the names of Bowman and Brücke. The muscle is known as the ciliary, and is the agent of accommodation (see Fig. 3, p. 5). The ciliary body consists of a congeries of fine vessels gathered into rolls to form the ciliary folds and processes, and of other vessels in great numbers. It is saturated with pigment, both free and in cells, and it possesses a very large number of nerves. All that go to the iris pass through it, and many are destined for its own function. The uncommon richness of this region in fine vessels is meant to furnish adequate blood-supply for secretion of the aqueous humor and nutritive material for the maintenance of the crystalline lens. The intricate plexus of nerves, mingled too with minute nervous ganglia, makes the ciliary body a region to which great importance attaches, both in health and in disease.

CYCLITIS AND IRIDOCYCLITIS.

It were superfluous to attempt to isolate inflammations of this region from those of the iris. We cannot see the structure, and we know its diseases by symptoms in adjacent parts. Much has been collected in the way of pathological description of cyclitis, because, when it is well developed, enucleation is often needful. An account of the varied morbid appearances is not necessary, except to say that while we have exudation and thickening of its structure, we also have exudation into the posterior chamber and behind the lens, the latter constituting varied phases of vitreous opacity. It may also happen that exudation comes forward and is visible in the anterior chamber. If serous exudation predominate, we may have increase of tension ; but in chronic conditions, softening of the globe, terminating in marked phthisis bulbi, is the issue. In a case of iridocyclitis we find discoloration of the iris ; it is often greenish in color, the tissue is atrophied, the periphery is retracted ; the outline of the lens is conspicuous, because of total adhesion of the iris ; the anterior chamber is deep, the cornea is small, the vision is extremely bad. But the signal symptom of cyclitis, whether or not associated with manifest iritis and hyperæmia, is exquisite sensitiveness of this region if it is touched ; pressing upon it with the tip of the finger, through the lid, or by a probe or pencil, causes the patient to shrink suddenly and spasmodically. The exhibition of pain is as if a nerve-twig were pinched with forceps, or a dentist had driven a peg into the pulp-cavity of your molar. This one symptom of painful sensibility has the gravest significance : it often dooms the eye to enucleation.

There is another kind of cyclitis—the gummy form—in which a mass of yellow exudation occurs at some point, and, infiltrating the sclera, renders it soft and yielding, and it pushes out as a tumor on the surface. It ultimately leaves a dark blue protuberance, which may be as large as the tip of the finger. This is usually syphilitic. Sometimes, as the result of more chronic inflammation, the ciliary region becomes distended by a series of such prominences or staphylomata, without there having been at any time a distinct yellowish exudation.

The *causes* of cyclitis are usually the same as those of inflammations of neighboring parts which have the same vascular system. Especially must we emphasize syphilis ; but we deal with it most often as the effect of injuries, either with or without the perforation of foreign bodies. We therefore must give special attention to

TRAUMATIC CYCLITIS.

It is through this zone that many missiles find entrance into the eye. Many times blows by blunt instruments, by sticks, or by the fist, if they rupture the eyeball, do so in this locality, and the line of solution of continuity is parallel to the cornea. Often the rupture is not through the whole thickness of the sclera, and shows itself merely as a dark line which may be ectatic ; there may be dislocation of the lens, tearing away of the iris from its periphery, and bleeding into the aqueous and vitreous. Sometimes the iris is folded back from its pupillary border and presents all the appearances of a clean iridectomy. If a foreign body, a bit of steel, or a birdshot go into the eye, its point of entrance may not be detected, and, after hitting the back of the globe, it may be reflected forward so as to lodge upon the ciliary region. It may also come forward to this part, as the effect of chronic intra-ocular changes. In whatever way a foreign body finds lodgement in the ciliary region, it excites destructive lesions and is the fruitful cause of sympathetic ophthalmia.

It is needless to say more about treatment of cyclitis in this place, because the preceding remarks upon iritis indicate the proper therapeusis ; but it must be added that, in all cases, a longer period of care will be needed in case the eye is capable of preservation, while in no small proportion the result must be enucleation. This must be resorted to either to rid the patient of a distressing malady, for which no cure is possible, within such time as the patient can afford, or with preservation of useful sight; or, on the other hand, the question is how to prevent harm to the fellow eye. I may say that, in some cases of irido-kerato-cyclitis, where the eye was thought to be condemned to removal, and had already become soft, a large iridectomy and good nursing during many weeks have saved an organ, not only to be correct in form, but of some value in vision. But large experience and suitable nursing facilities must be available to get this result, which is really of exceptionable possibility. One must be prepared to enucleate the eye if, after an iridectomy, the other eye show dangerous symptoms. The subject of traumatic cyclitis is of such paramount importance that it leads naturally to a larger consideration of the subject of traumatic eye-lesions, and we therefore, at the risk of some repetition, take up this topic.

The figure represents a condition often found in cyclitis, viz., great development of connective tissue about the ciliary body and in the vitreous. The result is shrinking and absorption of the vitreous. It adheres to the retina, and pulls it away from the choroid. The space between choroid and retina is occupied by serous effusion. The retina shrinks into a cord which runs from the optic nerve to the ciliary body. In the illustration the lens has, by the contraction of the cyclitic exudation, been drawn forward, and presses against the cornea.

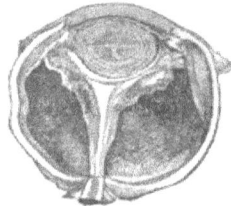

FIG. 83.

CHAPTER VIII.

WOUNDS AND INJURIES TO THE EYEBALL.

A few words as to foreign bodies on the cornea. If they adhere only loosely, they may be wiped off with the point of a folded bit of rag or by a small swab of cotton on a stick (Agnew). If they are imbedded, they must be removed by a cataract-needle, or a spud. To illuminate the eye, a bi-convex lens is necessary, and the patient should be near a window. If one do

not have an assistant, the lens may be carried upon the in-dex-finger, as the cut shows, by a suitable contrivance (see Fig. 84). In removing a for-eign body the surgeon should be behind and above the pa-tient—holding the upper lid by the index finger, which will press the tarsal border into the orbit, and, by the middle finger resting on the

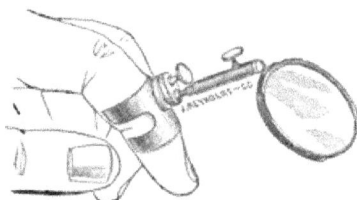

Fig. 84.

side of the globe, help keep it steady. If a bit of muslin be wrapped around the middle finger, its hold is better. With this finger the lower lid can also be pushed down. The manœuvre is illustrated in Fig. 85, the lens for illumination being on the forefinger of the left hand. The way to discover foreign bodies in the cornea is shown by Fig. 14, p. 21.

We have next to consider deeper injuries and the effects which they produce, and the possible presence of a foreign body. A penetrating wound without contusion or laceration becomes grave according to the region it involves and the depth to which it penetrates. If through the cornea alone, and small, no harm may occur, because the iris may not be disturbed. It amounts only to a paracentesis. If the wound be large, the iris will prolapse and remain entangled. It is rare in any case that effec-tual efforts of reposition will be possible. One may try with a small, smooth spatula of hard rubber or tortoise-shell, to gently push it back, or, what is more likely to succeed, is stroking the neighboring cornea to loosen it and excite the contractility of the iris, which may draw it back. It is useless or harmful to continue such efforts long, and resort must be had to excision of the imprisoned tissue. To do this demands a pair of toothed iris-forceps, which should be opened along the *length* of the wound to grasp the iris, and even be inserted into it to catch and drag it gently out. Then scissors laid flat upon the cornea cut off all that can be gath-ered up of the iris inside the part prolapsed. Sometimes the forceps must be used between the lips of the wound, to catch the iris on each side sepa-rately. The important thing is to grasp lengthwise, and not crosswise of the wound. After a prolapse has lasted a few days, there will be neoplastic tissue formed upon it; this need not prevent the attempt to excise the entangled membrane. In a case a week old, or even more, a prolapsed iris can be cut out without leaving a synechia, and it ought always to be attempted. Generally ether or chloroform will be necessary. The forceps

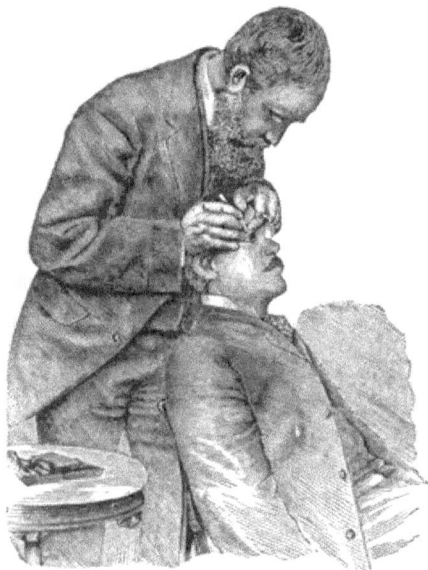

Fig. 85.

with projecting teeth are most serviceable (see Fig. 86), and they will also pick off bits of steel imbedded in the cornea.

The region of prolapse is of consequence. When near the pupil it is least dangerous, because traction is least; the nearer to the periphery of the iris, the greater the danger. This is true, even though the amount be small, if it be near the ciliary region. It has been observed that while cases of this sort may recover and for some time do well, they are prone to suppurative iridocyclitis, or to a chronic form of inflammation of danger-

. Fig. 86.

ous quality. A wound which involves prolapse of iris and rupture of lens-capsule has the added dangers coming from rapid swelling of the lens and cataract. Besides excising the prolapse, rigorous restraint must be enforced, confinement to bed, atropine used with energy, and cold-water lotions. If symptoms do not yield, but pain and hyperæmia continue and the lens press forward, it may be removed by linear extraction (vide infra). Sometimes its removal becomes imperative because of increased tension and pain, but oftener it is better to wait. A lacerated and contused wound of the cornea, such as is often inflicted by a stick of wood in chopping, may involve not only laceration of the lens-capsule, but displacement of the lens. The amount of hemorrhage and of laceration will determine the treatment. Unless there should be evident symptoms of trouble due to the swelling of the lens, or to its being in contact with the ciliary body, it is better to remove the prolapsed iris and to await developments.

If a wound have gone into the ciliary region, the future of the eye is very precarious, and so in some measure is that of the other one. Contused wounds or blows on the eye may rupture the cornea, and may tear away the iris from its periphery. The latter lesion may occur without the former. A common cause of such mischief is the unexpected pop of the cork from a soda-bottle. For such lacerations (called iridodialysis) nothing is to be done but to apply cold water and atropine, and good nursing. A total separation of the iris has happened a number of times, and I have seen it combined with a small wound of the cornea, which permitted the entire extraction of the iris. The lens and deeper parts were not disturbed, and moderate sight was preserved.

For injuries which cause hemorrhage into the anterior chamber and apparently no extensive lesion, only ordinary care is needed, and the absorption of blood will occur in from one to six days. But one must be guarded in prognosis, because laceration of the suspensory ligament may permit the lens to tilt back and forth, or the latter may be dislocated. If either of these lesions have happened, cataract may slowly come to pass, or secondary glaucoma with hyalitis, choroiditis, etc. It also happens that contusions of the eye, which exhibit only slight visible marks of injury, may be attended by laceration of the choroid, as will be referred to at a later time.

Incised wounds of the sclera, in whatever region, require no other dressing than a bandage or closure of the eye, and cold compresses. But sometimes they are so extensive as to gape, and then a suture in the sclera, including as little of the tunic as may suffice, will be proper. I have seen a flap of sclera, turned up by a bit of glass from the bursting of a soda-water bottle, lie well in place and the eye make a good recovery. If vitreous protrude, the eye will be quite softened, and manipulations with a suture will be liable to increase the prolapse. An eye so injured had better be left to itself; its preservation will be very doubtful, and the best chance is gained by putting the patient to bed and attempting to repress inflammation.

Injury of the eye by bits of iron, such as the heads of bolts or fragments struck off by a sledge-hammer, are apt to tear it so extensively as to be tantamount to its destruction. When the cornea and sclera present an irregular wound and the eye is sunken and filled with blood, the course most humane and surgical is to enucleate it immediately. A prolonged and distressing inflammation is thereby avoided, and an artificial eye may be worn within a fortnight. I would not always remove the whole globe, but might sometimes be willing to remove its anterior part with vitreous and choroid, and leave the posterior half of the sclera to shrink down. A longer period for healing must be expected, and a better stump is gained.

This whole subject of injuries of the eye requires careful consideration of the precise condition of the organ, and, while the utmost conservatism is to be practised, we are obliged to carry in our thoughts the possibilities of hurtful effects which may be communicated to the fellow eye in that subtle way which characterizes sympathetic ophthalmia. This possibility often makes us insist on removing an eye when we would gladly avoid the mutilation, and to this decision such circumstances contribute, as the patient's status in life and means of livelihood, and accessibility to competent advice, his occupation, etc.

Foreign bodies in the eye give rise to most serious complications. If large they may compel immediate extirpation. To detect the presence of a large foreign body supposed to be in the eye may justify the use of a probe, but this instrument is to be handled with the greatest circumspection. There must be a wound quite large, and evidence quite convincing that a large fragment has entered, to authorize one to put in an exploring instrument. It can happen that a fragment has entered through a linear wound behind the equator, not disturbing the iris and lens, and been stopped just inside the globe ; a probe would push it farther, and destroy the little chance which might have existed of saving the organ. To convince the patient or one's self that enucleation is unavoidable, a probe may be used to detect a foreign body in a badly lacerated globe.

The most embarrassing cases are those in which the foreign body is small, such as bits of percussion-caps, bits knocked from chisels or hammers, bird-shot, etc. The position they have reached is a chief consideration. A foreign body imbedded in the cornea may be taken out from the front by a forceps spoken of before (p. 214), or may be pressed forward by putting a broad needle behind it and then catching it with forceps or prying it with a spud. But if a foreign body have gone into the anterior chamber and lie loose at its lower edge, it may be difficult to discover. A drop of water pushed up on the edge of the lower lid, so as to rise to the level of the cornea, will sometimes bring it to view, as a penny in the bottom of an empty bowl, and hidden by its side, is brought to sight by pouring water into it. To get it out, a very peripheral wound must be made by a narrow Graefe's knife, puncture and counter-puncture being on either side of the foreign body, and no regard paid to its edge in cutting out at the angle of the chamber. It may force out the foreign body, or it will bring it within reach of a small, blunt hook, whose shank must be bent to proper shape. A loose body in the iris may sometimes be coaxed out by a blunt hook, through a wound in the cornea so placed that the inner edge shall be close to the foreign body, but shall allow the curved portion of the hook to pass. I have satisfactorily used a pair of forceps called Matthieu's, with blades bent at an angle of 45°, and not more than four millimetres long beyond the bend. They open by playing around the longitudinal axis of the handle (see Fig. 87). To such cases the magnet which Dr. Grüning has suggested will find application, and I have favorable expectations of its utility. Dr. Grüning has joined together a number of magnetized steel rods, separated from each other, but in close proximity, and fitted at their ends into iron caps, one of which is provided with a delicate point of malleable iron, 32 mm. long, 1 mm. wide, and 0.3 mm. thick. This point sustains with ease a weight of 225 grains. Chips of iron weighing from one to fifty centigrammes, in the vitreous of animals, are attracted from a dis-

FIG. 87.

tance of one to five millimetres, and easily withdrawn. The larger the piece, the more powerfully is it attracted, which is an unfortunate fact in the applications which we desire to make of the instrument. For portability and ease of manipulation in the vitreous, this is the best instrument. Dr. Bradford has constructed an electro-magnet, needing but one bichromate of potash cell, and having eight square inches of negative service. The tips in the induction coil, which is encased in hard rubber, are 3½ inches long by $\frac{3}{12}$, $\frac{4}{12}$, and $\frac{5}{12}$ of an inch in diameter. Their suspensive power is 11, 16, and 20 oz. respectively. The greater power of this instrument, fits it for acting on bits of iron inside the eye and drawing them toward the surface, while the coarseness of its points is against introduction inside the vitreous, although it may be passed into the aqueous chamber. An important suggestion is that the incision by which it is entered should not be straight, but T-shaped, to prevent the foreign body being stripped off the magnet as it comes out of the eye. For further knowledge of the use of magnets, see Dr. Bradford in *Boston Medical and Surgical Journal*, March 31, 1881. For Dr. Grüning's magnet, see New York *Medical Record*, May 1, 1880. Dr. Grüning gives likewise the history of the early uses of electro-magnets by McKeown of Belfast, and by Hirschberg of Berlin.

If a bit of iron stick fast in the iris and the magnet be not available, nothing will suffice but removal of a piece of it with the foreign matter included. The object will be to take out as little iris as may be necessary, and to place the coloboma in the least conspicuous locality. A foreign body in the lens will, except in a very few instances, be followed by cataract, and usually it is better to wait for opacification and some absorption to occur before interfering. If great swelling arise, one may be compelled to operate promptly, but considerable latitude must be allowed for various contingencies. A very small particle may be for a long time undisturbed, and the lens left to spontaneous absorption. Sometimes a discission may be done to hasten absorption, or the capsule be more freely opened, and the extrication of the foreign body be deferred to a final performance, when the capsule and foreign body may be brought out together. Usually iridectomy will have to be done, either at the time of lens-extraction or preceding it. At other times it may, because of the considerable size of the object, be necessary to do a regular extraction of the lens by a full section with iridectomy. On the other hand, minute particles entering the lens through the iris are never discovered, and are to be left alone. The cataract in time is gotten rid of and partial vision restored, while the intruder makes no sign of its presence. Years may pass without evil hap, and possibly it may never arrive; but there is no absolute security Cases of trouble after immunity for many years are on record. The case should be left to the indications which may arise.

When a foreign body of small size finds its way into the vitreous, we have a portentous case. Under certain favorable conditions it is capable

of extraction. For instance, it may have entered behind the crystalline, and be discovered in the vitreous by the ophthalmoscope. Dr. Knapp has by means of a grooved hook (see Fig. 88), removed a piece of percussion cap thus situated, first making a meridional wound in the sclera near the point of lodgement, and as a bead of vitreous appeared, the body came toward the wound; upon the second trial with the hook it was brought to view and removed. The patient recovered good sight. Subsequently a cyst was formed in the iris, which also was successfully operated upon.

Under such conditions, the magnet is the most effective appliance for pieces of iron or steel. Forceps cannot be depended on, nor does the attempt to operate with an ophthalmoscopic mirror before the eye make the performance more certain of success. Too often the foreign body is out of sight, and then no effort can be of use, except a chance endeavor with the magnet; while the additional lesion of the operation may, if fruitless, make the case worse. An eye enclosing a foreign body is usually doomed. A special tendency to rebound from the back of the eye forward to the ciliary region has been demonstrated by Dr. Berlin, and if it reach this locality fatal trouble will ensue. A long time may elapse before it comes, but it usually consists in development of hyalitis with formation of connective tissue in the vitreous, which contracts, becomes adherent to the retina and pulls it from the choroid, while a yellowish, serous fluid fills the subretinal space. The detachment of the retina in time becomes total, so that it stretches from the optic nerve to the ora serrata in a cord or funnel, with its base forward. Sometimes the retina will be pinned fast to the choroid at the place of impact of the body. The ciliary body will be infiltrated with connective tissue, the choroid atrophied or otherwise degenerated, the lens opaque. The globe may be reduced in tension, and **sometimes in** diameter markedly. The above sketch applies with variations to **many cases,** whether the intruding body be in one or another part of the vitreous chamber. The eye becomes sightless, and usually brings on the symptoms of sympathetic ophthalmia. The records of pathology are rich in descriptions of eyes enucleated because of such accidents.

The surgeon called to a case of foreign body in the vitreous, which he finds that he cannot remove, is obliged to give a judgment as to immediate extirpation. He has to satisfy the patient that it is in the eye. Sometimes this is not easy for a layman to believe, because a wound of entrance may, if through the sclera, be almost imperceptible, or it may become so in a few days. Fine bird-shot may enter the eye, and leave absolutely no trace of their passage. Dr. Williams, of Cincinnati, reported an eye in which, at its examination, one wound alone could be found, while three shots were lodged inside of it. Reduction of tension is often produced very early, sight will be very dim, and the ophthalmoscope may show blood in the vitreous, or possibly hemorrhage on some part of the fundus. An absolute scotoma or limitation of field may be found. Such symptoms are conclusive. Immediate enucleation is not often done, but the progress of the case observed. Should the reaction be great and with much pain, **the** operation may be insisted on, and will usually not be declined. But if, as often occurs, reaction be moderate, and the eye little troublesome, the patient will not willingly consent to part with it. Under certain conditions, enucleation need not be recommended. But a strong statement must be made as to the probable necessity for so doing before a long period, because the other eye will, in time, exhibit sympathetic symptoms. This warning must be most impressively and distinctly given. If the patient live near a competent surgeon, if he be watchful of his own condition, if intelligent, and if the eye be of nearly normal look, he may be permitted to retain it until the time of absolute necessity for its sacrifice shall come. Not a little consideration is to be given to the social position and sex and surroundings of the person. But if the person live at a distance from competent surgeons, if he be stupid and can ill afford loss of time, if the foreign body probably is large, if there be tenderness over the ciliary region, or notable softening of the eye, the operation should not be delayed. However distressing may be a firm insistance, it is the surgeon's duty. Wounds by bird-shot are most deceitful. It is possible for them to go entirely through the globe, yet more often they do not. In fact, it does not make much difference in the ultimate issue of such a case whether the shot lodge or go through the eye. The processes of hyalitis, choroiditis, etc., above sketched, are equally liable to be begun, although, perhaps, not equally soon, and the eye will have to come out. The danger to the patient lies in the subtle and unappreciated character of the early symptoms of sympathetic ophthalmia and in his false security. As will hereafter be shown, there is time during which interference may be effected, and this period of grace will not last long. Too often it falls to the surgeon to see a patient who has allowed the second eye to go too far on the road to ruin to admit of its being rescued.

It remains to state a rather uncommon lesion which can cause sympathetic ophthalmia, namely, burns of the eye. I have seen one case where caustic lime was the cause, and the eye had to be extirpated in two weeks, because of mischief in the other. Usually burns are not followed by any such misfortune, their injurious effects being confined to the surface of the globe.

CHAPTER IX.

SYMPATHETIC OPHTHALMIA.

UNDER this head, a large variety of lesions are included, and it seems fitting to discuss the subject in this place. The lesions are practically divided into sympathetic irritation and sympathetic inflammation.

Under the head of *irritation* we have the symptoms of weariness, shortening of the range of accommodation, photophobia, sometimes phosphenes, watering of the eyes, and, as a result of these troubles, inability to use them, and sometimes limitation of the field. The conspicuous sign is the photophobia and reluctance to use the eyes. As sympathetic *inflammation*, we have iritis serosa, irido-cyclitis, irido-cyclo-choroiditis (irido-choroiditis), retinitis, and neuritis; much less frequently have been recorded conjunctivitis, keratitis, and neuralgia. The dangerous and frequent inflammations are those of the uveal tract, especially irido-cyclitis. It is needless to repeat what has been said descriptive of this lesion, but some additional remarks may be useful. A faint, circumcorneal injection, perhaps a very narrow zone, the discovery of a few pupillary adhesions, and of a slight, possibly local discoloration of the iris, with a little loss of sight, betokens the onset of iritis serosa, and is to be esteemed a danger-signal of the highest importance. The same process, more advanced, gives deposits on the posterior surface of the cornea, foggy aqueous, turgid iris, more numerous adhesions, and deeper sinking of sight, with more decided ciliary injection. These symptoms have been already enumerated. The tension may be a little plus and the eye not painful, except on touching the ciliary zone.

Irido-cyclitis plastica does not often succeed the previously described symptoms, but may do so, while generally its advent is from the start marked by exudation in the tissue of the iris, thickening and befogging it, causing adhesions, not only at the pupil, but upon the posterior surface. Sometimes hypopyum appears, straining through the periphery of the iris, and not infrequently small hemorrhages are seen. The vitreous may become hazy, and exudation appear behind the lens. For a short period tension may be increased, but softening soon sets in, ciliary injection becomes intense and diffuse, the eye painful and very tender to the touch, and sight extremely reduced.

By statistics collected by Alt, in 110 cases of enucleation we learn that 73 per cent. had affections of the uveal tract and retina, 27 per cent. affections of the uveal tract alone; one case had affection of the retina alone. In 52 per cent. of these cases the symptoms of sympathetic irritation had occurred, and 98 per cent. of them had lesions of the uveal tract, while 90½ per cent. had also affections of the retina and optic nerve. This shows most signally how irritative symptoms pass into or concur with tissue-lesions of the retina and the uveal tract, and justify the urgency with which such symptoms must be regarded. Among Alt's cases are included 19 per cent. in which panophthalmitis had set in. This condition cannot any longer be held to be a preventive against sympathetic trouble. Both the ciliary nerves and the optic nerves are the seat of alterations, and may be the means of transmission (see *Arch. of Ophth.*, September, 1881, p. 277).

The causes of sympathetic ophthalmia have already been alluded to, but an enumeration taken from an author of extraordinary accuracy and large experience, Prof. Arlt of Vienna, which I can endorse from my own observation, will enforce the subject. 1st. Traction of the iris by an ectatic corneal cicatrix (partial or total staphyloma). 2d. Prolapse of iris, at the corneal margin (cystoid cicatrix). 3d. Sunken scars, with entanglement of the ciliary body caused by penetrating wounds. 4th. Ruptures of the globe, by contusions of the ciliary region. 5th. Operations for cataract, of which Becker, in 1875, reported after flap-extraction seven cases, and after Graefe's extraction, eleven cases, and four cases operated on in other ways, making a total of twenty-two cases. 6th. Operations of sclerotomy (probably when the iris is entangled). 7th. Perforating wounds of the lens. 8th. Luxations of the lens, under which head operations for reclination or couching make a large contingent. 9th. Calcified lenses. 10th. Intra-ocular cysticerci. 11th. Choroidal tumors. 12th. Foreign bodies in the eye. 13th. Ossifications in the choroid. 14th. The wearing of an artificial eye after it has become roughened by corrosion. The list is long enough, but does not include some rare causes. It will be noted that operations on the eye may be followed by sympathetic trouble. Critchett's operation of iridodesis (artificial prolapsus iridis) has on this account been almost abandoned. Iridectomy, when the angles of the cut are caught in the wound, even though cystoid cicatrix does not occur, has been followed by this result. Cataract extraction has the same liability. The flap method is less prone to it than Graefe's method, but the very large preponderance of the latter over the former renders comparison between the two methods impracticable. The number which Becker collected, viz., eleven, has since 1875 been largely increased ; but we know nothing yet, with accuracy, about its percentage.

The interval which may elapse before the trouble begins is usually between four and six weeks, while a few cases of much earlier outbreak are recorded (I once noted an interval of only two weeks), and the period of exemption may extend to twenty years. In these late cases the cause is ossification of the choroid, or what is often loosely called an inflamed stump.

Treatment.—The added experience of fifteen years has made the question of treatment more complicated than when the advantages and necessity for enucleation were first announced. We still have to declare that enucleation of the mischievous eye is the only trustworthy remedy. But we are embarrassed by two classes of facts : that, on the one hand, some cases of sympathetic serous iritis have been known to get well without enucleating the offending eye ; and second, that enucleation does not always protect. The first class of cases are extremely uncommon. Under the second head it is to be said that it is not claimed that enucleation will afford protection or bring about recovery, unless done before pronounced symptoms have appeared. Therefore it must be performed early, during the stage of irritation, or at the very onset of the inflammatory state. To enucleate when the first eye (the mischief-maker) is in acute inflammation is held to be dangerous, because the operative interference is added to the provocation under which the second eye suffers, and tends to bring about in it the dreaded so-called malignant irido-cyclitis. In this opinion I do not concur. It has also been shown that after enucleation the second eye may remain quiet for some days, and then break out in inflammation, and even be lost. This catastrophe clearly shows that the usual relations between the eyes could not in such a case exist. A number of weeks or months may elapse after enucleation, and the second eye may be attacked. In this case the cause may be the entanglement of a ciliary nerve or of the optic nerve in the orbital cicatrix. Opening the scar and excising the end of the optic, and also any tissue which may have been found on previous careful exploration to be tender upon pressure with a small instrument, has three times in my experience arrested the mischief in the remaining eye. Moreover, an artificial eye sometimes makes excessive irritation in the orbit, and this being dispensed with, relief of sympathetic symptoms follows. It is also to be said that there is much more chance of abatement of symptoms in the second eye if the patient be kept under stringent conditions of protection and nursing. It has been under these favoring circumstances that cases of sympathetic trouble, usually simple iritis serosa, have eventually recovered when enucleation has not been done. It is therefore not for the surgeon to hesitate in declaring the **necessity** for an operation on the offending eye, because, while there may be chances for the patient to escape, if he be contumacious, or be hindered by circumstances, he assumes great risks. We cannot yet say that our knowledge of these cases is perfect, but we must hold fast to such knowledge as shall insure the greatest probabilities of benefit.

Within three years a less severe alternative than enucleation has been proposed, viz., section of the optic and ciliary nerves of the offending eye. To this proceeding many otherwise recusant patients would consent. But the offending eye often has a certain degree of sight, or, by operation, such as the removal of cataract or by iridectomy, may become capable of sight. Its vision may be better than in the other eye, which has, perhaps, passed through its stages of sympathetic inflammation. Under these conditions no one would think of optico-ciliary neurectomy. This operation may involve great hemorrhage and sharp orbital cellulitis, and we do not yet know how much protection it can afford. It, however, has claims to consideration, and is being extensively tested. I have done it three times without mischance surgically, and have not had sympathetic trouble. The operated eye is liable to grow softer, **and** may even become considerably **smaller.**

It **is** not a trifling matter **to** deprive a person of **an** eye which, though sightless, may not be a notable deformity and may not be painful. The necessity for an artificial eye is not only a trouble, and to the poor an ex-**pense to** be considered, but after a number of years it often must be given up because of shrinking of the conjunctival sac. The personal deformity **becomes** a grief, and often makes **it** difficult for clerks, shop-girls and others **to** find employment. I therefore deliberate most carefully over a case in which enucleation is to be considered. If an iridectomy in the offending eye, or removal of a foreign body or of a lens, can allay its irritability, while assiduous care is given to the patient's surroundings, I refrain from enucleation so long as the symptoms of the fellow eye are mainly those of irritation. But if actual inflammation, as shown by pupillary adhesions or other signs, takes place and does not yield to atropia vigorously tried, or if neuro retinitis appear, enucleation is the safe and urgent thing. Optico-ciliary neurectomy may be tried in case there is no foreign body or growth in the eye and it have a satisfactory form and appearance, and there be no reason to expect great hemorrhage or reaction. If this operation fail to give relief promptly, **or even** within forty-eight hours, enucleation must be done.

When an eye has reached the acute stage of sympathetic inflammation, no operative interference upon it is to be done with any view of relief, **except** to remove pain. An iridectomy may be followed by relief of pressure, but by no gain of sight, because the gap is soon filled up by new tissue. An operation upon **an** eye which has gone through the acute stage and become quiet, **does no** good. To make a clear pupil is usually impossible, because the iris pulls off in shreds, or the uveal layer is left sticking. Even relief of pain is often temporary. It is best, when sympathetic irido-cyclitis is in full march, to treat it by medical means and assiduous nursing, **expecting** it to last for six to eight weeks or for months, and carry the patient through as may be possible. In time the interior of the eye may grow more clear, and through a very small and clouded pupil the patient may in twelve months gain useful sight. The temptation to operate in such cases should be resisted so long as the lens has clearness. But, when cataract comes, the operation must be such that lens and iris must be taken out together. This will be described hereafter. Even **under** these circumstances I have been greatly disappointed by the results of some cases in which a succession of operations to gain a clear pupil have been faithfully tried. The globe would become soft, and the pupil close, and though light perception remained, there was no gain in useful vision. Mr. Critchett, of London, has very lately suggested operating upon these cases by means of two needles, to procure absorption of the lens. The first discission to **be** very carefully done, and only to open the capsule ; the lens-matter appears and the eye is left to events. After a time the same careful operation is repeated, and even a third may be needed before all of the lens has been absorbed and a free hole gained through the membranous obstruction. The condition of success in these proceedings will be the avoidance of all **violence to escape the** tendency to reproduction of connective tissue and occlusion of the aperture. Months will pass before the result is gained. The mode of using the needles is represented on a later page.

A description of the operation of optico-ciliary neurectomy and of enucleation will be **given** hereafter.

CHAPTER X.

FUNCTIONAL TROUBLES OF THE IRIS—MYDRIASIS—MYOSIS—HIPPUS.

A VERY large pupil belongs to many myopes, because they do not use their accommodation. A large pupil is often the token of anæmia and debility ; but mydriasis means permanent enlargement of the pupil. Paralysis of the sphincter iridis occurs as the result of paralysis of the third (motor oculi) nerve most frequently, and is usually associated with double vision and loss of accommodation. It may occur without any such symptoms, and is then supposed to be due to deficient innervation in the ganglia contained in the iris, or to cerebral disease. It occurs from overtaxation of the eye, as I have seen in a distinguished artist who once painted miniatures. A most common cause is diphtheria, when not only will the pupil be enlarged, but the accommodation be paralyzed. Atropia is sometimes fraudulently used, and the mydriasis which it occasions differs from that caused by paralysis of the motor oculi in that, when the latter is the cause, the use of atropia makes the pupil still larger. A functional, but habitual enlargement of the pupil, is seen as an early token in some cases of insanity. A permanent mydriasis occurs in meningitis, hydrocephalus, and cerebral tumors ; it also occurs in absolute amaurosis, because the reflex action of the retina is destroyed.

Treatment.—The nature of the cause must be discovered, whether syphilis, rheumatism, orbital growths, diphtheria, cerebral causes, etc. These will determine the treatment in a general way. The local treatment is the use of electricity, or of sulphate of eserine ; the latter may be used in solution of gr. ss. to gr. ij. ad. ℥ j., as often as it requires to be renewed. Hydrochlorate of pilocarpine, gr. ij. ad. ℥ j., is less painful and sufficiently efficient. In fact, the local treatment is of minor value as compared with the constitutional.

MYOSIS.

Continuous contraction of the pupil is caused by irritation of the fifth nerve, as when a foreign body is on the cornea, or by whatever may cause ciliary neuralgia. A blow on the cornea will sometimes cause myosis, which will resist the most vigorous use of atropine for hours. Irritations of the brain at the origin of the third nerve, or beginning meningitis, have the same effect. The poisonous influence of opium acts in this way, while pilocarpine and eserine produce myosis by direct influence on the nerves of the iris.

Myosis from paralysis of the dilating fibres of the iris is the result of a lesion of nerve-centres, and especially of the cervical spinal cord, or it may be caused by lesions of peripheral fibres of the great sympathetic in the neck. Horner has called attention to a condition originating in paralysis of the sympathetic in the neck, in which there is flushing or sweating of half the face, slight protrusion of the globe, partial ptosis, and contraction of the pupil.

Treatment need not be specially discussed.

HIPPUS

is a term applied to rapid contraction and dilation of the pupil. **I have** never seen it, and simply give the definition. It is not the same as *tremulous iris*, which is often seen, and means that the anterior part of the vitreous is fluid, and that the suspensory ligament of the lens is torn or much relaxed. The same is often seen after extraction of cataract.

TUMORS OF THE IRIS

are rare. They are generally sarcomata, either white or pigmented. What is called granuloma is usually simply a mass of granulation-tissue, the result of healing after a wound or of extensive ulceration of the cornea.

Cysts of the iris are more frequent. They are the effect either of sacculation of the iris in consequence of a wound or operation, or they result from the proliferation of cells introduced into the anterior chamber. For instance, if cilia should fall in through a wound, cells from the sac of the hair have been assigned as the beginning of a cyst. In all cases the cyst is very transparent; it may be in the substance of the iris, or simply lift its epithelial layer. It is lined with cells, and expands gradually, pressing back the iris, and may disturb vision seriously. If not dealt with, it would set up internal inflammation and destroy the eye. The *treatment* is to excise the cyst; puncturing it is ineffectual. In its excision the cyst may be always expected to rupture. To make an incision of proper size at the limbus, a puncture should be made near the cyst with a Beers knife, large enough to enter a scissors-point, and by this, with successive clips, each of which shall not cut more than two millimetres, the wound may be made as large as needful. I have seen the wound completed without puncturing the cyst, and the same might possibly be done by a very narrow Graefe's knife. Iris-forceps then grasp the cyst, and with it the iris to which it is attached, and then the rupture takes place. Care must be taken to remove the whole sac, and not to leave any tissue lying in the wound. Healing occurs kindly. A relapse is sometimes observed. If the case be allowed to go too long, the eye may require enucleation, because the cyst may invade the ciliary **region.**

Coloboma of the iris **is** a congenital **defect. The fissure is** generally downward, **may** be partial or complete, and may **or may not** extend to the choroid. It may affect one or both eyes. It is sometimes accompanied by microphthalmus. Irideræmia, or total want of iris, is a rare congenital defect. In the fetal state the pupil is closed by a membrane. Shreds of this are sometimes discovered after birth.

OPERATIONS ON THE IRIS.

Iridectomy—Iridotomy—Irido-dialysis—Iridenclysis—Corylesis—Iridodesis.

The operation most frequently done upon the iris is iridectomy. It has a wide range of applications, and these have two intentions—therapeutic and optical. Iridectomy for therapeutic purposes is called for to diminish

15

intra-ocular tension, whether this be exhibited in staphylomata or in general hardness of the eye, as in glaucoma of various types; in cases of sclerokeratitis; also, it has application in cases of adhesions of the pupil, or of unyielding or recurrent iritis, or for rapidly increasing ulceration of cornea; sometimes for keratitis with hypopyon, for rapid swelling of the lens; it may be coupled with linear extraction, and with the removal of foreign bodies on the iris or in the lens; and also as preliminary to extraction of cataract. For optical purposes it is done for opacity of the cornea, for occlusion of the pupil, for central stationary opacity of the lens or its capsule, for luxation of the lens, and as part of the operation for extraction of cataract.

Before doing iridectomy for visual purposes, it must be ascertained that there is perception of light—enough to warrant its performance, and next, that the visual field is not too greatly encroached upon. The choice of place for the new pupil is in many cases determined by the region where the transparency of the tissues will give the best sight. If the opacity of the cornea or lens be central, the choice of place is at the lower part of the cornea, below the horizon. When an artificial pupil is made, chiefly for enlarging the field of sight, it is often best to place it outward. Should both eyes be operated on, the pupils ought to be symmetrical, i.e., both inward, or both outward, or both downward. Should the upper part of the cornea be selected, the pupil may be made more available by cutting the superior rectus tendon, provided the patient has but one eye; but, if he have two available eyes, a small piece of skin may be taken from the upper lid. The place of election in therapeutic iridectomy is determined by the lesion in many cases; but if the locality be optional, it should be made upward, that the upper lid may cover it as much as possible. The distance of the incision from the corneal margin will vary with the case. The rule may be thus stated: for an optical purpose it should be kept as near the centre, or the best portion of the cornea, as may be possible; for therapeutic purposes, it should be laid in the scleral edge, even one and one-half millimetre from the transparent cornea. A pupil for optical purposes should be small, i.e., from two to four millimetres wide at its base. For therapeutic purposes, it should be broad, from five to eight millimetres wide. The thickness of the cornea must always be remembered: that it gives an outer and an inner wound, and that the latter is always smaller, and that it regulates the size of the iridectomy. The instruments to be chosen are the following: a spring-speculum, fixation-forceps, an iris-knife, iris-forceps, scissors, and a curette or spatula; a bit of muslin with a mass of cotton upon it, to be used to check spasm or hemorrhage, is to be at hand in every operation which opens the eyeball. If anæsthetics be used, a lid-elevator of large size must be ready for instant use to pull forward the tongue by slipping it over its base and hooking it up. The patient may sit in an operating-chair, but I prefer to have him on his back on a table or hard bed. The kind of knife depends on the size and situation of the iridectomy. For a very narrow one, a broad needle, straight or bent, is the best. For a larger one, up to five millimetres, a lance-knife, straight or bent, is suitable, and it can be used in the transparent cornea or at the limbus. For very peripheral, and also for large incisions, a Graefe's knife is to be preferred with which to make puncture and counter-puncture. Lance-knives should be held perpendicular to the cornea until the point is seen to be inside; then depress the handle, that the blade may become parallel to the iris, and press steadily forward. Such knives penetrate hard, and a to-and-fro pendulum-like movement of the handle as-

sists their entrance until the point is within. For forceps we may use
straight or curved, according to need. Scissors should be curved on the
flat, and be sharp-pointed. Another form, with a bend at the joint in the
plane of their spread (Maunoir's), is sometimes useful.

In operating the performance is as follows : If anæsthetics be used,
two assistants are better than one. The pain is not great and the operation short, but the anæsthetic is quite as advantageous to the operator as
to the patient. Wait until muscular relaxation comes, then put in the
speculum with blades closed and held by the screw-head ; open the lids
by turning the screw, apply fixation-forceps opposite the point where the
incision is to be made, and close to the cornea. The operator will himself
hold the eye, and with the other hand make the incision. He then has
entire control, and his hands act consensually. Watch the point of the
knife, and go forward steadily to the depth required. Withdraw with
care, and avoid pressing on the wound, that aqueous may leak as little as
possible. Pass the fixation-forceps to the assistant by dropping it down
to him, and he must avoid all pulling on the eye, remembering that it is
to be rotated, not dragged. His hand must be light, and he must see that
he do not make the wound gape. If the iris has prolapsed, the iris-forceps
need not be placed within the eye, but will gently draw it out to the required extent. If it should not present in the wound, carry in the forceps
with closed blades, and when the margin of the pupil is reached, let them
spread to the full, and with a little backward pressure close the blades and
draw forth the iris-tissue. Be mindful not to let the traction be too sudden,
if there be adhesions. When all that is wanted comes outside, cut it off with
scissors. For a small iridectomy one clip suffices, pressing firmly against
the cornea. For a large incision two cuts are better, the points of the

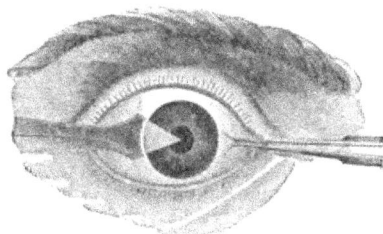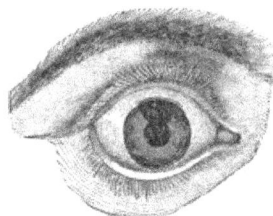

Fig. 89.   Fig. 90.

scissors being used, and the forceps drawing out the iris as it is cut.
Hemorrhage is now apt to occur. Its escape from the wound is favored
by keeping up slight pressure with fixation-forceps, and at the same time
gently pressing on the posterior lip of the wound with a curette or spatula.
If it do not fully flow out, a bit of muslin folded to make a corner or tip,
rubbed along the wound, will accomplish it. The greatest pains must be
given to the complete return of the iris to its normal position, if not adherent. To leave a little blood in the chamber matters nothing, but there
must be no approximation of the pillars of the coloboma toward the
wound. This rule is imperious. The whole operation is done in about
three minutes. Loosen the fixation-forceps, shut together the speculum
and slip it out gently, lay upon the eye the pad and cotton, holding it
gently with the hand. After a few moments open the lids and carefully

remove clots from the wound by iris-forceps, and from the conjunctiva by bits of muslin. Then cover both eyes with a bandage and put the patient to bed. Fig. 88 shows the place of entrance of the knife, for an iridectomy outward. Fig. 89 shows how the coloboma should appear after an iridectomy. The angles of the pupil are returned to their proper place, and the pupil has the correct " key-hole " shape.

Iridotomy, or, as sometimes called, iritomy, is required chiefly in cases of absence of the lens, when the pupil is closed and the iris is adherent to a false membrane or to the lens-capsule. It is therefore a sequence to an operation for cataract. It may be done by a knife or needle, with a single incision, or be done by a more formal operation, by Wecker's forceps-scissors (see Fig. 91). The simpler operation requires a knife, of which there are

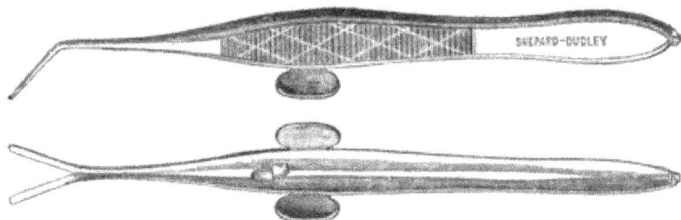

FIG. 91.

several forms, straight and bent, narrow and wide. The incision should be made at right angles with the direction of greatest traction in the iris. A knife is thrust through the cornea a little from its centre, and through the iris, and the cut in the latter made as large as possible by a to-and-fro movement, the corneal wound being the fixed point. A double-edged blade is usually to be preferred. When larger openings are required, or there is not tension enough in the tissues to make a proper opening, the method by Wecker's scissors is amply adequate. It is of recent introduction, and has been in my hands very satisfactory. The eye being exposed and held, the patient anæsthetized to the full degree, an incision four to five millimetres long is made in the cornea near the limbus by an iris-knife, while a stop-knife can be used which shall give the exact length. Push the blade forward to its shoulder, then withdraw about half way to let aqueous ooze out, and bring up the iris; then push the point again forward to pierce the iris and its lining false membrane. This manœuvre brings the wound in the cornea and that in the iris at a little distance from each other, although parallel. Next carry in the scissors flatwise with blades closed, slip one blade through the iris, opening and canting the instrument a little, let the other blade come in front of the iris, thrust down the blades as far as the cut is to go, shut them and withdraw. The cutting being done across the line of traction of the tissues, a good and clear pupil results. But if the tissues be loose and not retractile, two cuts in the shape of an ∧ must be made, and the pupil will become arrow-headed in form, thus, △, with certain variations. Loss of vitreous is not uncommon, but does not entail mischief. This operation takes the place of iridectomy after cataract operations, and gives clean results. It is also applicable to some cases of anterior synechia, for dividing a limb of prolapsed iris, and to some traumatic lesions in which a dense band sometimes traverses the anterior chamber it is invaluable.

Irido-dialysis, or *iridorhexis*, is not often done. It means tearing away the periphery of the iris when only the extreme margin can be availed of for sight. For most of these cases a peripheral iridectomy by Graefe's knife is better to be done, especially if the cornea be opaque. Desmarres did it for adhesion of the iris over its whole surface to the lens, where iridectomy often fails. A sharp hook is employed.

Iridavulsion denotes total removal of the iris, and tearing it from its periphery. It has been done for certain special cases, and I have found it to have a remarkable effect in two cases of hydrophthalmus. The whole iris may be drawn out with a hook or by forceps, or by both in succession. Of course care must be taken not to wound the lens, and the operation is done, not for restoring sight, but to improve the health of the eye by opening free communication between the anterior and vitreous chambers, and by releasing the lymph-channels and vessels at the margin of the anterior chamber.

Corelysis.—The separation of pupillary adhesions is done by Streatfield's hook or by a fine pair of toothless forceps. By the former method the adhesions are liable to be re-established and the lens injured. By the latter method the same trouble can occur, but less often. Consequently, and because the operation has not been fully approved by experience, it need not be described in detail. Streatfield's hook is to be slipped underneath the pupillary edge, close to an adhesion. By the forceps, risk of rupture of the capsule is avoided, and one simply needs a broad needle or narrow lance for an incision in the cornea near the adhesion, to permit entrance of the instrument.

Iridodesis, as a means of making a peripheral pupil for conical cornea, or for central opacity of the lens or cornea, has been already said to be objectionable, because it has led to cyclitis and sympathetic ophthalmia. It has been superseded by iridectomy, coupled with tattooing of the cornea at its middle, when needed for opacity ; and for conical cornea spherico-

Fig. 92.

cylindric lenses and Graefe's or Bowman's operation are to be substituted (see page 194). The proceeding consists of making a small wound at the limbus, through it to pass a blunt hook and catch the pupillary edge of the iris or its middle, to draw this into the wound, and leave it there to grow fast. The pupil is displaced, and may be made quite small. On behalf of Mr. Critchett it may be said that he punctures the cornea in its transparent part, while others have gone through the sclera, and he seizes the iris in its breadth, not by the pupillary edge.

A Tyrrell's blunt hook (see Fig. 92) is sometimes to be preferred to forceps in drawing out the iris for iridectomy, especially when a small pupil is desired. The amount withdrawn and excised of course limits the size of the pupil.

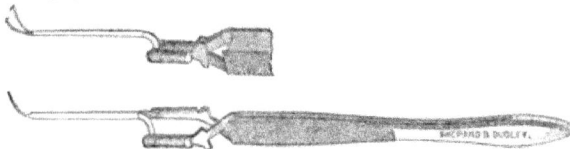

Matthew's Iris Forceps.

CHAPTER XI.

THE CRYSTALLINE LENS.

Anatomy.—The **crystalline** lens **is a** biconvex body whose equatorial diameter is from 8.1 mm. **to** 10.3 mm., and its axial thickness is from 3.6 mm. to 4.7 mm. Its edge is not in contact with the ciliary processes, while its front surface touches the iris over a considerable area. It is enclosed in a capsule, of which the anterior half has an endothelial lining on its posterior surface, while the posterior half is thinner, and has no cells. It is **very** elastic, and if divided the cut gapes. The lens is made up of long fibres or cells of a prismatic form, with serrated edges, which bend **upon** themselves at its margin, and their extremities terminate on certain **lines** arranged in radii in a peculiar manner. The result of the arrangement is to divide the **mass** into sectors, whose planes, seen from the front, **have** a star-like form, **and, seen** from behind, the limbs of the **star** are intermediate in position **with those in front.** The **lens is** also divisible into **concentric** lamellæ. **The fibres** have nuclei **near the** margin; they are

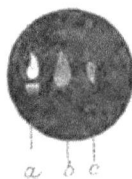

in **reality very** elongated **cells. The** middle portion is denser than **the exterior layers, and is** called the nucleus. The lens in youth is soft **and** easily changes form, its front becoming more convex in efforts of accommodation. Gradually, with years, **it** becomes harder and more flat and less transparent, assuming a smoky **or** amber tinge. This is manifest in focal illumination, and is a chief cause **of the** dull **sight** of old age, and **the** less black pupil. The lens

Fig. 93.—*a*, image from the cornea; *b*, image from front of lens; *c*, image from back of lens.

fits into a depression in front **of the** vitreous, known as the hyaline fossa, to which the posterior capsule closely adheres, and is supported by the suspensory ligament or zonula of Zinn. The space between the front and back layers of the suspensory ligament, formerly called the canal of Petit, is now asserted not to deserve the name canal, because intersected by numerous very fine and interlacing fibres (Gerlach). In its normal state an image can be reflected from the surface of the lens, if the pupil be of medium size, and a single candle in a dark room be employed as the object. The front image is upright and diffused; the rear image is small, sharp, and inverted. These images afford the means of measuring the curves of the lens during life (see Fig. 93).

In old **age the** lens sometimes becomes marked by short linear marginal opacities, which are known by the name **of** arcus senilis lentis. They are only visible with dilated pupil.

DISLOCATION.

The lens is liable to dislocation both from violence and from disease, and as a congenital condition. It may be tilted into an oblique position, and may swing back and forth, as on a hinge—the zonula being only partly torn ; or it can be pushed slightly out of place in a vertical plane, or in any direction. It may fall backward entirely out of the pupil into the vitreous, and it may be lodged in the anterior chamber. Sometimes, in old choroiditis, it passes forward and returns again backward through the pupil. In all these displacements the lens may be either transparent or opaque. Under severe violence it may be pushed partially or wholly through a rupture in the sclera, and either be entirely extruded from the globe or lodged under the conjunctiva. Displacement is most frequently caused by disease. The vitreous humor has become diffluent in its anterior part, and perhaps in all its substance. The suspensory ligament has become stretched and atrophied, and gives way in whole or in part. Extreme myopia is a condition which favors this occurrence, and it may present all the varieties of luxation. If the lens is pushed out of position so that a part of the pupil is uncovered, a person has monocular diplopia, or at least two different qualities of refraction in the same eye. A myope in this condition will sometimes seek a dim light, that the expanded pupil may give him the benefit of the part not occupied by the lens. Sometimes a patient uses both pupillary regions.

An eye not myopic is made greatly hyperopic by displacement of the lens, and also loses its power of accommodation. The iris sinks, the anterior chamber becomes deep, the pupil is generally small. Often the iris seems pulled back by some fibres, yet adherent to it. Disease of the choroid, in patches of atrophy, and diffuse opacity of the vitreous, sometimes its entire obscuration by floating bodies, is to be seen. Chronic irido-cyclitis is sometimes the beginning of the process, but more generally it starts from the ciliary body, and from it advances to the iris. Diagnosis of luxation of the lens is sometimes very obvious, and again close attention is required. In a partial luxation, if the rim come to the pupil, it will be recognized by the dark border which it always presents, because of the total reflection of light. Again, if part of the pupil be unoccupied by the lens, the ophthalmoscope will give a double view of the fundus ; one may even see the optic nerve in two places at once—one image seen through the lens and the other seen beyond the lens. This, of course, is only possible by the indirect method. The difference in refraction is at once recognized, according to the part of the pupil which is utilized. To determine the entire absence of the lens sometimes requires skilful observation. We may assure ourselves of the fact by looking for the images reflected from it in its normal state, this being what is called the catoptric test, which was once of value in diagnosis of cataract. Purkinje observed the images, and Sanson utilized their diagnosis. The figure, page 230, shows what they are. The flame to the left is from the cornea, and both the other two are from the lens ; the larger is from the anterior, and the smaller from the posterior surface. The candle is held to one side of the eye, in a dark room, and moved about. The brightest image is from the cornea, and upright ; the smallest is from the lens, and inverted. If all or two of these images are visible, the lens is *in situ*.

Treatment of luxations.—When in the anterior chamber, the lens should be extracted by a peripheral cut. The wound can be made by a very narrow Graefe's knife, and, if necessary, can be completed by scissors. I have in some cases been obliged to transfix the lens by a needle passed into it from behind the iris, to prevent its slipping back through the pupil. The capsule should not be ruptured. A small space will usually remain, through which the knife may pass without piercing the lens ; this will be above, if the lens is left to itself; while, if it be impaled on a needle, it may be lifted up and a cut made below, as is the most desirable. There is much liability to loss of vitreous. When luxated under the conjunctiva, removal is very simple.

Partial luxation should not be meddled with so long as the lens remains clear. The impaired and astigmatic sight may be aided by a cylindric glass. Total dislocation into the vitreous had better be left untouched. It is the common event of partial dislocations that though the **lens for** some time may remain transparent, it at length becomes opaque, and the loss of sight then demands an operation, provided it do not lie in the depth of the vitreous. It is also frequent that cyclo-choroiditis, hyalitis, or detached retina should ensue after traumatic luxations, and these grave complications must be met as the exigencies demand. Besides the surgical treatment, it will be needful to employ medical treatment, to modify the lesions which the deep structures are apt to undergo; but, besides a hygienic regimen, there is little that can be useful. If, however, the **lens be** partially displaced, a cylindric glass will often aid the astigmatism thus occasioned, provided the eye be in a sufficiently healthy state. For complete displacement, convex glasses, as after cataract operations, will be used. In congenital cases the eye is generally otherwise imperfect; it may be microphthalmic, or have a partial or complete cataract, or be myopic. Usually, if the media are clear, there is decided amblyopia. It is always doubtful how much help can be afforded. Usually both eyes **are** affected, and, sometimes several children in the same family have this condition. Sometimes the subjects are albinos.

CATARACT.

This name is applied to opacity of the lens, or of its capsule. The opacity of the capsule consists of proliferation of its endothelium, while its structureless membrane remains clear. In certain cases of anterior polar cataract, the result of perforation of the cornea, the capsule has a dense opacity, and the same may be seen in some cases of posterior polar cataract. In advanced cases of lenticular cataract, there may be specks of deposit on the capsule by cell-proliferation; or, if the lens have undergone great degeneration, its detritus may be precipitated on the capsule. In former days, stress was laid upon the distinction between capsular and lenticular and capsulo-lenticular cataract; but this is a matter of small consequence. The value of recognizing deposits on the capsule is to give an idea of the stage to which an opaque lens has arrived, and of possible lesions of the deeper tissues of the globe. Capsular, or as it is often called, membranous cataract, will be considered under the head of secondary cataract, it being generally the sequel of removal of the lens.

Lenticular cataract is to be classified in a variety of ways, namely, according to the extent to which the lens is invaded, into partial and total; according to its cause, into traumatic and spontaneous; according to its density, into soft and hard and fluid. We also have congenital and acquired cataract; we have it unattended by any other disease of the eye, so far as we can discover, viz., simple; and we have it as the result of many internal ocular diseases, viz., complicated cataract. We also have cases in which it depends on constitutional disease. The proximate *cause* of cataract in traumatic cases is the degeneration of the lens-fibres through imbibition of aqueous humor, which, when the capsule is unbroken, cannot in undue quantity pass through its epithelium. In congenital cataract it is assumed to be an arrest of development. In senile cataract the fibres have undergone sclerosis and molecular changes, and sometimes fatty degeneration, cholesterine being frequent. There is also a cleavage of the laminæ, especially at the margin, and an accumulation of fluid between the fibres. The pathology of cataract is not yet fully known, and there is manifestly great variety in different forms. It will not be proper to say much on this subject. In diabetes mellitus the density of the eye-fluids is regarded as the agent of the change, and this is rendered probable by experiments on frogs in which cataract was produced by injections of sugar and salt solutions into the blood, and these substances were subsequently found in the aqueous. So-called inflammatory cataract, which occurs in iritis, choroiditis, glaucoma, etc., is the result of perverted nutrition. Why it should then take place is easily comprehended, when it is remembered that the lens is very far removed from the circulation of the blood, and that on the health of neighboring structures it depends for its supply of nutriment.

For practical purposes we consider cataract in the following phases: A. As to the extent and locality of opacity, we have : 1, anterior polar, and 2, posterior polar ; 3, laminated ; 4, nuclear ; 5, cortical. There are also endless varieties in the appearance of a cataract, so far as the opacities may be in striæ, in molecules, in meshes, diffused, concentrated, near the surface, or in its centre. B. Having regard to consistence, we have : 1, soft ; 2, milky ; 3, mixed ; 4, hard. C. As to the stage, we have : 1, immature, or incipient, or unripe ; 2, mature, or ripe ; 3, over-ripe, or degenerated, or Morgagnian. D. We have : 1, stationary, and 2, progressive. E. 1, simple, and 2, complicated. F. Traumatic and idiopathic. No age is exempt, but the spontaneous and senile are most frequent.

Symptoms.—They are subjective and objective. The subjective consist in slowly developed dimness of sight, which induces the person to hold small objects very close, and if glasses be worn, they will be thought to be at fault. Distant sight is also reduced, and there is fogginess in the air and over all objects. The eyes feel uncomfortable, there is uneasiness or irritation of the lids ; there may be phosphenes. If, as happens sometimes in the incipient stage, the lens swell, the symptoms of uneasiness, and of muscæ and of phosphenes, may be quite annoying. It is often the case that the vitreous is hazy, and this causes the foggy look of the air. Frequently an atrophic choroiditis is setting in, and slight cataractous striæ, with general haziness of the vitreous, are its accompaniments.

Sometimes myopia is caused by the swelling lens, and a concave glass will greatly aid the sight. It is a common circumstance that sight improves to a noticeable degree when the first few months of cataract have been passed, and remains better for quite a period. This is because the vitreous grows clearer, and the hyperæmia of the choroid is not so great. Moreover, the lens begins to shrink, and this is quieting to the eye. The degree of dimness of sight, which shall be the effect of a given amount of lens opacity, can never be predicted. One is sometimes astonished to find how much sight remains when the lens is densely covered with striæ. Sometimes astigmatism, both of the regular and of the irregular type, is a notable feature. The pupil is apt to be small, and its behavior is important to be observed. If it do not dilate well by mydriatics, or dilate slowly, the indication is bad, while a quickly acting pupil, which may be fully dilated by atropia, is satisfactory. In some cases the tension of the eye is plus, and this would be likely to go with the sluggish pupil. There should be no limitation of the field of view, and the light-perception be quick. Patients with cataract are annoyed by strong light, and they instinctively put their back to the window when they wish to get the best view. Sometimes they find blue or smoked glasses helpful. Objective symptoms will be included under the next head, viz.:

Diagnosis.—This includes the recognition (A) of the presence of cataract and its extent and quality, and whether it be ripe or unripe, or overripe; also (B) whether it be simple or complicated; also (C) what is the state of light-perception and of the visual field.

A. With *dense opacity* recognition is easy, because the varying size of the pupil will render its visible extent larger or smaller, and by looking sidewise at the eye the opacity is seen to lie behind the pupil. For *incipient* cataract the visible appearance may be wholly undecisive, although often some opaque parts may be seen projecting into the pupil. If there be doubt, a weak solution of atropia, gr. ss. ad. ʒ j., must be instilled. Then oblique or focal illumination may clear up the case. But a possible and frequent error is found in the smoky hue which belongs to the lens of old age, and which may be so dense as to be pronounced cataract. True, this is a loss of transparency, but not in the sense of cataract. To settle all doubt, use the ophthalmoscopic mirror. Let the light be feeble. Examine from twelve inches' distance, and from various points make the light play across the pupil by slight movements of the mirror, and cause the patient to look in all directions successively. Any real opacity will be certain to be revealed. It will be remarked that, as the patient's eye changes direction, striæ which were visible at certain spots have disappeared, and again they come to view as the first position is resumed. Or, as the observer changes his point of view, a given sector of the lens will seem to be now opaque and again to be clear. This proves that cataract consists largely in irregular refraction of light, as well as in its obstruction by the lens-fibres. Now put a three-inch convex lens (13 D.) behind the mirror, and inspect the lens at short range from various points of view, as before. It will be hardly possible to fail to detect opacities. Their relation to the iris and to the pupil fix their location.

We must call attention to certain special types of cataract. In the diagnosis of posterior polar cataract, a special difficulty exists. When there is a small spot on the back of the lens at its centre, as occurs in choroiditis, in hyalitis, and in extreme myopia, particular attention should be given to the behavior of the opacity during the movements of the globe, whether it seems stationary as the cornea moves or performs a somewhat extensive excursion. The centre of motion of the globe is about four millimetres behind the posterior surface of the lens; hence, an opacity on this surface and axial will remain almost motionless as the eye turns, while the farther in front of this point opacities may be, the greater will be their extent of motion. Of course the pupillary edge will have a greater range of motion than any deep-lying lens opacities, and by this circumstance we largely form our judgment. This is called observing the parallax of opacities. Again, one may find a clearly defined opacity in the axis of the lens, with possibly a few knobs or rays in its edge, and oblique inspection and comparison of its parallax with that of the pupil will locate it in the middle substance of the lens. This will be a laminated or zonular cataract, and it will be understood by the figure, which shows several varieties (see Fig. 94). It is congenital, and stationary at least for a long time. Some cases ultimately develop complete opacification. This is one of the most common varieties of strictly congenital cataract. It often fails to be noticed by the naked eye, and parents and patients are often surprised to be told that this condition exists. A gray blur in the pupil is to be observed by close attention. These patients generally suppose themselves only near-sighted, and they hold objects close, and their distant vision is bad. Often there may be some other fault in the development of the eye, and possibilities of improvement are to be judged by the activity of the pupil and by the quickness of light-perception, as well as by the actual visual acuity. I have seen such a subject to be also truly myopic. The anterior polar cata-

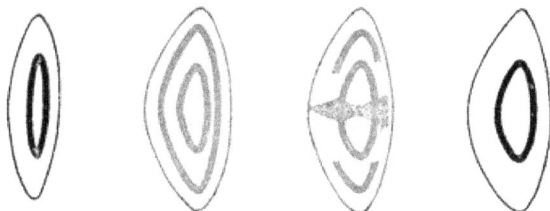

Fig. 94.

ract resulting from perforation of the cornea is usually only a minute white dot in the centre of the pupil. It is not so harmful to sight as is the faint corneal scar often to be discovered in front of it. Sometimes a thread runs from the cornea to the lens-spot. If the actual mass in width bear considerable ratio to the diameter of the pupil, its effect on sight will be serious. The farther in front of the nodal point an opacity may lie, the more it obstructs the entering rays. Sometimes a little cone rises above the capsule, and an inverted cone may penetrate into the lens. These opacities are stationary. Usually, nothing is done for them, except rarely a very small iridectomy.

As to the *quality* of cataract, we are to regard its consistency as well as its maturity. A hard cataract is known by its yellowish or amber or smoky hue seen by focal illumination. The nucleus may have this amber tint and the overlying layers be whitish, and this indicates the common condition of soft cortex and hard nucleus. When broad, radiating striæ are seen, the lens is likely to be soft; when the striæ are few and narrow, it is probably hard. A white, bluish, milky mass, especially if large and pressing the iris forward, signifies a soft and pulpy cataract. Sometimes the milky-looking mass is really fluid, and the points to be observed are the want of striæ and of all appearance of texture, while nothing but particles will be detectable, and these diffused uniformly through the whole. A hypermature cataract is one which has lasted for years. Its outer layers have degenerated into a semi-fluid granular substance consisting of fat drops or cholesterine and detritus. Sometimes this liquefaction involves half the bulk or even more. Its nucleus, which is yellow, small, and hard, falls downward and sometimes changes position as the head is inclined to one side or bent forward for five minutes. Usually one can see the rim of the nucleus lying across the pupil. The surface will show no striæ or organized form, but a pasty or creamy or gruel-like look.

Special attention must be given to the *size* of the cataract, as has already been mentioned, by looking to the depth of the anterior chamber and the plane of the iris, whether it be flat or convex, and by noting how much space exists at the margin of the anterior chamber. Traumatic cataracts in their beginning are large and often irregular. Portions often push out in front of the iris or press upon it, and may break off and lie floating in the aqueous humor. They are always soft. They are apt to be bluish or like watered milk in color, and they exhibit no traces of the proper lens-structure in either striæ or marks of regular arrangement. It must be noted whether the cornea have its usual diameter, viz., about twelve millimetres, whether it have normal sensibility to the touch of a hair, whether the arcus senilis be very great or complete. The fact of an arcus is not serious in its **effect** on the healing of a wound, but it has a bearing often upon the opinion as to the general health. A very wide arcus may obscure the place of entrance and emergence of the knife to an uncomfortable degree. In immature cataract we find portions of lens remain translucent enough for light to be reflected through them from the fundus. A *mature* cataract—that is, one in which the whole lens has opacified, is known, when no light is reflected through it, and when the pupillary edge is seen to lie immediately upon the opaque lens-surface. If the pupillary edge cast a shadow on the cataract, or a space seem to exist under the pupil, into which one's eye can penetrate, the surface layers of the lens are not yet opaque; the cataract is immature. Sometimes a cataract is wholly ripe, and yet a patient will count fingers at one foot with good light. It will, in these cases, be noted that the lens is very dark, of a reddish or mahogany, or a deep yellow hue, that it has been very slow in development, that its opacity is quite uniform, that the whole is homogeneous, and the color comes out toward the surface. Again, it will be noted that its bulk is small as discovered by the depth of the anterior chamber and the flatness of the iris. Such cases are most favorable for operation. An unripe cataract is more bulky than when mature. A cataract formed rapidly is usually large; one of slow formation has been drained of its watery contents and is reduced in size. A large cataract presses forward the iris, and the pupil is not quick to react.

B. Cataract is *simple* when no other lesion of the eye can be discovered, although it is true that an otherwise perfectly healthy eye would not have cataract. *Complicated* cataract is one in connection with which we have other recognizable diseases of the eye. Abnormities in the cornea will readily be seen. So, too, adhesions of the pupil will not escape notice. Discoloration of the iris, atrophy of its structures, softness of the globe, tenderness about the ciliary region, tell of cyclitis or choroiditis. A very yellow or chalky white lens, or one in which cholesterine is seen, or with many spots of distinct capsular opacity, signifies more or less serious degree of deep-seated trouble. A shrunken or deformed lens, or one reduced to a wafer, indicates extensive choroidal and vitreous disease. If the iris and lens flutter, this is a warning of fluidity of the anterior part of the vitreous, or loosening of the suspensory ligament. An eye with symptoms of absolute glaucoma will often have cataract, and, without stating the symptoms in extenso, attention is called to the need of testing the tension of the globe. If extreme, a most unfavorable condition exists; if moderately increased, the case is to be regarded with suspicion, even though it be susceptible of operation.

C. Special attention must always be given to the degree of *light-perception*, because it is the most important factor in the case. It may be premised that qualitative perception of light, that is, ability to see objects or to count the fingers—not his own, but of another—precludes the idea of operating. The patient should wait, unless it should be seen that the lens is of the deep, smoky variety mentioned on page 234, in which case, the patient whose sight has for a long time been much reduced may continue to count fingers within six to twelve inches, for years, and should not be refused an operation.

On the other hand, in simple cataract a patient should recognize a lighted candle in a dark room at more than forty feet with perfect ease. A more ready way is to use the ophthalmoscopic mirror with feeble light, and hold it two to six feet away to make its illumination extremely faint; this the patient should promptly recognize, and also be able to indicate the direction from which the light comes; this tests what is called the projection of the retina, as well as its perceptive power. Any abatement of this quick response to light, and inability to recognize the situation from which it comes, awakens suspicion of internal lesion of the eye which clouds the prognosis. Marked limitations of the field or of projecting power indicate detachment of the retina, or atrophy of the choroid, or disease of the retina or of the optic nerve, or turbid vitreous, etc. When these facts are made known, one may feel justified or bound to operate, but the probability of success and the gain in vision are put at a level which ought to be appreciated both by the surgeon and the patient. It is sometimes desired to have a cataract removed, when in only one eye, to get rid of the blemish, without regard to improvement of sight. This is appropriate, if not contraindicated by extreme tension or tendency to hemorrhage or presence of a foreign body, or other serious intra-ocular lesion, and often succeeds. I have done it with success, knowing that there was partial detachment of the retina, and to the advantage of sight.

For good prognosis, one ought to have an active pupil, as well as a mature cataract and the anterior chamber of normal depth. A quiet temper and a hopeful and obedient disposition are scarcely less requisite. As to the general prognosis of results of operations, this cannot be stated specifically, unless the peculiar quality of the case in all its bearings be borne in mind. We may distinguish between degree of sight gained and the successful removal of the lens with a clear pupil, but without sight. The former, if satisfactory, is functional and also a complete success; the latter may be merely a surgical success, and of no use to the patient except in removing a blemish. For congenital cataracts it is rare to have surgically bad results, but the gain of sight is uncertain, because so many cases are amblyopic. For complicated cataracts the same remark applies as to sight, while the surgical dangers are greater. For persons with constitutional diseases, or in feeble health, the outlook is always grave. For traumatic cataract, much depends on the nature of the injury and the possible presence of a foreign body.

But, for the cases of hard and senile cataract, very great pains have been taken to construct statistical tables of results. To get at fair conclusions, even with the limitation to cases considered favorable, strict logic requires that we have many elements for calculating the results which the tables do not afford. Most tables are compiled to show the advantages of a particular method of operating. Of course, each method must be judged by itself. Each operator ought also to be judged by himself, for he is the most important factor in the problem. But each operator can never use his best skill on all occasions, nor will his surroundings always be the same. A multitude of causes, needless to be mentioned, come in to modify the mere technique of the proceeding in the hands of the same surgeon. The innumerable circumstances which combine to make patients dissimilar and eyes dissimilar, are all to be weighed. Therefore it follows that not less than a thousand cases, by the same operator and by the same method, are to be considered a sufficient basis for drawing trustworthy conclusions. In estimating results of vision, the rule has been adopted that cases which gain $V = \frac{1}{5}$ and better are counted as full successes; qualitative vision less than this is called partial success, and quantitative vision (only light-perception) and worse results are called failures. The requirements, it is seen, are lax enough, but as strict as we can at present work up to. The result of the most generally practised operation, viz., that first known as Græfe's, is about six per cent. of failure among **the best operators.** This is an inference drawn from a compilation which **I have made of 11,000** cases. See "Trans. Am. Oph. Soc.," 1880.

Duration and progress.—Soft cataracts progress more rapidly than hard. Traumatic cataracts develop with speed proportionate to the degree to which the capsule is opened. Very small openings in **the** capsule have been known to heal, and the lens to recover transparency **at** the affected spot; but the contrary is the rule in the large majority of cases. Congenital cataracts do not always appear until some months after birth, or at least do not always become complete. For this reason the class of juvenile cataracts has been set apart by recent authors. Hard or senile cataract is an evidence of involution of tissues, and is apt to be hereditary. Its rate of development is excessively variable. The better the health, the less rapid its progress. Cataract accompanying myopia is of exceedingly slow growth. I have watched one case eight years, with very moderate abatement in sight. If cataract develop quickly, it is accompanied by symptoms of retinal irritation due to its swelling. The iris may be seen to be pressed forward and the tension somewhat increased, as already remarked. Another feature of senile cataract is that the index of refraction of the lens becomes higher; the effect of this may be to bring about a degree of myopia. In case the pupil, as commonly is the fact, is quite small, and the striæ few and well defined, vision for near objects becomes better than for years previous. At least, the use of convex glasses in reading becomes superfluous, and the individual gains what is called his "second sight." I have one patient who, at sixty-five years of age, has complete cataract in one eye, in the other partial cataract with a myopia of one-twelfth (3 D), which has been acquired within a few years, and who is able with a concave glass to see distant objects reasonably well. No period can be fixed during which the cataract shall complete its development and become mature. From a few months to twenty years are limits recognized.

Spontaneous cure does not take place, except by absorption of the lens, as in traumatic cases, and in some other soft cataracts, or by spontaneous luxation, which can happen in hard cataracts and in cataracts of myopic eyes when the vitreous is fluid. Sometimes a blow does this suddenly, to the great joy and astonishment of the patient.

Treatment.—While we have no means for bringing about absorption of opacities in the lens, and only a very few reliable cases of spontaneous disappearance of these opacities are reported ("Bericht über Augenklinik Wien," 1867, where some such cures in traumatic cases are recorded), it remains true that in the early stages of the disease something can be done to mitigate accessory symptoms. The principal matter is strict attention to general health, removing all causes of indigestion or of feeble nutrition, adopting a simple and nourishing diet, keeping the bowels regular, insuring sufficient sleep, perhaps giving mild saline waters, requiring moderate daily exercise in the open air when the weather permits, ordering sometimes change of scene, and securing, so far as possible, cheerful surroundings. I have, when health seemed good, and when I saw the vitreous to be hazy, given iodide of potassium, gr. v. ter in die, for several months. Whether *post hoc vel propter hoc,* I have found the sight better. Sal rochelle might perhaps have been as good. So, too, mild tonics and bitters, strychnia (iron often disagrees with old people), phosphoric acid, acid phosphates, and such remedies, have fit use. These measures may be aided by soothing lotions to the eye to allay conjunctival irritation, viz., borax and camphor-water, etc. A very weak solution of atropia, gr. $\frac{1}{16}$ ad. $\overline{3}$ j., will be / of use to some cases, but strong solutions are unpleasant, because of the great glare of light. Moreover, encouraging words as to the future mean much to a blind person.

Outward applications of a stimulating kind, and digital massage, have had popular repute, but no real success. Moreover, galvanism and every variety of current has been used in former and in later days. Some asserted cures have been published within three years. They have been exploited in newspapers, and had large currency. They have no real foundation in fact, and I have had experience of two of these cases.

It remains to take up the *surgical* treatment of cataract. But some preliminary matters are to be considered. Shall the operation be done upon one eye while the other is perfectly good or has useful vision? For soft, including traumatic cataract, the answer would be yes, because it involves little risk; for hard cataract there is a difference of judgment, but my own practice is in favor of it. I admit that some have been followed by squint, and have complained of confusion of sight, but the much larger number have greatly enjoyed the benefit conferred by the restored eye. The gain is in the enlargement of the field of vision, in the stronger mental impression, because of a greater impulse of light-stimulus to the central ganglia, and because, notwithstanding no correcting-glass was worn by the operated eye, a degree of stereoscopic vision was secured which the patients many times found of great advantage. One man about forty years of age, with monocular cataract and the other eye good, $V=\frac{2}{9}$, was so much distressed by losing binocular sight, that he demanded an operation, and acquired in this eye, when corrected, $V=\frac{2}{9}$. Without wearing a glass he recovered his former ability to locate objects, could grasp correctly, go up and down stairs with assurance, and was loud in praise of the benefit he received.

Should both eyes be operated on at once? To this my experience leads me to say no. In one case of a German woman for whom I extracted both cataracts by a flap (Beers' method), I found that intra-ocular hemorrhage happened in both eyes during the following night, and the clot was pressing out through the wounds. In other cases I have seldom found both eyes do equally well; and when the possibilities of loss of one by Graefe's method, and the possible entailment of sympathetic ophthalmia in the other, are considered, the argument to me is imperious. In cases where the first eye does well and the second is fit for operation, the latter may, in urgent cases, be dealt with in eight to ten days after the first. Usually, several months are allowed to pass.

The variety of operations performed for cataract-extraction is too numerous to catalogue. For many years flap-extraction and reclination were the only alternatives. The latter has been consigned to oblivion, because it leaves in the eye a foreign body, which will, in the end, work destructive mischief. I have been twice induced to do it in cases of hard cataract in the very old, who were too feeble to expect recovery from flap-extraction, which was then the only operation known, and in both cases the result was failure.

Flap-extraction was done in the cornea and by a large wound, and with a pupil undisturbed, except by the stretching of the extruded lens. When it succeeded its results were brilliant, but almost all the men who learned it and gloried in it have given it up for the modern method, first introduced by Graefe. The one surgeon who performs it regularly is Hasner, of Prague. It requires more than ordinary skill, and is specially exposed to accidents in its performance, and to suppuration in the healing. We retain the knife with which it was performed, and which is known as Beers' knife.

Passing by many operations which my judgment does not approve, the following are to be considered : 1st, we operate to obtain absorption of the cataract ; 2d, we operate by the extraction of cataract, and of this proceeding there are several varieties, suited to different conditions.

For *absorption* of cataract we are confined to lenticular cataracts, because the capsule never is absorbed ; and also to those which are soft, and occur in persons whose youth shall insure sufficient absorptive activity. The operation is done almost exclusively on those who are under fifteen years of age. Sometimes older persons are thus treated, but absorption is very slow, and the lens-fragments are more likely to cause irritation than in younger eyes ; especially is this remark true of those beyond thirty. The object of the operation is to open the capsule, not to break up the lens. A needle which is double-edged and sharp, and cuts for at least two millimetres from the point, is the proper instrument. A sickle-shaped needle will also do the work well. An anæsthetic is not usually required. The full effect of atropine is to be obtained. For congenital cataract, the operation may be done as early as the age of three months. Care must be taken to find out that the case is one suitable for absorption, *i.e.*, not a circumscribed laminated cataract. For a person under fifteen years of age and whose pupil dilates easily, simple discission will do well. If the pupil dilate imperfectly, and in persons over twenty, iridectomy should precede discission by eight weeks, as Graefe advised. If the lens is more or less fluid, a broad needle may be used to evacuate some of it. The lids are separated by a stop-speculum, and the globe steadied by one finger ; the needle pierces the cornea on its horizontal meridian, just outside of the dilated pupil, in a direction nearly perpendicular, and going to the opposite side of the pupil, cuts down into the capsule by lifting the handle and depressing the point of the instrument, using the cornea as a fulcrum. If this cut do not go down too deeply into the lens, a second cut at right angles with it will fully open the capsule. Care must be taken to avoid displacing the lens *en masse*. Sometimes, in congenital cases, the lens is so small that this necessarily happens ; but then it is not so likely to touch the ciliary processes injuriously. If the lens be reduced in bulk by partial absorption, a different mode may be used. Dilate the pupil by atropine, give anæsthetics if the patient be not very steady, put in a speculum, let one assistant hold the globe by fixation-forceps, and the operator taking two straight discission needles, one in each hand, enters one on either side of the pupil, aiming for the middle of the lens ; and when he has pierced it through, he inclines the handles of the instruments to each other, thus separating the cutting points, and opens a chasm in lens and capsule, which makes a black pupil, and sight is gained at once. This proceeding will often be the second after the (see Fig. 95) one previously described. In young children the eye will bear more disturbance of the lens than it will in youths. Atropine is to be dropped into the eye, sol. gr. iv. ad. $\bar{3}$ j.; a bandage over both eyes, and the patient is put to bed. If, within twelve hours, pain occur, look to see whether the lens is swelling inordinately. If this be so, the iris will be pushed forward, the lens-substance may be welling up through the pupil, and hyperæmia may appear around the cornea. For twelve hours use atropine every hour or two, and apply cold compresses. If this do not control the trouble, resort to paracentesis by a lance-knife,

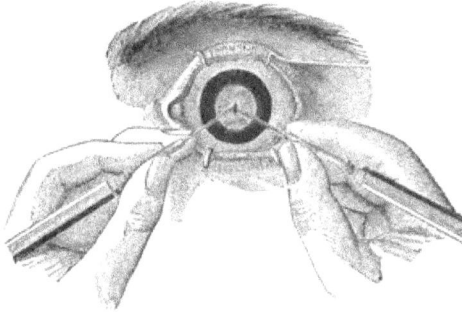

Fig. 96.

and attempt to get out some lens-matter as the aqueous gushes, holding the wound open with the tip of the lance. This proceeding is of the utmost value. It must be done without hesitation if there be much pain, if the aqueous be muddy, the iris discolored, flocculi of lens-matter floating free, and decided ciliary hyperæmia with slight chemosis. To do it an anæsthetic will be needed, while an assistant is not essential, although a good one will be convenient.

For cases which are not attended by any mishap, two to four months will easily procure absorption ; but there are many exceptions even with infants. In some the capsule is dense, and does not permit free use of the needle. In all cases it is best to err on the side of prudence. It is important to plan the incisions so as to leave a clear central opening in the capsule at last. Until absorption is complete, atropine should be used.

For certain cases of nuclear (zonular) congenital cataract, iridectomy has been advised in preference to solution of the cataract. Its theory is the supposed preference to be given to vision through the transparent margin of the lens, to the vision possible to a lensless eye. I grant this to be true in case the central opacity is very small, i.e., not greater than one-fourth the diameter of the lens ; but, if it equal one-third or one-half the lens in diameter, I decidedly favor solution. For further remarks, see below.

In the operation just described, it was necessary to introduce the proceeding of evacuating both the aqueous humor and the softened lens-substance, to control possible reaction occasioned by excessive swelling of the cataract. We now speak formally of that proceeding under the name of

SIMPLE LINEAR EXTRACTION.

This, in reality, is a form of paracentesis, and it presupposes a very soft lens—perhaps mushy is the word to designate it. The case may be spon-

16

taneous or traumatic, and may be in a youth or older person. It may also apply to what may be called milky cataract, which I have a few times seen in persons under twenty, in whom a soft cataract had degenerated into a fluid emulsion like a saccule of cream. The capsule needed only to be pricked, and all its contents gushed out. I have found these cases liable to be followed by violent reaction, because the contents of the sac appeared to be an irritant to the iris, and I have seen one eye destroyed. It is better to make a broad wound, to get the matter out of the globe as quickly as possible. A hypodermic syringe might be suggested for such cases, but it is very doubtful if the material could be conveyed out without coming in contact with the walls of the anterior chamber.

Simple linear extraction is applicable to any age, and has regard only to the consistency of the lens. But by persons of middle life its manipulations are not so well borne as by those under thirty, and it may in them be substituted by peripheral extraction (Graefe's method). The operation may be done either without or with iridectomy, and the iris operation may be coincident or precedent. If no iris be removed, the amount of lens-matter must be small and very soft. Under such conditions (using anæsthesia, a speculum, and fixation-forceps, which are to be applied at a point diametrically opposite the intended place of puncture), a narrow lance-knife (see Fig. 96) is thrust almost vertically through the temporal side of

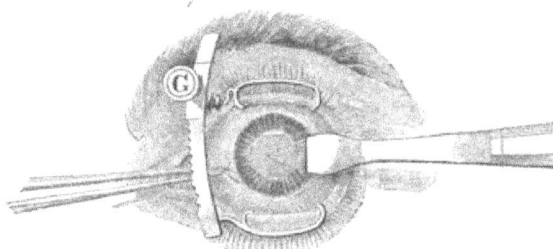

Fig. 96.

the cornea, just at the apparent edge of the fully dilated pupil. It is carried into the lens so as to open the capsule freely. The operator may increase the capsular wound by lateral, lever-like movements of the handle. Then withdraw the blade until its point just projects within the wound, and open it by pressure backward; the lens-matter comes out, and its exit may be aided by slight counterpressure with the fixation-forceps. Let the exit go on slowly to avoid prolapse of vitreous, and keep up pressure very gently, until the pupil clears. At the first sight of vitreous, pull out the knife and close the eye.

Another way is to withdraw the knife quickly after it has made the wound in the cornea, and to put in a sharp cystitome to open the capsule, then evacuate the lens by the point of a curette laid at the wound to press it open. The fixation-forceps may be taken off, and counter-pressure made by a rubber scoop, which shall be passed over the cornea, squeezing out and following up the lens-matter (see Fig. 97).

If this operation be done with iridectomy, the wound is made .½ mm. inside the limbus, the knife going in parallel to the base of the cornea, and being a lance of medium size. With it make a wound about five millimetres long, introduce forceps to draw out the iris, and cut it close to the cornea. Now with a sharp cystitome open the capsule as freely as possible, for this is a most essential step. The expulsion of the lens-matter may be done by the curette, aided by counter-pressure with the rubber spoon. The fixation-forceps may be taken off before expulsion, or, if the patient be very quiet, left loosely in place. There is danger of loss of vitreous, because the inner wound lies near the circumlental space. This is the danger to be guarded against. Sometimes lens-matter must be left be-

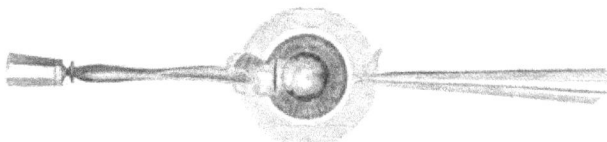

FIG. 97.

hind. It may not make any trouble. Atropia, gr. iv. ad. ℥ j., is put in, a bandage is put over both eyes, and the patient put to bed. The eye will not be examined at the first dressing, which will be after twenty-four hours, except in the most cursory way, the outside of the lids gently washed with boracic acid solution, and the bandage reapplied and changed once daily, the eyes kept covered for three or four days, and if all go well then use the double shade. Atropine used twice daily after the first twenty-four hours, and later, when the bandage is off, three times daily. Keep the patient in quietness for seven to twelve days.

Laminated cataract requires careful study of its extent and of its relations to the pupil. If the pupil will dilate wide enough to extend beyond the opacity, and afford sight, a weak solution of atropia may be used once in three days, and an operation may be postponed, if the subject be very young, to the age of four or five years. At this age the best proceeding is an iritomy by Wecker's scissors. A wound of four or five millimetres is made opposite the edge of the dilated pupil and one blade of the scissors slipped behind the iris, and the other in front, to cut two-thirds of its breadth with one slit. Of course care is needed not to rupture or bruise the capsule, nor do violence to the iris. This operation requires much delicacy and skill. If the central opacity be too large to leave rim enough in the lens to be useful, discission will be proper. Often the central opacity has a thick, white centre, and thinner gray surrounding rings. There are several varieties (see Fig. 94). Most cases require discission. An iridectomy may be done in some instances, but the rim of the lens is highly astigmatic, and the existing lens-opacity causes dispersion of light, so that, if the opacity be more than three millimetres in apparent diameter, discission and subsequent use of cataract glasses will give the best vision.

We come now to the consideration of the operation which is the capital achievement of ophthalmic surgery, viz., the

EXTRACTION OF HARD CATARACT.

The methods are as follows: flap-extraction in the cornea upward or downward; flap-extraction in the sclero-corneal junction, upward or downward; either of these may or may not be accompanied by iridectomy.

In this operation the cut traverses a circle of latitude, and forms a flap which does not easily withstand intra-ocular pressure, and causes various mishaps because of the gaping of the wound. Graefe conceived the thought that if a wound be made in or very near a meridian of the globe, this cut could not very easily be forced open, and would be a great safe-guard against bad coaptation, with resulting suppuration, prolapse of iris, etc. To form such a cut, he contrived the narrow knife which bears his name, although it is variously modified. He laid the wound deep in the sclera and beyond the visible margin of the cornea. He restricted it to a length of 4½''' Paris (about ten millimetres) and added a very peripheral iridectomy. In this position it is adequate to the expulsion of most lenses. After opening the capsule he used considerable pressure, by various in-struments, to expel the lens. By this proceeding he obtained a great gain upon the previous percentage of successful results. The operation com-mended itself by its greater ease of performance and the less irksome after-treatment, and by the quicker healing. But it was found that harm arose from the proximity of the cut to the ciliary processes, and a chronic cyclitis, lighted up by the wound, caused many so-called half results, which were practical failures ; and in the sixteen years since its introduc-tion, not a few cases of sympathetic inflammation of the other eye have been reported, and the unreported ones are probably more numerous. Moreover, it is sometimes very difficult to get the lens out through so small an aperture, because it may have very little soft cortex, or may be of unusual size. Cataract will vary from seven to ten millimetres in diameter. Hence, the wound was by many operators brought down nearer the limbus, and increased in length to eleven or twelve millimetres, if needful. Such a peripheral character is given to the wound that sufficient coaptation is maintained by the help of a pressure-bandage, which, too, was Graefe's in-vention. The iridectomy is retained. The method of opening the capsule is now attended to more carefully than by Graefe, and the expulsion of the lens is effected more readily. Instruments passed into the eye are dis-carded, except there be loss of vitreous. Very many minor modifications of the operation have been published, but these must be regarded as the "personal equation" of the operator. It may be done at the lower edge of the cornea or at the upper ; the latter is to be preferred.

The patient is to be prepared for the operation by securing a move-ment of the bowels and good sleep the night before. The best anodyne is bromide of sodium ℈ij. and chloral ℈j. unless it be known that morphia agrees well. Some operators put in eserine, gr. iv. ad. ℥ j., an hour before the operation, to keep the pupil well in the anterior chamber. One never should use atropia within a few hours of the operation, except in a feeble dose, say gr. ¼ ad. ℥ j., and this simply to learn the peculiarities of the cata-ract. The question of employing an anæsthetic depends on the habit of the operator and the wishes of the patient. If a patient be intractable or timid, it must be given ; if the operator distrust himself, he had better use it. If the patient is very easily nauseated, he should not take it. Subse-quent vomiting, while often immaterial, is sometimes disastrous. Ether is less pleasant than chloroform, and for the timid operator is to be preferred. For the timid patient I would choose chloroform. If ether be chosen, profound relaxation must be secured, and be kept up through the whole proceeding. If chloroform be chosen, it can be given to the procurement of the first stage of anæsthesia, annulling pain during the first two and painful steps of the operation, and then be abandoned, and the patient al-lowed to recover his senses. Ether cannot so well be managed to gain this

effect, because its early stage is more exciting than is that of chloroform.
I leave out of view any discussion of the applicability of one or the other
anæsthetic as a general surgical question. In Europe anæsthetics are
rarely used ; in this country the practice is variable. For myself, I avoid
them as much as possible, but frequently have to resort to them. A
patient who will squeeze the lids, or holds the eyes rigid, must have an
anæsthetic. I have given full doses of bromide of sodium and chloral,
within the last year, the night before, and also an hour before the operation, to great advantage.

Conditions unfavorable to an operation are : extreme depth of the eye
in the orbit, and also decided prominence of the eye. The former makes
the operation difficult, the latter is less favorable for healing. Chronic
catarrh of the lachrymal sac, pterygium, and all diseases of the lids, are
detrimental, because the secretion is apt to get into the wound and cause
suppuration. All conditions which hinder the patient's remaining quiet
in bed are bad, viz.: cough, asthma, bladder trouble, etc.

I prefer that the knife should be very narrow and sharp (see Fig. 98).
The best blades are ground hollow, like a razor, and their width is not more
than one and one-half to two millimetres. The point must be very sharp,
the back thick enough to be stiff. It is intended to make a wound
around nearly the upper third of the cornea, which shall be eleven to

Fig. 98.

twelve millimetres across its extremities, which shall be in the sclera one
millimetre from the limbus. The patient is on his back, the lids parted
by a spring-speculum, the globe seized by spring fixation-forceps applied
near the cornea, below and on the vertical meridian, and in the operator's
hand. The operator will stand behind for the patient's right eye, and be
in front for the patient's left eye. It is not worth while to suppose the
operator to be ambidextrous to the degree requisite for cataract-operating,
although the possession of this quality is of eminent value.

First.—The point of the knife is thrust into the sclera at a spot 3 mm.
below a line laid horizontally across the upper edge of the cornea, and is
pushed down to the middle of the pupil ; then it is brought up to aim at a
spot diametrically opposite to the place of its entrance. If counter-puncture
should not be exactly correct, the knife may be withdrawn and the effort
repeated. In this manœuvre the fixation-forceps are made to counteract the
tendency of the eye to rotate on its axis, which the advancing knife causes.
During transfixion the back of the knife has been pressed a little down, and
when the point appears on the nasal side, it is thrust rapidly around and
upward, so as to form as much of the inner end of the wound as possible by
the full thrust of the blade ; meanwhile the heel is not allowed to cut, and
no leakage of aqueous occurs here (see Fig. 99). When the blade has gone
to the full length possible, the heel is lifted up as the edge is drawn backward, and the cutting proceeds at the outer end. Finally, only the centre
remains to be cut, and here the edge may have to be turned a little forward,
to avoid going too deep in the sclera. The manœuvre thus described in
detail I have found to be a sufficient protection against falling of the iris
over the knife, because the flat of the blade comes upon it before it can
fall forward. If one fail in the endeavor thus to make the cut, this accident

may happen. Some conjunctiva may be lifted, but, unlike some operators, I do not desire it. It promotes bleeding, and is in the way.

Secondly.—We proceed to the *iridectomy;* but if bleeding take place into the chamber, be careful to coax it all out by wiping the wound with a soft bit of muslin, and at the same time press with the fixation-forceps. A light and dextrous hand is very needful in this performance. The assistant now takes the fixation-forceps, and this transfer is delicate business. The iris sometimes presents in the wound; if it should not,

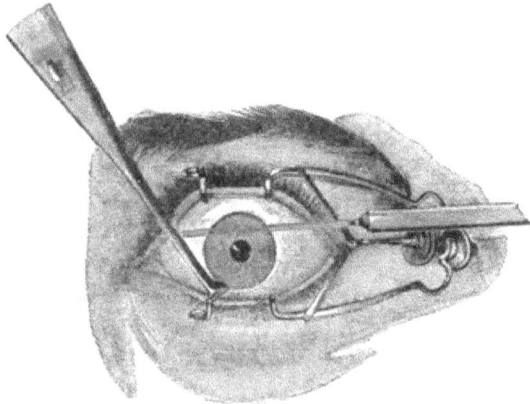

Fig. 99.—The operator is supposed to be behind and using his left hand. The knife is represented too narrow.

catch it near the pupil at the right-hand side of the wound; gently draw it out and cut with scissors pressed against the sclera. Make a second cut at the middle, a third at the distal end of the wound. The eye must be turned down, chiefly by the patient's effort. Much care is needed to prevent imprisonment of the iris in the angles; it is to be softly coaxed back by stroking the limbus with a rubber spoon. If the iris do not return to its place after a little pressure at the angles of the wound, use forceps to catch it, and excise it. The operator then resumes the fixation-forceps, and is vigilant to wipe off clots, and coax out by slight pressure what gathers in the chamber.

Third.— Cystotomy.—Take a cystitome, sickle-shaped, or with point projecting at an obtuse angle, as devised by Dr. Knapp (see Fig. 100); its shank may need to be bent, if the eye is deep. The point must be sharp

Fig. 100.

and able to make a clean cut vertically from below the lower pupillary edge to the top of the lens. To cut the capsule properly, the lens must be crowded forward by a little pressure of the fixation-forceps, and the patient must look as far down as possible; there is great advantage gained by his intelligent co-operation. If the lens have been displaced a little out of the hyaline fossa, it may be lifted up to place by the point of the cystitome before making the cut. Having succeeded so far without accident, the patient is exhorted to keep looking down while the next step is taken, viz., the expulsion of the lens, which now rests against the cornea.

Fourth step, *delivery.*—If the openings in the **cornea,** iris and capsule are all of suitable size, no trouble will be experienced, unless the lens have been dislocated, or unless vitreous have escaped. The fixation-forceps may often **be** laid aside, but, if still in situ, press gently backward with their **point so** as to tilt the upper edge of the lens into the wound. If **it come to** view, change the direction of the pressure so as to follow its movement **upward,** and lay **a curette** gently along **the posterior** lip of **the corneal wound. The lens keeps rising, and** as it gets midway into the wound place **the curette or rubber spoon on the** lower border of the cornea to crowd **up soft matter and squeeze it out in a compact mass,** following it up with **the** spoon **on the cornea, not coming higher than its** centre until the pupil seems clear. **Wet the** spoon, **and** be extremely gentle. It is now that the quality of the **hand** appears in knowing how to apportion pressure to **re**sistance, and **not to** cause loss of vitreous. When the lens is out, take out the speculum, close the lids, and apply a pad of cotton with moderate pressure.

Finally, the pupil is cleared of all fragments, and this may be done by means of the edge of the lower lid, which the operator pushes with his thumb over the cornea, or by a spoon wet in sol. acid. boracic. gently strok**ing the** cornea and making the wound gape. No instrument should be **put into** the eye if it is possible to avoid it. Wash out the conjunctiva **with sol. acid.** boracic., and let the patient try to count fingers before putting on the padded bandage. Do not put in atropine, because this will favor the entanglement **of** the iris in the wound. The eserine, **which some op**erators **use before** the operation, is for the purpose of keeping **the pupil well in the wound. I** do not approve of its use after the operation, **but a correct** situation of the pupil must be secured ; and if nothing else can be **done, the** angles of the protruding iris must be snipped off Next apply the **flannel** bandage, which shall be two inches wide and two and one-half yards **long.** Lay a square of muslin over each eye, and on it picked lint or **cotton** wadding. The padding must be carefully packed about the eyeballs. Let the mass be enough to make a broad and elastic wad, level with the forehead. The bandage must lie smoothly, evenly, and without painful pressure. The patient then goes to bed. He is not to eat solid food for twentyfour hours, but to take milk, soup, and nutritious food which does not require to be chewed. He is to take a position in bed **which** he can keep quietly for several hours, whether on the back **or on** the side; he must not rise out of bed for the **relief of bowels or of** bladder, if it can be avoided. The rule is that the minimum of disturbance, local and general, and the maximum of comfort and of healing power, are to be secured.

After-treatment.—Make sure that good sleep is obtained, and, as a rule, **order** a competent anodyne. Some smarting follows **the operation,** but in two or three hours the eye feels well. If there should be pain **after** six hours, give a hypodermic of morphia sulph., gr. one-fourth. If there be no decided discomfort within eighteen hours, do not disturb the patient at all ; **nor should the** bandage be taken off until after thirty **hours,** if **there be no** complaint. **Then** it may be removed, and a glance at **the secretion on the** muslin patch will tell how matters go on ; if it be scanty, it augurs **well.** Wash off the outside of the lids gently with warm sol. acid. boracic. four per cent., and renew the dressing of patch, cotton, and bandage. At the end of another twenty-four hours do the **same,** and the edges of the lids may be pulled asunder **to get out secretion. Do not put** in atropia until the third day, and then the solution will be gr. ij. ad. ℈ j., one to thirty, only two drops. From this time on, the bandage may be changed from once to twice daily. In favorable cases, the after-treatment is tame and unvaried, and the patient complains only of being in bed. It may be stated broadly that a normal operation, unattended by accidents, and in which the eye is not handled harshly, will be followed by normal and prompt healing. The general nutrition is of less importance than a perfect operation and gentle manipulation.

For three days the patient will remain in bed, and his diet will be simple, but not meagre. During this period the bowels may be left to themselves. The patient may then sit up a short time each day, and gradually be allowed more liberty. The bandage will remain on both eyes for a week or ten days, then· to be succeeded by a shade, and the room to be moderately dark. While wearing a bandage the room may have light enough to permit every attention to the patient. From twelve to twenty days is required, on the average, for after-treatment.

It is proper to speak of the accidents which may happen at the operation, and what is to be done in view of them. If the iris fall over the knife in a large fold, it is best to withdraw the blade and treat the case as a wound ; another attempt may be made in ten days or two weeks. If the piece caught be small, it may be simply cut away and the operation continued.

FIG. 101.

FIG. 102.

If prolapse of vitreous occur before the iridectomy, tie up the eye and treat the case as a wound. Another attempt must be deferred for one to three months. If vitreous prolapse, after or during iridectomy, make the attempt to extract the lens without capsulotomy, by a traction instrument, such as may seem best suited to the case. The loop set with teeth is very reliable, while sometimes a hook answers better (see Figs. 101 and 102). In any event the manipulation is delicate, and more vitreous will escape. For loss of vitreous after opening the capsule, the same method must be used, but will be easier of execution. No pressure is to be exerted on the globe by fixation-forceps.

If the lens come out unwillingly, the wounds are too small either in the cornea, which is to be enlarged by the iris-scissors, or in the capsule, which is to be cut again by the cystitome. If the lens be dislocated before the capsule is opened, it may sometimes be safe to try to get out lens and capsule together, because the suspensory ligament is to some extent torn, and must be atrophied. Nothing but gentle pressure is to be tried, and if this do not succeed, catch the anterior surface of the lens by a small, sharp hook, and with this or a proper cystitome cut the capsule, and deliver in the normal way. After all is over, beware of the patient suddenly nipping his eye or curiously trying whether he can see. He may squeeze out vitreous. The harm of this accident lies in the rude disturbance of the deep structures of the eye, in preventing proper removal of cortex, in causing entanglement of iris in the wound, and in the interposition into the wound of a substance which hinders close coaptation. It may not be followed by disaster, but it is always dangerous. The occlusion of the eye after this accident must be more careful and persistent, and undue reaction is probable.

In after-treatment, should undue reaction occur, it usually starts from the wound, because either its edges were bruised, or the iris, or a tip of capsule or lens-matter have been caught in it. If mild, it will abate under use of warm fomentations, begun so soon as the pain and the swelling about the wound denote its onset. It may extend and become iritis, or inflammation of the capsule, with proliferation of its cells, or it may go to the ciliary body and choroid, and then real mischief is brewing. The best remedies are atropine and hot fomentations; for pain, apply two or four leeches to the temple, and inject morphia. Often a patient feels better by being allowed to sit up some of the day. The case may go on for three or four weeks.

Iritis or capsulitis may occur without irritation of the wound by the third, or tenth, or twentieth day, and is generally due to some imprudent use or exposure, or blow, or pressure. For exudation in the anterior chamber, I have many times practised its removal by iris-forceps, and if the corneal edges have not been decidedly involved in suppuration, I have found the proceeding helpful, and thus saved some eyes; but often such cases go to a bad end. Cyclitis is liable to be the mode of destruction, and softening of the globe. Reaction in the wound, with purulent infiltration of the edge of the cornea and in the pupil, is the great mischance after this operation. The treatment is hot water for many hours daily, and atropine, and keeping up the patient's condition.

It remains to speak of some other ways of operating on hard cataract. We have two modes, in which the wound is laid in the cornea, viz., the operation of Liebreich and of Lebrun. In both these the section is in the cornea, either below or above the pupil, at the junction of the middle with the upper or lower third. The cut begins in the sclera, one or two millimetres behind the cornea, and goes in a great circle to the same point opposite (see Fig. 103). If the iris prolapse, it must be cut off; the capsule is opened by a transverse cut, and the lens slipped out as in the Graefe method. For the cut downward (Liebreich's) a speculum may be omitted, and also fixation-forceps sometimes, but both make the method easier. Before cystotomy the speculum may be taken out, and the operator's fingers will control the lids, and evacuate

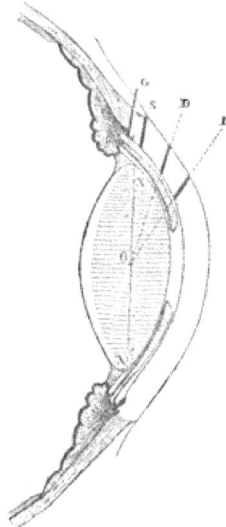

Fig. 103.—O, centre of lens; X, X', equatorial plane of lens; G, place at which Graefe originally laid the incision; S, place at which most operators now make the incision; D, place of section in the discarded flap extraction; L, place of section in the Belgian, or Lebrun's, operation; if made at the lower part of the cornea in a corresponding place, it becomes Liebreich's section.

the lens, aided by a spoon pressing on the lower edge of the cornea. Prolapse of iris is very liable after the lens is out, and if not well replaced it should be excised.

Lebrun's, or the Belgian method, is similar and a little more difficult. To both these operations the objection lies of adhesions of the iris to the wound, especially at its angles, and irregularity in the cornea from pouting of the cicatrix. These methods are good substitutes for Graefe's operation when the latter may be unusually difficult because of great depth of the eye, or where the latter has been tried with bad result. The figure shows where the section of the cornea is made in the various operations.

The objection to the two last is that if, immediately after the operation, the iris seem to be quite out of the wound, the stump of it is almost certain, after a time, to become adherent to the cornea, and make a peripheral synechia, which is a perpetual threat. Astigmatism, after these operations, is apt to be great. They are more easy to do than Graefe's method.

Weber's operation is done by a very broad lance-knife, but it has not found many adherents.

Pagenstecher and others have advocated removal of lens and capsule together, but only in exceptioanl cases can it be done without imperilling the eye, through loss of vitreous and rough handling, caused by the introduction of traction instruments or by difficulty in delivering the lens. In all extractions it is a bad sign to see the cornea collapse after its section, or to see a bubble of air enter the chamber. It comes of the rigidity of the tunics of the eye, which prevents that contraction of all its diameters which the loss of the aqueous necessitates. Extreme hemorrhage also suggests sclerosis of vessels and impaired healing power.

SECONDARY OR MEMBRANOUS CATARACT.

In a large proportion of cases, some membraniform obstruction appears in the pupil after the operation. We have every grade and may speak of the simple and complicated secondary cataract. The simple may be extremely thin and nearly transparent, or quite thick and opaque. The condition of the capsule, after escape of the lens, is as follows : the epithelium remains in a normal state, the torn anterior capsule is thrown into folds, adheres to the posterior capsule, and within the pocket thus formed, lens matter and amorphous substance are shut up. Soon, by the peripheral cells, a mass of irregular and imperfectly developed lens fibres are formed, which are transparent, and mingled with them is more or less cortical substance. For this reason there is a thick rim, while the centre may be very thin, even hard to see, or be broken by openings, or be thicker (see Fig. 104). It is not adherent to the iris, but to the ciliary processes ; the pupil acts freely, and has a keyhole shape.

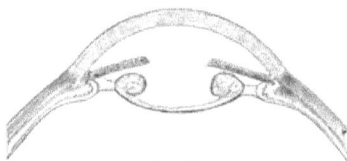

FIG. 104.

Sometimes a secondary cataract may not appear until months or a year after the operation. Peripheral section of the capsule produces it as a frequent immediate result. The complicated varieties of membranous cataract come immediately. The rule with them is, that they thicken and contract, and draw the adherent iris up to the wound, and may obliterate the pupil. Consecutive changes in secondary cataract may be severe, and the eye go into atrophy with detachment of the retina, the vitreous grow fluid, and the changes which cyclitis causes ensue. From this may follow also the irritation which sets up sympathetic ophthalmia. The time when secondary cataracts may be operated on is not easily determined, but is seldom less than within six weeks or three months. Never until all irritability of the eye has ceased, and this may be after six months or a year. Again, a patient with a normal operation must be prohibited from having glasses for a month after his cure, and atropine must not be **given** up too soon, say for two or three weeks after discharge ; the contrary course tends to produce secondary cataract.

The operations for simple secondary cataracts are as **follows** : discission by a single straight needle or by a sickle-shaped needle **drawn** horizontally across from the inner to the outer side, being entered **in the cornea, near** the limbus, on the outer side. Mr. Bowman. for thicker **opacities,** uses two needles, as the picture **shows** (see Fig. 95, page 241), **and this** method pertains to all cases where **care** must be taken not to drag **on the** ciliary region. For similar cases, Graefe's iridotomy knife is used, being put into the cornea perpendicularly and near the level of the dilated pupil. Piercing a thick capsule by a broad needle or a narrow iridotomy knife, or cutting a hole as large as needful by Wecker's scissors, have given me **excellent** results.

For complicated secondary cataracts, the difficulties are sometimes great. We have the capsule, and behind it and the iris a newly formed membrane, composed of connective tissue, pigment, and blood-vessels. These structures are matted together, and sometimes adherent to the **wound and to the** ciliary processes. Frequently the **membrane** is under **unusual tension.** The operation **must** avoid traction, **direct** or indirect, **on the ciliary region.** The following are methods from **which** a choice can be made : **If the** wound be above, make an iridectomy **below** ; and, if the **space be not clear** operate afterward on the membrane in **some** suitable **way. If** there be no great thickness or toughness in the membrane, and simply a few adhesions of iris, Dr. Agnew's method is good. A broad needle pierces the cornea and capsule at the upper edge, a wound is made below near the limbus corneæ, through this a small, sharp hook is carried, and its point engaged in the capsule wound. Getting a good hold, the **hook** tears down the tissue and is resisted by the needle in the operator's other hand, which also defends the ciliary region from traction. As much tissue is drawn out of the wound as possible and cut off with scissors.

For cases of greater difficulty, I devised the following method, published in 1869, in "Royal London Oph. Hosp. Reports :" A Graefe's knife pierces the temporal edge of the cornea, and emerges at the limbus diametrically opposite, making a wound on both sides 2 mm. long ; withdrawing it, the point is plunged into the middle of the membrane, and then it is entirely taken out of the eye. The operator takes two small blunt hooks, whose shanks must be bent to the proper shape, and puts one of them into each of the corneal wounds and conducts them to the aperture in the membrane. By pulling in opposite directions, the hole is enlarged, and whatever tissue comes out is cut off with scissors by an assistant.

Since the introduction of Wecker's scissors I have done most of this class of operations by their aid. A small iris knife, which may be furnished with a stop, makes a wound in the cornea 1 mm. or 2 mm. from the margin and 4 mm. long. After the knife is entered to the needful depth, it is slightly withdrawn to allow escape of a drop or more of aqueous, and make the iris bulge ; its point is then thrust into the middle of the membrane and quickly withdrawn, leaving as much aqueous as possible. The scissors are now passed, with one blade behind and the other in front of the iris, and one cut made, which, if it gapes well, is enough, but if this do not occur, a second cut at an angle to the first is made, as if trying to form an inverted V. The included tongue retracts, and gives a pupil of arrow-head form.

The cut is often made above, but the peculiarities of the case will decide this, and a suggestion of Dr. Green, of St. Louis, to place the corneal **wound in** such a way that the scissors shall cut across the direction of greatest traction, is to be borne in mind. The scissors are figured on page 228 (see Fig. 91).

In cases still more difficult a piece of iris can be cut out, **as Bowman** and Weber have devised, by means of the same forceps-scissors.

GLASSES FOR APHAKIAL EYES.

In about a month after a patient has been dismissed and has meantime used the atropine twice daily, he may get glasses and begin to use his eyes. The test for glasses ought not to be made sooner, although in practice this rule cannot always be observed. He will at this time by no means have the full degree of sight which he may expect. The vitreous will grow clearer, and pupillary obstructions will disappear. Or at this time it may be proper to tear any delicate membranous obstruction in the pupil. The refractive error will be hyperopia, which in a previously emmetropic eye will reach about $+ \frac{1}{4\frac{1}{2}}$ to $+ \frac{1}{4}$, or $+ 9$ D to $+ 10$ D. But astigmatism is very likely to occur, and will be from $+ \frac{1}{36}$, $+ 1$ D to $+ \frac{1}{12}$, $+ 3$ D. **For** reading it is right to add about $\frac{1}{8}$ or $\frac{1}{10}$, viz., $+ 5$ D or $+ 4$ D to the above, to bring the distance to eight or ten inches, and any required cylindric correction will be included.

For cases where only one eye is to be used, the glasses may, if the patient like, be set in reversible frames, which will save some trouble. Glasses so thick and heavy are uncomfortable. Dr. Loring proposed to make the glasses very small, not more than 20 mm. across, and circular, and cement them on a piece of plain glass. This leaves them equally correct and lighter. The frames must be strong, and gold is the best, if expense is not an objection. Patients may be taught to get an artificial accommodation from the glasses by putting them away from the eyes, by which the distance glasses are made to serve for shorter ranges, and so, too, the reading glasses will better bring out fine print. Some patients like a third pair, viz., for use at table or for two feet distance.

This variety of glasses is necessitated by the total want of accommodation. A small pupil is an advantage, and a large one is disturbing, while faint pupillary opacities are seriously depreciating to sight.

CHAPTER XII.

THE VITREOUS BODY.—CORPUS VITREUM.

Anatomy.—This structure occupies the space between the crystalline lens and the retina. It is, therefore, a flattened spheroid with a depression in front, called the lenticular fossa or fossa patellaris. It adheres to the optic nerve and to the ciliary body, while with other parts it has no attachment. It is enclosed in a glassy membrane called the hyaloid, while some authorities assign this to the retina. I have, in the dog's eye, seen the lenticular fossa lined by posterior lens **capsule** and another similar membrane which were perfectly separable. **The hyaloid** is plicated into the ridges and depressions of the ciliary body, and **forms** the walls of the so-called canal of Petit (suspensory ligament of the **lens**). **The** vitreous is **transparent and** jelly-like, and its intimate structure has been much disputed, and is not yet fully agreed upon. Its middle portion is said to be arranged in concentric layers, while its outer parts are divided into sectors by membraniform partitions. There are found in it cells of very diverse forms, most of them in its outer parts. A canal runs through its centre, between the crystalline and the optic nerve, which is the remains of the hyaline artery of fœtal life. Sometimes the stump of it is visible by **the** ophthalmoscope. The corpus vitreum has neither blood-vessels nor nerves, yet it must be recognized to be an organized structure because of the cells which it always contains.

Diseases of the vitreous.—There are numerous morbid conditions to be seen in the vitreous, consisting of opacities of various kinds, some mobile and clear, others fixed, and which may either be seen as individuals or be so diffused as to render it hazy, or deeply opaque, and sometimes it is absolutely black. We find distinct membranes, loops of blood-vessels, and sometimes pus. The substance is liable to be rendered perfectly fluid, either at its anterior or posterior part, or as a whole. Sometimes it contains cholesterine, or tyroline and phosphates. When fluid, its state is called synchisis, and if there be cholesterine in addition, the name is given of synchisis scintillans.

If we look at a bright surface through a pin-hole in a card, numerous semi-transparent bodies of rounded, elongated, or beaded shape, will float **before** the eye in space. They are muscæ volitantes. They consist in part of mucus and epithelium, falling down over the cornea, as the movement of the **lids** will demonstrate, and they consist also of specks in the aqueous **or vitreous** humors. The vitreous is liable to be detached from the retina, **and the space to be** filled with fluid, yet it may be hard to distinguish between this and liquefaction of the posterior part. It may, too, contain connective tissue, which may adhere to the retina and undergo contraction, and may draw **the** retina from the choroid. If greatly occupied by fibrous tissue, it shrinks and **forms** a small mass, such as we find in phthisis bulbi. We have hemorrhage **into the** vitreous from the ciliary body, which will be found close to the lens ; or, coming from some other source, it may occupy any portion of the mass, or the whole of it. Sometimes hemorrhage occurs in the "canal of Petit" alone. Finally the vitreous is said to show, in very rare cases, ossific deposit.

Many of the lesions above enumerated may occur in connection with inflammation of the iris, ciliary body, choroid, and retina. But modern study has shown that morbid processes are capable of originating in the vitreous itself, and this accords with the existence of cells in its substance, despite the absence of vessels. The same reason holds good as in the cornea. But the enormous proportion of water in the vitreous makes the inflammatory process peculiar. The doctrine of diapedesis (immigration of white-blood corpuscles and their transformations) further teaches us how to understand the lesions we find. Still the subject is obscure, and we speak of its morbid processes only clinically. To Klein ("Lehrbuch der Augenheilkunde," Wien, 1880) or Schnabel (*Arch. of Oph. and Otol.*, vol. v., p. 171, 1876) we may refer as giving the latest views.

Diagnosis.—Floating bodies in the vitreous are often overlooked, both by patient and observer. The former will see them if they lie near the visual axis, and will often be much annoyed or frightened about them. In myopic eyes they are very common, and may have serious meaning, especially if in the deeper parts. When not near the axis, patients may be unconscious of their existence. An ophthalmoscopic observer will not see small vitreous opacities from a distance of twelve inches, but will easily see larger ones. When close to the eye, he will only see them by a correct optical adjustment for that position. For such as lie in the front part he will use a +4 (+10 D) glass behind the mirror. By telling the patient to look up and down, and using very feeble illumination, he will catch sight of fibres, specks, and curls dancing about. For the middle of the vitreous, a weaker convex glass, say +8 (or +5 D), will serve, while for the posterior part the refractive state of the fundus will govern the choice. In general turbidity of the vitreous, whether slight or dense, the diagnosis rests on the impossibility of getting a clear view of the fundus by any correcting-glass, and by feeble light the minute specks may possibly be seen. To the inverted image, the optic nerve will be visible, and it always looks red, as does the sun in a foggy atmosphere. Cases of this kind are mistakable for retinitis or neuritis. With inverted image a strong light is needed, and the retinal-vessels may be seen, and possibly a hemorrhage be detected. A hemorrhage into the vitreous will make a strong light needful for both direct and indirect ophthalmoscopy. In cases of extreme opacity, we may find the vitreous filled with black-looking membranes and strings rolling about in obscurity, or find it so impermeable to light as to give absolutely no reflex. Such a condition can occur in secondary syphilis, and I have seen complete recovery, with no discernible changes in choroid, or retina, and vision $\frac{2}{9}\frac{0}{0}$ after the lapse of six months. A condition of total obscurity may also be caused by hemorrhage or by a new growth.

Treatment and prognosis.—For single or a few opacities, no treatment is to be given, while if they appear in quantity in a myopic eye, which likewise has considerable choroidal change, they call for admonition as to strain of eyes and general hygiene. They will not, in any case, really disappear, but may float into some other position, where, being eccentric, they will not give annoyance. Of course they imply that the vitreous in which they float is liquefied.

For diffused haziness one may be in doubt whether it depend on disease of the adjacent structures. In iritis we know it to be so dependent, and so, too, in many cases of choroiditis and retinitis. The primary disease will be the subject of treatment, but the attendant hyalitis will call for a longer continuance of remedies than would otherwise be adopted. In the **early** stage of cataract it is a common experience to have opacity of the **vitreous.** I usually order iodid. potas., gr. v., ter in die **for two** or three **months,** and have felt impressed with the conviction that **thereby its** transparency was promoted. I certainly have seen the vitreous **grow clear,** and vision manifestly improve. In case any other indications for treatment could be found, I would take it, viz., tonics, phosphoric acids, etc.; **but** in default of other indications I give iodid. potass. There are cases **of soft** cataract in persons above twenty-five which will be attended by **a great** amount of hyalitis. For them extraction by Graefe's method is the **best** treatment, and in time the vitreous opacities may clear up. I have witnessed both their persistence and considerable disappearance.

For hemorrhage, the artificial leech may be applied to the temple **from** one to three times at intervals of five days, and salines given either **as mild** purgatives or as diuretics. In any event, weeks or months will be required for absorption. I have seen **a** hemorrhage in the front of the vitreous, and which also showed in the anterior chamber, and therefore came from the ciliary body, entirely disappear in two weeks. The patient must, during the first ten days or two weeks, if not longer, be kept **in** moderate darkness, **in** bed, **or** in his **room.** It may be advisable **to use a** bandage.

For the dense opacities which occur in secondary syphilis, whether **with or** without choroiditis, iod. potas and ung. hydrarg. i.e., mixed treatment **is to** be steadily kept up and supported by all that the patient needs **to** aid **his** nutrition. In such cases, though at first discouraging, the prognosis is **at** least measurably good and may be entirely favorable.

Foreign bodies in the vitreous have already been spoken of. If recent, they should **be** extracted if this be feasible, by a hook or by **a** magnet. If suppuration have begun about them, the attempt at extraction may **be admissible,** but if they **cannot** be withdrawn, the eye must be enucleated **or** eviscerated.

Suppuration occurs in the vitreous commonly because of blows or wounds or of lesions of adjacent tissues, and is part of the state of panophthalmitis. It may also occur in **more** limited inflammations, and may be the result of metastatic choroiditis, or of meningitis or of relapsing fever, and may exhibit a picture **which** is almost identical with glioma of the retina. Such cases were formerly called **scrofulous** choroiditis. The pupil may be enlarged, or small and adherent, **while a** whitish yellow mass occupies the vitreous and may **come** up **to the lens. In one** case of a girl two years old, which followed meningitis, **it seemed to me that** the diagnosis from glioma could be supported by the densely **white** look **of the** periphery of the mass, while its middle had **a** degree **of dull** transparency like wetted ground **glass.** The eye was in a hyperæmic state at the beginning of the disease, **and this** subsided, leaving the **vitreous** opacity. For such cases the use **of iodide** of potash with mild mercurial inunction is the treatment, and not **the** impetuous resort to enucleation. Of course, the case must be vigilantly **watched,** lest **a** true **glioma** go too far to admit of protective enucleation. **In the case I have in** view, a preceding meningitis and sudden blindness, **followed** soon by ciliary injection and the discovery of the intraocular mass, gave **strong** grounds for a diagnosis of an inflammatory deposit, and justified the **effort to** save the eyeball, which proved successful. Schnabel (*Arch. of Oph. and Otol.,* vol. v., p. 172), reports a case of suppurative hyalitis without choroidal lesion, which he examined anatomically.

When distinct membranes appear there is usually no probability of improved sight, and this is the more likely when accompanying choroiditis can be seen. If the tension of the globe begin to grow less, the outlook is certainly bad. On the other hand, I have watched a case which might be equally well described under another head, where membranes in the vitreous were an early symptom and were adherent to the retina near the optic nerve. Ultimately, spots of choroidal atrophy came, and the vitreous opacity continued but little altered.

Blood-vessels in the vitreous may appear **in** connection **with membranes or as** prolongations from vessels of the retina. Several **such cases** have I seen, **and** one of very notable character. They have **remained unchanged** for **the months** during which I observed them.

It follows, of course, that in all cases of vitreous opacity the vision will suffer. For mild cases the complaint will be of muscæ volitantes, or, sometimes, of muscæ fixatæ. There may be imperfectly defined scotomota, but usually no limitation of the field, and no pain, and no change in the pupil, except **in** the suppurative cases, or in those complicated with other diseases. Tension of the eye is reduced in purulent cases, not changed in **others.** While we have cases of total or partial recovery in so-called plastic (as distinguished from purulent) hyalitis, we also find them inclined to relapses. A striking picture of vitreous opacities is found in Jaeger's "Ophthalmoscop. Bilder."

Sometimes the arteria hyaloidea is found permanent for a considerable length, and in a case exhibited to the New York Ophthalmological Society by Dr. Kipp, of Newark, a complete hyaloid artery was visible in each eye. The cysticercus cellulosæ is found in the vitreous among populations whose food is liable to contain the germ of the parasite. We read of it in German treatises, **and some** cases have been reported in this country.

CHAPTER XIII.

CHOROIDEA.

Anatomy.—This membrane reaches from the ciliary body to the optic nerve, and lies in contact with the sclera. The optic nerve passes through it. Loose connective tissue unites the choroid and sclera, which, as the two are separated, appears in shreds. As it remains on the sclera, it is called the "lamina fusca," while what remains on the choroid is called the supra-choroidea, and it is charged with pigment. The two, where united together, constitute a lymph space of large size, in which some ganglion-cells exist. The sclera is obliquely perforated at its posterior third by the short posterior ciliary arteries, veins, and nerves, which go to the choroid and to the remainder of the uveal tract.

The choroid consists of blood-vessels, of which the coarser are exterior and the finer interior, and are called the chorio-capillaris. These two layers, among which stellate pigment-cells with pale nuclei, and round cells, and ganglion-cells, are interspersed, compose the stroma of the membrane. Upon them is a transparent, structureless membrane, called lamina elastica, or lamina vitrea, and most internal is a layer of irregular, but mostly hexagonal cells, charged with pigment, and having round, pale nuclei. Some cells have no pigment. The whole compose a beautiful tesselated pavement of one or two layers, which after death adheres to the choroid, and has hitherto been assigned to it, but recent studies have given it place with the retina, and it is now so classified. Some call it a membrane by itself, and it seems to have the function of secreting the visual purple of Boll, as well as of damping the light by its dark color. Its color is deep brown in Anglo-Saxons, but varies greatly in intensity in persons of different complexions. In albinos it has no perceptible color, and through it the choroidal vessels are distinctly seen. Its tint is deepest about the centre of the eye, while in all persons the choroidal vessels are visible near the equator.

DISEASES OF THE CHOROID.

In a preceding chapter the frequent association of choroidal disease with troubles of the iris and ciliary body has been noticed. It is not necessary to repeat what then was said. Acute affections of the choroid, which attain considerable intensity, must, of necessity, implicate these structures and give rise to opacities in the aqueous humor, to adhesions and obstruction of the pupil, as well as to effusions into the vitreous. It therefore becomes impossible to view the depths of the eye, and we can give no clinical account of the choroidal lesions. In the case of serous iritis (Descemitis), mention has been made of the fact that the choroid participates, and that a dim view of the interior of the eye can sometimes be ob-

17

tained, but nothing more can be discerned than opacity of the vitreous and some connective-tissue fibres, while the choroidal tissue is invisible. Acute glaucoma is, by some writers, spoken of as acute serous choroiditis ; certainly effusions come from this structure and cloud the vitreous ; but all the vascular tissues of the globe suffer the shock of the attack, and we have no accurate knowledge as to the place from which it starts. It is not fitting to discuss this condition in this place.

We are therefore obliged to limit our considerations to those lesions which we can find with the ophthalmoscope. Of outward signs of trouble we have none. The cornea and iris and the lens, except in special cases, are free from change, and ofttimes the vitreous is clear. The lens does lose transparency in part or completely in certain advanced cases, and the vitreous may present various alterations in formation of connective tissue, of membranes, of fibres, and molecules, and in some cases we find blood-vessels developed in it. But in this connection we are often unable to separate lesions of the choroid, of the retina, and of the vitreous from each other as to the origin of the morbid process. Liquefaction of the vitreous is a common result or concomitant of choroidal diseases (synchisis). It must be remembered that the anterior part of the choroid is not within our inspection. The part which we can see will vary greatly in both healthy and morbid appearance, according to the quantity of pigment naturally present. The tawny look of the negro's eye contrasts strongly with the light red hue of the blonde. In some cases the granular contents of the epithelium can be noted. A condition, formerly called "pigment maceration," does not in reality have any existence in the sense attributed to it. Most of those cases which once I set down under this head, I am convinced were probably errors of refraction or cases of amblyopia. For many years that diagnosis has ceased to appear among the records of clinics.

There is, however, a lesion which, in a few rare cases occurs, which consists in the apparently complete removal of the hexagonal epithelium over a considerable area. When this occurs, the vessels of the choroidal stroma are disclosed distinctly, and resemble angle-worms in a bait-box, while all about this spot no such vessels appear because covered by the usual dark granular layer of cells. Jaeger gives a picture of such a patch of epithelial absorption, occupying the centre of the fundus and extending over an area equal to three nerve-diameters. Around its edges there is very little pigment deposit. I remember several similar cases. They might, according to modern classification, be set down as cases of retinal disease, but it is convenient to mention them here, and we do not know whether to impute their beginning to the choroid or to the retina. That the latter is not seriously involved is manifest from the moderate degree to which vision suffers.

The most characteristic fact about diseases of the choroid is the accumulation of pigment in spots and masses. We have spots of plastic exudation, and spots of atrophy more or less extensive ; but the result of the former and the usual accompaniment of the latter is the presence of black blotches. It would seem that under most conditions, whatever happens to other elements of the tissue, the pigment is not destroyed, but simply displaced to the edge of the atrophic spot, or on its surface. If this be not so, it is produced in abnormal quantities and remains indelibly.

Many chronic diseases of the choroid evidently take a more serious hold upon the retina than they do upon the former membrane, as evidenced by the state of vision. Of course, we may be in error as to the true seat of the malady, but the cases which I now have in mind are some in which there will be marked dimness of sight, which has been of slow development; there will be nothing abnormal in the optic nerve and no changes visible in the retina, while somewhere a dot of pigment will be found—perhaps at the periphery, perhaps near the centre of the fundus; several may be found, but sometimes only a single speck will appear to give us any clue to the disease. With it there may be neither exudation nor atrophy. It is not rare to find that in these cases atrophy of the optic nerve may ensue. These cases insensibly grade into those where abundant pigmentation appears, chiefly about the periphery of the fundus, showing irregular figures and resembling what is called retinitis pigmentosa. The latter disease differs in microscopic features from these choroidal lesions, but clinically they are difficult to be discriminated until atrophy of the tissue appears and the sclera is brought to view.

Choroiditis disseminata is a well-recognized form of disease and the most frequent of those we now discuss. It may attack any part of the membrane, but is, perhaps, oftenest seen at the periphery. Many times the person is unaware of the disease, and its presence would not be shown except by careful examination of the periphery of the field of vision, or by inspection with the ophthalmoscope. Take the following case: I was asked to see a gentleman in consultation with my friend, Dr. R. H. Derby, who had received a sharp blow on the eye by a racket-ball while playing the game. Dr. Derby, on examination, found near the margin of the fundus a patch where choroidal tissue had disappeared and exhibited the white sclera, and at its margin was an inky pigment border; some traces of choroidal stroma remained within the space. The point was raised whether we might have here a laceration of the choroid. The pigment deposit, the remains of choroidal structure, and the absence of ecchymosis decided in favor of a spot of choroiditis disseminata. It had doubtless been produced many years before, and without the knowledge of the patient.

Such lesions are apt to appear in childhood, to be arrested, and not to be recognized when their location is eccentric. On the other hand, if they fall upon the region of the macula lutea, they obliterate central vision and cause a positive scotoma. A lady about sixty-five years old came to me recently for advice respecting a slight conjunctivitis of the left eye. The other eye, she said, was unable to see, and had been in this case since childhood. The late Dr. Wallace of this city had given his opinion that some fault of accommodation was the cause. In his day the ophthalmoscope was unknown, and he had given much study to the function of accommodation. I found that at the centre of the fundus there was a large patch of choroidal atrophy with moderate pigment deposits. To see any object the lady had to turn the eye to one side so as to use the adjacent retina.

The disease begins in the choroidal stroma as masses of cells and amorphous exudation forming faint yellow spots. The superjacent retina and vitreous also become hazy, and if the lesion be near the optic nerve it will be reddened. After a time atrophy succeeds to exudation, the walls of the vessels become sclerosed and they also disappear, while pigment becomes deposited. I, for a long time, watched a case, which gave all the appearances of chronic neuroretinitis, but at length patches of atrophy appeared in the choroid near the nerve, and there were remarkable membraniform products in the vitreous, springing from the retina and the nerve. The choroidal patches were streaked with red lines of vessels not yet destroyed, while the sclera shone through interstices from which the vessels and other tissues had faded away. In this case, choroiditis disseminata was added to retinitis proliferans. The various elements of morbid change which have thus been noticed may be combined in the most varying proportions, and the clinical picture will exhibit the most extraordinary variety. In some cases the fundus looks as if it had been corroded by an acid, so large a part of the choroidal tissue has been swept away; in other cases isolated patches are scattered about, and in some, one large spot shines out. Usually, if retinal vessels lie in the track of choroidal atrophic spots they pass unaffected, but sometimes they have pigment adhering to them. The figure (105) from Klein represents a crescent of choroidal atrophy next to the optic nerve, which is found in many cases of myopia, combined with spots of atrophy in other places caused by choroiditis disseminata.

In some respects similar to the lesion thus described is the change which occurs about the yellow spot in some cases of advanced myopia.

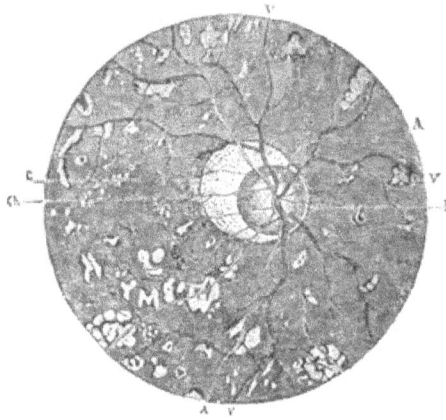

Fig. 105.—A, arteries ; V, veins ; B, optic nerve ; Ca, edge of choroidal crescent ; C, pigment deposit.

The process is in essence an atrophy due to attenuation of the membranes by the stretching of the eye. It is not preceded by exudation, nor is pigment-deposit conspicuous; there may also sometimes be hemorrhage. One sees a spot where the uniformity of surface gives place to irregular yellowish lines and dim pigmentations, the appearance of half-absorbed tissue, which suggests a resemblance to a worm-eaten surface. Vision is always impaired, and sometimes that peculiar distortion of objects takes place which is known as metamorphopsia. By this is meant that straight and parallel lines do not appear to be straight, but have a bend at the spot looked at. If parallel lines be ruled close together both horizontally and vertically, the person will be able to trace out the region over which the lines appear crooked, and the form which they take. The reason for this is because the bacilli have been irregularly stretched asunder, and, being dislocated, the visual projection of each one is changed, and the result for the disturbed group is metamorphopsia. The same result in the retina may be caused by exudation among the bacilli. Analogous to this anomaly are the conditions known as megalopsia and micropsia. In the former case, bacilli, being crowded together, a greater number than normal are covered by an image of a given size, and the object, therefore, is estimated to be larger than in reality. In micropsia fewer bacilli receive an image, because they are spread apart, and therefore the object is estimated to be smaller than in reality.

A form of exudative choroiditis occurs, which presents small yellow dots, not bigger than a pin's head apparently (upright image), scattered about the fundus. There may be a little redness of the nerve or none ; there may be no haziness of the vitreous or retina, and no pigment deposits. The exudations lie below the level of the retinal vessels, and may be very numerous. Vision will be reduced, and there will be a knowledge of when the trouble began. When it disappears no traces of mischief may remain, but the case may be otherwise. This is commonly a syphilitic trouble.

A form of choroiditis was described by Förster in 1862 as choroiditis areolaris, because it is confined to a limited locality, and in this respect it is like choroiditis disseminata, and the resemblance between them also extends to the essential quality of the pathological changes. But in clinical features they are unlike. The form now referred to is accompanied by redness of the optic disc, by haziness of the circum-optical retina, by exudations along the retinal vessels, especially along the veins. A marked feature is vitreous cloudiness, which comes and disappears suddenly. To mark the choroidal nature of the ailment we find in many cases spots of exudation, in other cases pigment deposits and patches of atrophy. They may occur at any part of the fundus, and be few or many and be large or small. They will be separated by intervals of healthy tissue, or enormous surfaces may have been ravaged. This disease is syphilitic, and comes in the late secondary or early tertiary stage. It is remarkable for its tendency to repeated relapses.

There remain some other choroidal lesions, which are much less important than those above mentioned, but which should be stated. We find minute glistening points quite often about the centre of the fundus, perhaps one or several, which evidently lie rather deep in the membranes, and are not accompanied by any other lesions, or by any loss of vision. They are also found in other parts, and are said to be often seen in anatomical examinations about the periphery. They are said to be minute exudations on the lamina elastica, which separate from their base and push forward into the retina, perhaps even to the optic nerve layer. Their peculiar brilliancy attracts notice, while they are really unimportant.

Another lesion which may be more extensive, and is closely allied to the preceding, is the so-called colloid excrescences from the elastic lamina (Drüsen). They were anatomically investigated by Donders and Müller many years ago, and have been repeatedly referred to in pathological examinations. They appear more in old than in young persons, yet have been found by Alt in a boy twelve years of age. They may be numerous or few. They displace and cause absorption of the retinal pigment, but do no harm to the remaining retinal structure. They may form simple nodules, or several may be piled in a heap. The choroidal stroma does not suffer, nor is vision in the least degree impaired. To the ophthalmoscope (by direct examination) they appear like mustard-seeds, or dull yellowish bodies, and have no pigment about them, except in rare instances. They seldom reach one millimetre in diameter. I have seen one case which I regarded as of this nature which was remarkable for the extent of the deposit. It was in a woman sixty-five years old, and the fundus was covered by bright patches like those seen in albuminuric retinitis. They caused no loss of sight, $v. = \frac{6}{6}$, color perception was perfect, the urine was normal; the patches were most extensive about the centre of the eye, and elsewhere were dots sprinkled about; there were no signs of inflammatory change in either the retina or optic nerve. Both eyes were affected. The case was seen during four months, and no change took place.

Prognosis of choroidal inflammations varies with the nature of the process. The first essential is to have a correct idea of the character of the lesion, and whether it be in the non-progressive stage. For many cases of choroiditis disseminata, this will be true, or if the disease be increasing its rate of advance is to be measured only by years. In some cases the association of vitreous opacity and of optic nerve congestion, will clearly indicate the activity of the process. For the atrophic varieties, such, for example, as occur in myopia, little more need be said. For other forms in which sometimes great absorption of tissue occurs, little is gained by treatment. A singular fact which I have several times noticed is that persons having extensive choroidal atrophy have become insane.

Treatment.—For certain forms we can trace in children some constitutional dyscrasia, perhaps hereditary syphilis—or scrofula, which is often the same disease farther removed from its origin. In these cases, moderate use of mercurials—and especial reliance on cod-liver oil, healthful diet, and correct mode of living in all particulars, are the chief dependence. For the more distinctly syphilitic types of disease, viz., choroiditis areolaris, a full mercurial and iodine treatment is to be adopted. Mercurial inunction, mercurial vapor-baths, corrosive sublimate, and iodide of potassium, with such adaptations as each case may require, are the suitable remedies. Local depletion is of no special use. The eyes must be guarded from brilliant light by smoked glasses. They may not be used in near work. While the existing attack may pass with little injury to sight, relapses, as already said, are very common. If much vitreous opacity occur, it will remain. There is danger in cases of metamorphopsia that this condition will continue. It follows, therefore, that a patient should be enjoined to continue the specific treatment after the disease seems to be gone, because the constitutional cause cannot be known to be eradicated by this fact alone. Other forms of mercurial may be adopted, but the essence of treatment is to be maintained for weeks or months.

Choroiditis suppurativa — Panophthalmitis, choroiditis metastatica, vel embolica.—From a clinical standpoint it is unnecessary to expatiate on the ordinary cases of suppurative choroiditis which follow injuries or operations or are consecutive to lesions of other parts of the eye. We denote such cases as panophthalmitis, and there is no occasion to single out the choroid for special regard. But metastatic or embolic choroiditis is likewise of the suppurative type, and an affection of the utmost gravity. It has already been alluded to in the chapter on the vitreous (see p. 204), but a few additional words may be in place, especially upon etiology. It is one of the accidents occurring during the puerperal state, or during low fevers, or after surgical operations, or from caries of the cranial bones, or from meningitis, or cerebro-spinal meningitis, or mumps, etc. In short, whatever originates embolism, or phlebitic inflammation may set up thrombosis in the choroidal vessels and suppuration.

The symptoms may begin quietly, as in a case reported on p. 204, loss of sight and a yellow reflex from the fundus being the chief symptom (besides moderate external hyperæmia)—or its onset may be violent with swelling of lids, hypopyum, chemosis, violent pain, extreme tension, etc., in other words, speedy development of acute panophthalmitis.

Treatment is to be confined in most instances to relief of pain. The mild cases will get on with atropine and warm water. The severe cases will need leeches in the outset, morphia, and an early incision into the eyeball to evacuate the lens and part of the vitreous. If suppuration be fully established an incision may not confer adequate relief, and the interior of the globe may be wiped out with a sponge. Enucleation is sometimes practised, but it is attended with profuse bleeding, unusual reaction, and some degree of danger; a few fatal cases having been reported. To obtain a good nidus for an artificial eye and early relief from suffering is what is obtainable. The former is best secured when as much of the eye as can be saved is retained.

Tubercles of the choroid.—Such deposits take place in the choroid either as minute specks or as diffuse infiltration of the stroma in isolated prominences. The little specks occur by preference in the region of the optic nerve, and project more toward the sclera than toward the retina. For a long time the hexagonal epithelium is undisturbed, and if the nodule do not exceed one-fourth millimetre in height it may escape the search of an expert ophthalmoscopist. When they do break through the epithelium they have no defined border, no pigment about them, are pale yellow or rose yellow in color, and therefore quite inconspicuous. There may be only a few or fifty. In size they may reach three millimetres in thickness. They affect both eyes; they do not injure sight. In rare cases they cause local choroiditis and then may exhibit pigment. Their anatomical character is identical with that of miliary tubercles. Of course they are liable to be confounded with warty (colloid) excrescences. But the points of difference are their situation near the optic nerve, their indistinct color, and that they lift both the retina and choroid in their elevation; also that there is no aggregation, but a diminution of pigment near and about them, and that when multiple they differ notably from each other in size. At best the diagnosis is not easy. They occur most commonly in cases of tubercular meningitis and in general miliary tuberculosis. One case is on record where the choroidal tubercle preceded by six months all symptoms of general tubercular deposit. (Fränkel.) Jaeger, who was the first to describe the disease, gives a good plate (No. 72 in his "Atlas"). The lesion in the choroid has been experimentally demonstrated in the guinea-pig, in which tuberculosis had been developed by inoculation. The subject has greatly interested pathologists, and has an interest in a clinical point of view.

Lacerations of the choroid.—Injuries, such as blows, which make no external mark on the eye, and seem not to have caused serious injury, sometimes cause laceration of the choroid. The place of its occurrence may be near the equator, but is much oftener about the posterior pole. It may surround the optic nerve in a crescentic form; it may be at the macula, and a straight rent. There may be a succession of two or more, concentric with each other. Usually the retinal vessels are not torn, even if they traverse the rent. There is at first hemorrhage about the spot, and vision is affected in some degree, but of course most seriously when the rupture is near the macula. The effused blood begins to absorb and the sclera then appears to view, showing that a disruptive lesion has occurred. After a time the blood, which may have been either light or very dark in hue, disappears, and then pigmentation occurs in and about the tear. I have a drawing of a case of long standing in which the deposit of pigment is extraordinary, and the rupture is not less than ten discs in length. There is no treatment for these cases except rest and protection of the eyes, and abstinence from use until signs of irritation have disappeared.

Tumors of the choroid.—The greater number of choroidal tumors are pigmented sarcomata; some are wholly black (melano-sarcoma), others have less pigment, and a few are white. According to Nettleship, some are spindle-celled or mixed, others are composed of round cells; some are alveolar, in many there is very little connective-tissue stroma and no very defined arrangement of cells; some are very vascular and liable to hemorrhage. They generally have a broad base, and may start from any situation. They usually grow slowly, and if peripherally seated may not attract the patient's attention, although a defect in the visual field will exist. They appear most often about middle life; are not very rare after twenty-five, but before then are exceptional to a high degree. I have a specimen which occurred in a boy fourteen years of age—a Jew. Its interior is white, its surface black and covered by the choroid.

The utmost importance attaches to the early discovery of these growths. If they get beyond the globe, constitutional infection and return, either *in loco* or in some distant organ (preferably in the liver), is almost certain. Exemption from infection is not secured even while they are intra-ocular, because as has been said (see Nettleship), the cells of a small tumor may pass out along the sheaths of the perforating blood-vessels and produce extra-ocular growths. The lymphatic glands do not enlarge.

The proportion of the disease as between men and women is as two and a half for men to one for women. Prognosis is made relatively favorable by early removal, while its natural course is to death. A careful examination of the nature of the tumor is essential to a just prognosis. Pure melanoma may be twenty years in growing, and is comparatively innocent. It has been shown that a considerable proportion of the tumors are fibromata; nevertheless, the larger number are cases of melano-sarcoma, and mixed forms also appear, having a degree of carcinomatous quality whose result is most disastrous.

Course and symptoms.—After remaining latent for an uncertain period, viz., months or years, a notable loss of vision appears, and then the ophthalmoscope discovers the growth. At this (first) stage there may be detached retina, and if the latter be clear and the subretinal fluid also clear, the tumor will be detected. If the contrary is the fact respecting the retina and its subjacent fluid, the diagnosis is made difficult, perhaps impossible. Then searching inquiry must be instituted about a loss of peripheral sight of slow development, followed by a sudden aggravation of the loss—the former symptom belongs to the tumor, the latter to the detachment of the retina. Sometimes the retina lies close upon the tumor, and then little difficulty is experienced. Again, there may have been a hemorrhage into the vitreous, and here the cross-examination as to the behaviour of sight during successive months, and recently, must be most rigorous. Again, if detached retina exists while no tumor is to be seen, and at the same time the globe, instead of being a little minus in tension, is or soon passes into a state of plus-tension, the diagnosis of choroidal tumor is certain. This diagnostic token applies in some measure to the complication of hemorrhage into the vitreous. But ere long further developments occur (second stage); acute inflammation, with great pain, and aggravated tension (acute glaucoma) sets in. The iris may be greatly or slightly pressed forward, the pupil may be large or, being adherent, be small. All signs point to acute glaucoma, and the suspicion of tumor may not arise. A careful study of the previous history will be the chief aid, because at this time no view of the fundus will be possible, or can be sufficient for diagnosis. If iridectomy be done, the symptoms may abate and relief ensue, as happened in a case which I treated last winter, and in which my suspicions of tumor were expressed before I did the operation. The person was sixty-five years of age, and such experiences are said to happen oftenest in old subjects. But the inflammatory reaction is again set up after a brief lull, and enucleation must be done.

But a patient may not come for medical aid during this **stage**, and through panophthalmitis the globe may rupture and subside **into** a phthisical stump. The probability of an intra-ocular tumor will be suggested by the peculiar form of the bulb—in its not being reduced equally in all dimensions, but having the equatorial measurement nearly normal, while the front of the globe is much flattened. It will also be hard to the touch. Such stumps should be enucleated.

In their natural growth, tumors in time make their way through the tunics of the eye and appear as external black prominences (the third stage). Sometimes, if they be near the equator, their presence is indicated by dislocation of the lens, by propulsion of some part of the iris, and cataract may appear. When they exhibit themselves on the surface, it may be by continuity of growth or by the formation of separate growths. Should the neoplasm advance to the front, the normal textures of the globe break down and disappear, while an offensive, nodulated black mass pushes between the lids, and, if left to extreme development, becomes most unsightly and repugnant, with putrescent and disgusting discharge. On the other hand, the growth may increase backward into the orbit, may spread out over the back of the sclera, or develop a special tumor in the orbital tissue. Then the globe is likely to be protuberant and its range **or manner** of movement peculiar.

During all these phases of growth, unless from pressure or from inflammatory attacks, the patient suffers no material pain. The cause **is** often inscrutable, sometimes an injury may be ascertained.

Treatment consists in extirpation at the earliest possible moment. There should be no delay or hesitation. Pains must be taken to excise as much of the nerve as possible, and it is easy to remove 8 or 10 mm. The mode of doing it is as follows: after dividing the conjunctiva and recti tendons, to press the globe forward, and with blunt-pointed scissors to go deep into the apex of the orbit and to sever it. I have aided myself in this kind of nerve-section by passing the forefinger of my left hand behind the globe, and then using the scissors; but it is more effective to catch the nerve next the globe by a strabismus hook, which drags it forward, and the scissors, which are curved on the flat, are easily introduced behind it and almost to the apex of the orbit. The cut end of the nerve must be closely inspected to see if traces of the growth can be discovered. In this event, the distal nerve stump is to be found and another piece excised. To do this, the finger in the orbital tissues is the best finder, and the hard, cord-like nerve may be caught with spring fixation forceps and the division performed. To attempt to dissect out the nerve in the blood-infiltrated orbital tissue is an almost impossible task. Should a growth be found in the orbit, the cavity must be absolutely cleaned out, including the periosteum, and all that admits of removal at the apex be gotten out bit by bit. Should the bony wall at any spot feel rough, the actual cautery, or chloride of zinc paste, should be used. It cannot be asserted that such thoroughness will prevent a relapse *in loco*, but it offers the best security. A return in some other organ is to be anticipated, and a consequent fatal issue.

Ossification of the choroid.—This occurs in old phthisical eyes, and while specially interesting to pathologists, it appeals to the practitioner in that such degenerations are liable to lead to sympathetic ophthalmia. Such a stump has a remarkable hardness when seized between the thumb and finger, and rough points may be discovered in it. They should be extirpated. To accomplish this with ease, catch the bulb with a sharp hook, **and** then proceed to sever the retaining parts while the eye is pulled forward. Unless this device is adopted, the enucleation is quite **troublesome.**

Apoplexy of the choroid.—Such a hemorrhage may occur, and be confined to the stroma of the membrane, or tear through the retina into the vitreous, or find vent into the supra-choroidea. In all cases the vitreous is rendered to some degree turbid, which beclouds sight, while, if the blood is near the posterior pole, the reduction of sight increases in proportion. Hemorrhage into the choroid differs from similar effusions into the retina, in not being elongated and in streaks, but rather in irregular and more extensive patches. Sometimes detachment of the retina accompanies it. Causes are traumatic, such as contusions of the eye or of its neighborhood. **Again,** this may attend acute choroiditis, and **be found in** posterior staphyloma (extreme myopia with choroidal atrophy), or it may attend upon lesions of the general circulation, such as disease of the heart, sclerosis of the vessels, dysmenorrhœa, struma, pertussis, dyspnœa, etc.

Treatment relates solely to the general indications and to hygienic care. Absorption takes place slowly, during weeks or months, and the blood passes through changes of color, growing of a lighter red, then yellowish, and at length leaves a white spot of choroidal atrophy exposing the sclera, and bordered, as well as speckled, with pigment.

Albinism of the choroid is a congenital defect, and it concerns the whole **uveal** tract, although the iris **is** likely to be more supplied with pigment than the choroid. Often there is arrest of development of the lens, and some degree of nystagmus, often extreme, is to be expected. The choroid presents a brilliant picture to the ophthalmoscope, showing the fine vessels and the venæ vorticosæ. Around the centre of the fundus, if any pigment exist, it will be found in greatest quantity. Persons with this defect always exclude their eyes from light, and usually wear dark glasses with dark sidepieces. They may also have refractive errors. They always bring objects close to their eyes, partly to get larger images, and partly, by straining the adductive muscles, to hold their eyes more steady. Their vision is always defective for distance, yet I have known two persons who were strongly albinotic, able to use their eyes in severe professional pursuits.

Coloboma of the choroid.—This congenital defect may or may not be attended with cleft of the iris; it may be more or less extensive; it may be bridged over at some spot by normal tissue. It always appears on the inferior part of the eye, running forward from the optic nerve. It presents a wide white or yellowish spot, with irregular distribution of vessels, some of which are choroidal and others are retinal. Sometimes the cleft runs into the sheath of the optic nerve, and the disc is usually of an abnormal look, and the retinal vessels may be singularly distributed. Vision is always below standard, partly because there may be imperfect development in the **retina, and** partly because the glaring reflection from the patch of denuded **sclera** overcasts the retinal image. Sometimes such persons have other evidence of arrest of the development of the eye, and often have nystagmus. Careful search should be made for refractive errors, especially for astigmatism, and if no glass should be found available or needful for this error, lightly tinted smoked glasses will sometimes be acceptable, and even corrective glasses may well have a shaded tint.

CHAPTER XIV.

GLAUCOMA.

This word designates a morbid condition which has one signal peculiarity, namely, an increase in the hardness of the globe. The name comes down from olden times, and was applied, because in certain advanced cases the pupil acquires a greenish hue (glaucus, green). Graefe recognized that unnatural hardness was a fact characteristic of the malady, and since his time all affections in which the tension of the eye rises above the normal degree, are called glaucomatous. The standard of ocular tension varies in its physiological limits. In women it is normally less than in men, in children than in adults. An average of twenty-five millimetres of mercury, or twelve inches of water, is normal. For exact measurements special instruments are used, called tonometers, of which several have been made, but they are not resorted to in practice. Dependence is placed upon the sense of touch, and one finger of each hand is to be lightly pressed upon the eye as when feeling for fluctuation in an abscess. The sense of resistance can also be appreciated, although less accurately, by a single finger pressing the globe through the closed lid. Mr. Bowman suggested a notation and classification as follows : for normal eyes, T ; for those with increased tension, $T_+?$, T_+, T_{+2}, T_{+3} ; for those with diminished tension, $T_-?$, T_-, T_{-2}, T_{-3}. When the sign ? is put behind T, it indicates a doubt as to the real status, whether greater or less. Mr. Priestley Smith has given values for these expressions in columns of mercury, but they are, of course, but approximate, viz. : that while $T=25$ mm. Hg, $T_+=50$ mm., $T_{+2}=75$ mm., $T_{+3}=100$ to 125 mm. Different observers might estimate the same case variously, and the sense of touch must be educated.

The following subdivisions of the disease are recognized : glaucoma simplex ; glaucoma with inflammation, and this may be acute, subacute, or chronic : glaucoma hemorrhagicum ; and secondary glaucoma.

Glaucoma simplex is the most frequent affection. It is an insidious and very slowly progressive condition. It arises most often after middle-age, and in hypermetropic eyes. It has been seen in young subjects, and a considerable contingent is found among myopes. The symptoms are as follows : In external appearance the eye may be normal, except a notable whiteness of the sclera, with a few conspicuous and tortuous vessels coming from the recti muscles. The anterior chamber and iris may be normal, and the pupil a little sluggish but contractile. Visual acuteness may or may not be reduced. The field of vision will be restricted in the nasal side to a greater or less degree, and perhaps be curtailed in the remainder of the periphery. It can also happen that scotomata are to be found in other portions of the field, and they may lie near to the macula-lutea. The tension of the eye will be increased. By the ophthalmoscope the optic nerve will be found hollowed into a cup or excavation : it will have a white or grayish hue ; its arteries will not form continuous lines, but some of them will be broken as they pass from their place of entrance to the surface of the retina. In extreme cases of optic excavation they seem to spring from one side of the nerve, and the faults in their continuity will be conspicuous. They spontaneously pulsate, or do so under slight pressure. The veins will be large and dark. By the inverted image a great degree of parallax is given to the arteries on the disc by to and fro movements of the objective lens; with the upright image the difference of level between the edge of the nerve and the bottom of the pit will sometimes amount to several dioptries. The excavation extends over the whole of the disc and at its edges is undercut, which explains the partial disappearance of the arteries as they climb, in a somewhat oblique direction, up its sides (see Figs. 106 and 107). If there has been a previous physiological

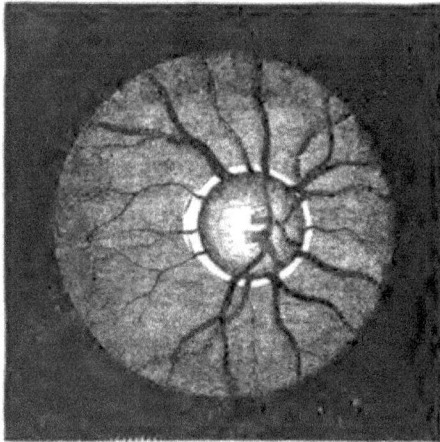

Fig. 106.

excavation, this will be added to the pressure-effect, and if the case chance to be one of myopia, with adjacent choroidal atrophy, a slope of the floor of the pit in this direction will be observed. It will be seen that the conspicuous facts of glaucoma simplex are increased tension, possible reduction of central vision, impairment of the field, especially on the nasal side, a general excavation and pallid look of the optic nerve, and spontaneous, or easily excited pulsation of the arteries on the disc.

There are subjective symptoms which are liable to occur, yet are uncertain. Almost always there is a notable diminution in the range of accommodation, that is, a retirement of the near point has taken place. By it a resort to uncommonly strong glasses for reading has been necessitated, or glasses have had to be adopted prematurely. Attacks of sudden obscurity of sight have taken place, when, for some minutes, everything became dark, even in good daylight. Some painful sensations have been noted about the eyes or brows. It can happen that one eye shall have become almost blind, with pronounced glaucomatous symptoms, without having excited the patient's suspicions of his loss. It is usual for one eye to be affected for some time before the other. The disease may occupy as much as five to fifteen years for its full development. Often it is mistaken in old persons for senile cataract, especially because the lens of old age has a smoky hue. This explanation of failing sight is considered adequate, and a resort to skilled advice is discouraged. Under this mistake the patient is told to wait until the cataract shall be ripe before going for relief, while the delay is simply affording time for total loss of all chances of recovery.

The full development of glaucoma may be reached by simple continuance of the foregoing symptoms until the so-called chronic or absolute glaucoma is established. There will then be augmented hardness of the globe up to T₃, a very shallow anterior chamber, with the pupil widely and unequally dilated and fixed, the iris discolored, the cornea more **or less** anæsthetic, not resenting the touch of a twisted thread, the lens may or may not be cataractous, the pupil will have a dusky or even greenish hue, the surface of the globe will be marked by tortuous and enlarged anterior ciliary arteries, which dip suddenly into the sclera, and sometimes a marked plexiform arrangement of vessels is to be seen about the front of the sclera. Should the media be clear enough to permit ophthalmoscopy, deep excavation of the nerve, with absence of capillaries and reduction of the arteries to slender threads, with marked pulsation and turgid veins, denote the extreme stage of pressure. Often the choroid, immediately about the nerve, undergoes atrophy, and presents by exposure of the sclera a marked ring. The patient will have lost vision, both direct and indirect, almost or quite completely, and have considerable pain in and about the eye.

Glaucoma with inflammation.—Inflammatory attacks may supervene upon glaucoma simplex and may be of various degrees; on the other hand, an attack of so-called acute glaucoma may take place in eyes which have had no objective and, perhaps, few or no subjective signs of the disease just described. It is not rare that a person in whom one eye has passed to **an advanced** degree of glaucoma should appear in the other to be entirely free from all morbid symptoms, and yet have occur in the unmolested eye an outbreak of acute glaucoma. Moreover, both eyes may be simultaneously attacked by such an inflammation without noticeable **warning** or lesion.

The symptoms are as follows: the person is apt to be taken during the night with severe pain in the eye and forehead; congestion of the scleral and conjunctival vessels rapidly develops, with subconjunctival œdema and swelling of the lids; within a few hours chemosis and tumefaction of the lids become extreme, and pain in the head **and** eye intense. There may be serious constitutional symptoms in **rise of** temperature and pulse, in vomiting, and tokens of a so-called bilious attack. On inspection of the eye the cornea may be hazy, although it may be clear, as I have verified; the aqueous will be turbid; the iris and lens pressed toward the cornea, reducing greatly the depth of the anterior chamber; the pupil will be obscured and dilated; the iris discolored, and diminished to a narrow ring. It is not easy to feel the tension through the swollen lids and chemosis, but after pressing through the boggy tissues the globe will offer great resistance to the touch, and the pressure will be painful. A view of the fundus is often impracticable, and always imperfect, because of the turbidity of the media; but the optic nerve is often red, and usually not excavated (if the attack be the first one); pulsation of arteries on the disc can seldom be made out, but is to be expected. The inflammatory effusions are chiefly serous, but adhesions of the iris may occur, although not so marked as in ordinary iritis. Purulent exudation and decidedly plastic exudations do not appear. Of course vision is greatly impaired, and sometimes is wholly annulled within a few hours. It is not to be expected that, under such circumstances, a careful examination of the field can **be made.**

Subacute glaucoma is a phase in which vascular engorgement in the **ciliary** region is more or less pronounced, the iris altered in texture and appearance, sometimes displaying an erratic and conspicuous vessel upon its surface, the anterior chamber shallow, the pupil enlarged and sluggish, **tension** augmented, the **vitreous** usually clear, the pit in the nerve and the pallor and displacement of vessels decided. The case simply exhibits moderate degree of **vascular** obstruction and hyperæmia.

Chronic glaucoma presents a picture already detailed as the ultimate state of glaucoma simplex. We may or may not have conspicuous vascular symptoms in the final stage, but we shall **seldom fail** to find in the vessels of the ciliary region great fulness and tortuosity **of the** principal branches, and the abrupt way in which they spring from the sclera, or are convoluted, will arrest attention. Added to these will be all the tokens of intra-ocular pressure above stated.

Glaucoma hemorrhagicum.—Under this name **is** grouped **a class of cases** which might in strictness be set down among those **of secondary glaucoma.** There is effusion of blood **in** the retina or optic **nerve, and** sometimes in the vitreous. It is generally true that no preliminary signs of glaucoma have occurred, but a sudden loss of sight is followed in **a** little time by pain and inflammatory symptoms. The point of distinction between this lesion and acute glaucoma is that the loss of sight precedes by some interval the acute inflammatory attack. On examination, the eye shows increased tension and turbid media, while, if the fundus be visible, patches of extravasation, more or less extensive, will be discerned. The pupil will be sluggish, and greater or less ciliary hyperæmia exist, but great variety may be found in the symptoms. The point to be dwelt on is that in these cases the hemorrhage is the occasion of the outbreak of acute symptoms, and there may have been no occasion for glaucomatous signs previously. One eye alone is usually affected and the other is free, while in the forms previously mentioned both eyes partake in succession. It has seemed to me quite certain in some cases that the beginning of the morbid process was embolism of retinal arteries with consecutive hemorrhage, and that the glaucomatous outbreak ensued because in such persons the vessels were atheromatous, and could not readily adjust themselves to the disturbance of the circulation.

Glaucoma secundarium.—By this is meant increased intra-ocular tension, consecutive to some other recognized disease. It ensues after ulcerations of the cornea when staphyloma appears; after wounds with incarceration of the iris; after total occlusion of the pupil, *i.e.,* complete posterior synechia; after operations for cataract, with false membrane filling the pupil; as the result of wounds of the lens, causing it to rapidly swell; after dislocations of the lens forward in front of the iris, or backward into the vitreous (illustrations of the latter condition are found in the after-history of reclination of cataract); it follows upon the growth of intra-ocular **tumors;** and the conditions called buphthalmus and hydrophthalmus, which may or may not be congenital, are specimens of secondary glaucoma. It may be well to describe these last-mentioned states, although mention of them has already been made. The whole globe is enlarged, the front more than the rest; the cornea expanded into a bulbous form, and very thin—it may be bluish and semi-transparent, or gray; the boundary between it and the sclera cannot be defined; the ciliary region is widened and bluish, and may project in nodules. The anterior chamber is of great depth and width, the iris thinned and discolored, in spots its tissue may have sprung apart and left holes. If the pupil be adherent to the lens, as is commonly the case, the iris will be pressed forward at the periphery, while the pupillary edge, being held down, will be at the bottom of a little pit. The pupillary area may be filled with exudation, and the lens be opaque. Generally, no view can be had of the deeper parts of the eye, and sight is wanting, or reduced to perception of light. I have seen a traumatic case which was caused by an entanglement of the whole pupillary margin in the central wound of the cornea, and the surface of the iris was applied to the posterior surface of the cornea. The posterior chamber was enormously developed, as the anterior chamber usually is.

Diagnosis.—No words are necessary to set forth the distinctive features of a well-marked case of glaucoma of any type, in addition to what has already been stated. But there are some conditions under which the recognition of the disease may require careful attention. The first to be mentioned is during the early period of glaucoma simplex. No dependence can be placed upon the external signs, because the eye looks healthy. The state of vision may be normal, or nearly so, and subjective symptoms may be obscured or have been unheeded. Tension may be doubtful. We are then shut up to two symptoms, viz., the appearance of the optic nerve, and a careful scrutiny of the visual field. As to the nerve, the excavation may be partial, involving only the temporal half, and the nasal side may be of normal level and red. There may be a vertical dip between the two levels, and the vessels climb over the steep edge partially concealed. The depressed side of the nerve will be white, dotted, and resemble atrophy, the lamina cribrosa being distinct. There may be spontaneous venous pulse, and the arterial pulse be easily evoked. To get this symptom to the best advantage, use the upright image; hold the ophthalmoscope in the left hand if examining the right eye, and use the thumb or forefinger of the other hand to both keep up the lid and make pressure. Much importance attaches to the facility with which arterial pulsation can be produced. In mapping the field, use a small bit of chalk or card about 10 mm. square, and avoid strong light. The perimeter is indispensable, because the peripheral part is the suspicious region. Limitations occur most often on the supra-nasal side, and any region of the nasal side is more likely to suffer than the temporal. But this rule is far from being of general application. The most diverse kinds of encroachment will be found in this disease. Scotomata may exist and no peripheral limitation. These can be best determined upon a black-board, and intelligent patients will sometimes describe them better than the physician. The liability to mistake in the cases now considered, is in regard to atrophy of the optic nerve. This may further be enhanced by the presence of cerebral symptoms, such as dizziness and headache. The lesion in both cases is undoubtedly in the optic nerve, while for glaucoma the symptom to be chiefly valued is ready pulsation of the retinal arteries—of course supposing that increased tension cannot with certainty be felt; the limitation of field, if discovered, will not be decisive.

Chronic glaucoma ought not to be confounded with senile cataract, yet this too often happens. Real cataract may coexist with glaucoma, but will be seen by oblique illumination, and certainly be discovered if the ophthalmoscopic mirror is used. Touching the eye will easily prove that the globe has become hard, and the pupil will generally be dilated. No reliance is to be placed on the smoky hue of the lens, because this belongs to its senile condition. There will be cause for suspicion in the defective power of light-perception, and in probable impairments of the field, neither of which signs exist in simple cataract.

Acute inflammatory glaucoma may be confounded with iritis. The distinctive features will be the state of tension, the sudden and remarkable reduction of sight, the vehemence of the hyperæmia and œdema, with a dilated pupil and the absence of adhesion. I have seen a case of subacute glaucoma with posterior synechiæ, which had arisen during an acute attack not long precedent. Such a concurrence is uncommon, and indicates that **care must be given** to determination of the tension as the important feature. Under such conditions the ophthalmoscope will not help us.

Prognosis.—The disease tends, with more or less rapidity, to total loss of sight. The cases of glaucoma simplex progress for years, with slowly creeping failure of sight. If an acute inflammatory attack take place, sight for the time may be wholly abolished, and it may, when the storm is over, return, but in reduced quantity. It may never return. Both eyes will eventually become concerned. For hemorrhagic glaucoma no good result for vision can be expected, and the eyeball will often have to be removed. For secondary glaucoma, every case must be judged by itself; but the tendency, as in all other forms, is to a bad issue.

Etiology.—We enter here upon the field of speculation, and at the present time opinions vary more than twenty years ago, when Graefe almost gave the law to ophthalmic opinion. For inflammatory glaucoma, **on** which he chiefly dwelt, he assumed a serous choroiditis. Donders **called** attention sharply to simple glaucoma, to which the choroidal theory **would** not fit, and invoked the influence of perverted nervous action, while experimentally Grunhagen and Hippel demonstrated that irritation of the fifth nerve would cause glaucomatous tension in the eye of rabbits. Within five years, anatomical study has shown that an almost constant lesion in chronic glaucoma is adhesion of the periphery of the iris to the border of the sclero-corneal junction, thereby choking up the apertures in the so-called space of Fontana leading to the canal of Schlamm. Knies, of Vienna, first, and, at nearly the same time, Weber, of Darmstadt, stated this theory. Weber accounts for this peripheral iris-adhesion on the supposition of swelling of the ciliary processes. Brailey, of London, found, as a constant lesion, atrophy of the ciliary muscle, and thinks its retraction diminishes the egress of fluid at the angle of the anterior chamber, by causing the adhesion of the iris. More recently, Priestley Smith, in an extended and admirable essay, and in subsequent papers, thinks the cause of glaucoma lies in diminution of the space between the border of the lens and the ciliary processes, whereby the transit of fluid from the vitreous to the anterior chamber is retarded, the periphery of the iris is pressed into the angle at Fontana's space, and increased tension results.

All of these theories assume the primal step and the essence of glaucoma **to** be increased tension. By it we have the nerves of the cornea stretched, and its surface rendered anæsthetic—for a similar reason the nerve-twigs to the sphincter pupillæ are paralyzed, the excess of pressure in the vitreous pushes forward the lens and iris, and reduces the anterior chamber; the circulation through the ciliary and choroidal vessels is hindered, and the anterior ciliary vessels become supplementally distended. The function of the ciliary muscle is also impaired; the retina and optic nerve lose their function, and the latter undergoes the change of level known as pressure-excavation. As part of this process, the choroid is drawn away from the edge of the papilla, and gives the scleral ring.

In opposition to these views, Jaeger, of Vienna, contends for a primary lesion in the optic nerve, and Mauthner, of Vienna, while holding to an optic-nerve lesion, also calls in the aid of a peculiar form of choroiditis. The latter author quite underrates the value of increased tension as a factor in the symptoms.

It remains true, however, that the only effective treatment of glaucoma is such as will reduce the tension, and, therefore, while we admit many difficulties in the way of a satisfactory explanation of all the symptoms, and recognize the occurrence of not a few puzzling cases, we hold to the fact of supra-normal tension as the key to the situation. Just how this is brought about does not yet clearly appear, but the cause is doubtless mechanical, as Weber insists, and may perhaps be found in the narrowed "circum-lental" space of Priestley Smith, or in the swollen ciliary processes of Weber. Just how atrophy of the ciliary muscle can be effective, or what incites it, is somewhat hard to understand. The increase of tension is behind the lens, and if this do not in itself fully account for the excavation of the nerve, it promotes it. This lesion of the nerve will probably be found to have indirect relations to plus tension which are not yet understood, and perhaps the circulation at the posterior choroidal circle

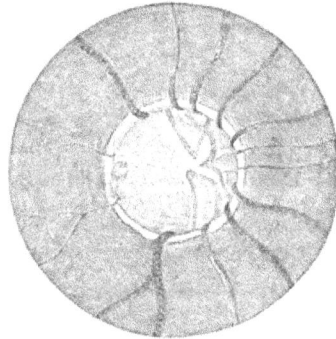

FIG. 107.—Shows a deep, glaucomatous excavation of the optic nerve. The vessels on the nerve-surface are pale, because out of focus; they are not continuous with the vessels on the rim of the nerve. The scleral ring is strongly marked.

about the papilla will prove to be the region at fault. It is certain that choroidal lesions in this vicinity are not infrequent, as my notes of cases relate.

Another condition, which was formerly assumed to have causative relations with glaucoma, was supposed *rigidity of the sclera*, but this has not been verified by pathological anatomy. The fact that many times hyperopia coexists with glaucoma, has attracted attention, but in what way the two conditions can become related is not known. Certainly the state of the ciliary muscle in H, which is one of hypertrophy, seems antagonistic to Brailey's theory of its atrophy as a cause. Yet a hypertrophied ciliary muscle might be supposed to readily occasion swelling of the ciliary processes, and thus favor Weber's theory, and likewise that of Priestley Smith. Glaucoma in myopia is liable not to have the sharply excavated optic pit, but one which slopes toward the region of circum-papillary choroidal atrophy, and it is not always easily recognized.

Treatment.—The merit of having discovered that iridectomy is capable of curing glaucoma, will stamp the name of Graefe with undying honor. Up to his time no remedy was known, and now no remedy, except an operation, is of positive value. The experience of twenty-three years has indicated some of the limitations of the operation, and has abated some of the hopes which were first cherished. The mode in which alone it can be effective was pointed out by Graefe, viz., that the incision must be in the extreme limit of the anterior chamber, and, therefore, not less than one millimetre behind the transparent edge of the cornea. It may require to be two millimetres distant from it. It must not be oblique, but as

18

nearly perpendicular to the surface as avoidance of the iris will permit. It should be from five to seven millimetres in length. Great care must be taken to prevent the angles of the iris coloboma from being caught in the wound, and the iris must be excised close to the surface. Generally, two strokes of the scissors are needful, and the iris must be well drawn out before being cut. The incision may be made with a lance-knife, or by a narrow Graefe cataract-knife.

Another operation which is now on trial, is sclerotomy. There are two ways of doing it: one is to make an incision with a lance-knife, but this is not to be recommended, because, if large enough to be effective, prolapse of iris is almost unavoidable, and is disastrous. The suitable manner is that of Wecker. The pupil is brought to decided contraction by eserine. A Graefe cataract-knife is entered at the angle of the chamber, about three millimetres above or below the horizontal meridian, is pushed across, and emerges on the opposite side at a corresponding point. The incision is carried upward by a to-and-fro motion, until the summit of the arc is almost reached. Then, by turning the edge of the knife forward, all the aqueous is allowed to escape, and the knife cautiously withdrawn. An undivided bridge of limbus is left, and, by careful management, the iris is neither wounded nor prolapsed. With a very shallow anterior chamber, it might be impossible to carry out this operation. The use of stop lance-knives, one bent and the other straight, has been proposed for such cases as a method of sclerotomy (Roeder). The efficacy of the proceeding, however performed, is not fully established, although some have written strongly in its favor. My own experience with it is small, and leads me to favor iridectomy in critical cases, where, for instance, but one eye remains.

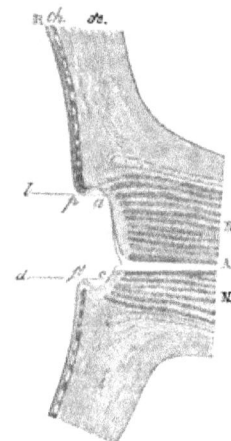

FIG. 108.—Shows a glaucomatous excavation in profile: *a, c,* the surface of the nerve; *p, p'*, the edges of the pit, and behind the vertical lines *d, c,* and *b, a,* the hollow is seen along which the vessels pass and are concealed from view; N, optic nerve trunk; A, sheath of vessels; R, retina; *ch,* choroid; *sc,* sclera.

The curative value of an operation is greatest in acute and subacute inflammatory glaucoma. In these cases the result may be perfect, even though the operation be not strictly in accord with rules. The longer the delay, the less favorable the prognosis. A week, or even two weeks, has not proved too long for restoration of sight. But the gravest uncertainty attends delay, and Graefe has said that after the third day the restitution is sometimes very imperfect. It is true that an attack may pass off and sight be retained if no operation be done. In chronic glaucoma the measure of benefit will depend on the stage of the disease. If no central vision remains, but little good will be gained. If there be no perception of light, and the eye be painful, an iridectomy may be unavoidably attended by loss of vitreous, or by intraocular hemorrhage, or by other accident, and lead to aggravated inflammation. Then nothing but enucleation will remain. In very painful cases of absolute glaucoma, enucleation is often done. But, as a less afflictive proceeding, sclerotomy may be attempted, and a few cases of optico-ciliary neurectomy have been reported with success.

For glaucoma simplex, iridectomy is the remedy which has the weight of authority. But while some cases yield a perfect result, others startle us by exhibiting worse vision after the operation than before it. This is measurably explained by the enlargement of the pupil and by the astigmatism which is very liable to be produced. On these accounts the iridectomy should be done upward, that the drooping lid may cover some of the coloboma. But the anatomical features of the eye and the lid sometimes forbid this mitigation, and a few cases are recorded in which a great reduction of sight ensued, which could be explained only by the direct mischief of the operation. If limitation of the field encroach almost upon the macula, central vision is pretty sure to be lost by the operation. To the cases of glaucoma simplex, late experiences show that sclerotomy is better suited than iridectomy. Hemorrhages into the retina and nerve are liable to occur in all forms of glaucoma after iridectomy, and the loss of sight thus occasioned will be regained in from four to eight weeks. The exceptional cases now alluded to are few, and cannot weigh against the well-known evil tendencies of glaucoma when a clear case is made out. But for the cases of doubtful diagnosis where signs are not positive, and vision, either centric or eccentric, is not distinctly implicated, the deplorable possibility of mischief thus hinted at should be squarely considered before advising iridectomy. Such a case should be subjected to rigid examination every few months, if at the time it is thought best not to operate.

It remains to state, as an additional drawback to iridectomy, that its performance upon one eye may be the occasion of an acute outbreak in the other and hitherto perhaps unimpaired eye. Of course the second eye must then be operated on. Cases of this kind are infrequent—I have seen but one instance. But, after all, it must be emphatically declared that glaucoma knows no cure but by an operation, and that the most favorable results are gained when it is done at an early period. We find that, as an immediate effect, the tension is not always reduced to the normal, and the longer it continues plus, the less satisfactory will be the issue. The excavation of the nerve will rarely be found much lessened. Sometimes the globe will in a few weeks regain its former state of increased tension, and for such a condition a second iridectomy should be done at a point directly opposite the first, or sclerotomy may be done. In a very few cases the operation fails to make any good impression on the tension and function of the eye, even when the case has not reached the stage of glaucoma absolutum. Graefe called this malignant glaucoma. Weber has offered an explanation and a remedy for this critical plight. He assumes that the lens, which by the loss of aqueous at the iridectomy comes forward, remains in this position, and by its edge continues to obstruct the angle of Fontana. He proposes to press the lens back, after the eye has recovered from the operation, by doing paracentesis of the vitreous, and, while the wound in the sclera is held open by rotating the Graefe knife, the finger is to be pressed on the cornea through the half-shut lid, and the pressure is to be made to bear especially over the coloboma. By simultaneously reducing the vitreous tension, and forcing the lens back, the angle of the chamber is liberated for a better performance of filtration, and he declares that permanent abatement of tension follows.

One of the not uncommon effects of iridectomy is that the cicatrix remains distended. The wound closes, but after a time the ocular tension rises, and the scar being weaker than any other part of the outer coat of the eye, bulges into a cystoid form. It has been assumed that this cystoid scar has much to do with abating intra-ocular tension, by permitting filtration of fluid through its attenuated walls. As an actual fact this is not necessary to the proper fulfilment of the purpose of the operation, but is to be looked upon as an evidence that the iridectomy has not accomplished to the full degree its purpose. For not a few cases of cystoid scar, the explanation is that the iris has been caught in the wound, either at the sphincter, or by a surface attachment. Sometimes this imperfection in the operation cannot be prevented, because the loss of aqueous does not lower the tension enough to permit the iris to be completely returned ; hence conditions favorable to cystoid scar are prepared.

In my own experience, I find that in studying 37 eyes, of twenty-four patients, 26 may be put down as having undergone abatement of tension by the iridectomy; while in 3, of three persons, no reduction was effected ; and in 9, belonging to six patients, the tension was aggravated. It cannot be affirmed of the 26, that in all the tension was brought to the normal, but an approximate idea is gained of what the operation actually accomplishes. The aggravation of tension occurred in both eyes of three persons having glaucoma simplex whose vision was $\frac{20}{20}$, $\frac{20}{20}$, and in the third was unrecorded. In the two persons with good sight this was not changed, while the sight of the third was not improved in one eye and was impaired in the other. In another case, both eyes of the patient with chronic glaucoma underwent reduction of vision as a result of iridectomy not followed by diminution of tension.

The explanation of how iridectomy effects reduction of ocular tension is not perfectly satisfactory, but we are to look for it in the case of acute glaucoma, in the sudden loss of aqueous and of blood, and the relief of pressure, by which the engorged circulation is enabled to regain its equilibrium. The increased effusions into the globe are permitted to drain away through the wound during some hours of its patency, and this suffices to bring about restoration in an eye whose proper filtration channels have not been extensively obstructed. But in chronic glaucoma we have more difficulty in understanding how any good is effected by an operation which directly concerns the filtration region for a space not more than one-sixth of its whole extent. We have to assume that a very slight improvement in the local conditions, by tearing off the iris at or near its periphery, enables the fluids to get out with increased facility. This is, moreover, despite the fact that no operation ever goes through the canal of Schlemm or the angle of Fontana, and that a stump of iris is always left. Further investigation is needed, and to such work as has been done by Priestley Smith, are we entitled to turn with hope of future elucidation.

The effect on the field is often negative ; in fortunate cases an extension of its boundaries may be discovered, but frequently this is not obtained when central vision has been improved.

Sometimes the operation is followed by severe reaction—leeches may be needful ; if the pupil tend to adhesions, atropine should be used, as in ordinary iritis—cold or warm water may be used as expedient. Should there be free hemorrhage into the anterior chamber, this will delay the recovery. Should severe vitreous hemorrhage occur, enucleation may be necessitated. Sometimes the anterior chamber is not re-established for a number of days, or even for two weeks. We must wait, and a fair result may be attained, although this circumstance is ominous. Should loss of vitreous occur at the operation, the effect will be bad and may be panophthalmitis. Should the lens-capsule be touched, the resulting cataract will neutralize the value of the operation. Cataract may ensue when the knife has not really touched the capsule, but when aqueous has been allowed to escape with a gush, or when the pressure of the lance-knife has been excessive before it enters the chamber, or if vitreous have escaped. In these cases the suspensory ligament has been probably torn, and this accident favors the slow production of cataract.

It must be remembered that cases of glaucoma somewhat advanced, and with shallow anterior chamber, require skilful hands for their proper treatment, and all that has been said regarding beneficial effects implies the ability to do the operation correctly. For acute glaucoma, an imperfect operation may suffice, but for other cases, crude surgery is at the peril of the patient.

Some stress has been laid upon the unfavorable influence of atropine upon a glaucomatous eye. There is no doubt that aggravation of symptoms has been thus produced. I have, myself, distinctly observed it in one case. But such is not a constant result. There is no reason why it should be used even for ophthalmoscopic examination, and the occasional cases in which mischief occurs precludes it entirely. On the other hand, eserine has temporary good effects, and may be freely used as a preliminary to an operation, or to tide over a period of unavoidable delay in seeking competent aid. It may be used in solution of sulphate of eserine, gr. iv., aquæ, ℥ j., every two hours during the day. But a few cases give us caution that it too will sometimes excite irritation, and it must then be withdrawn.

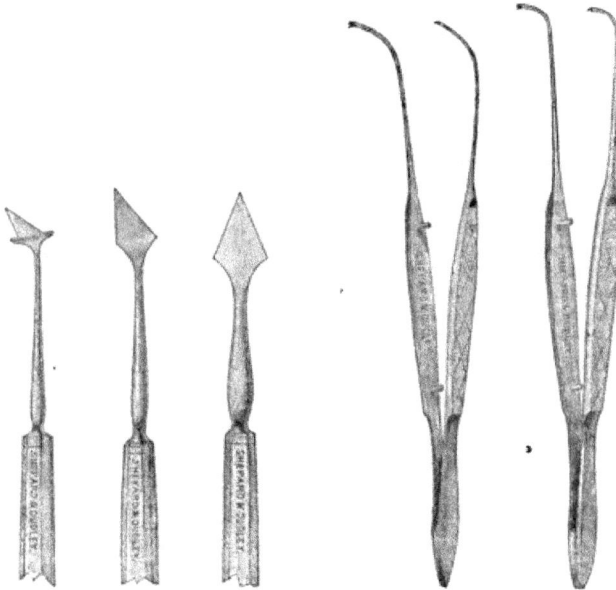

Lance-knives for Operations on the Iris.—The first is a stop-knife; of the others, one is bent and the other straight.

Iris-Forceps of Different Curves.

CHAPTER XV.

THE OPTIC NERVE AND RETINA.

Anatomy **and** *physiology.*—The optic nerve may be considered in three separate regions : its cranial, its orbital, and its ocular portions.   Within the skull we find the optic tracts winding around the peduncles of the cerebrum, and uniting into the optic commissure, from which the optic nerves proceed to the orbits.   The optic tracts are flattened bands, which have three roots, two of them white and one gray.   These roots come in connection with the corpora geniculata externa and interna, with the tubercula quadrigemina, with the **optic** thalami, and, finally, fibres are sent **off,** by which they reach the surface of the brain at the occipital region, and also enter into the spinal **cord.   It is not** expedient to trace out in **detail** these relations ; but the **parts mentioned** have important relations to **vision, and the most recent discoveries show that the** gray matter of the **occipital lobe, not far from the gyrus angularis, is the** ultimate origin of the optic-nerve **fibres.   It is also believed, from** physiological experiments and deductions **of** Charcot, **that the optic tracts** from **the** two sides come into relation with each other **in the tubercular quadrigemina,** and that, in conjunction, they pass **to** the occipital lobes of **each side.**   By this arrangement a destructive lesion in the occipital lobe **can** cause total blindness, or amblyopia, of the opposite eye, with preservation of **the** corresponding eye. The optic tracts unite to form the commissure, **or** chiasm, and emerge under the name of the optic nerves.   In the chiasm, the fibres interlace.   The chiasm rests upon the body of the sphenoid bone in the sella turcica, and the crossing fibres are said by Gudden to be upon the lower surface, the direct fibres upon the upper surface.   The anterior and posterior commissural fibres of the chiasm, which were formerly described, are proven not **to** be nervous, but connective tissue.   Part of **each** tract, and the lesser quantity, goes to the eye of the same side, and **the** remainder to the eye of the opposite side.   The *partial* decussation of **the** fibres has been within a few years vigorously controverted, but general opinion is settling down to a belief in its correctness.   The tract of the right **side** supplies the right half **of** each eye, the division being at the macula lutea ; the tract of the left side supplies the left half of each **eye.**   It follows that the inner or nasal halves of each eye are supplied **by the crossed** fibres, and the outer halves by the corresponding fibres of **each tract.   As** the outer parts (temporal) of the visual fields **are** much more important than the inner (nasal) parts, it is clear that **it is** fitting that **the** larger portion **of** the fibres of the tracts should be those which pass through the chiasm.   Each optic nerve reaches the orbit by the foramen opticum, or canalis opticus, and adheres closely to the edge of the canal by its external sheath.   This attachment is very firm, **a** strong pull will not loosen **it.**   (It may be well to call attention to the close proximity of the canalis opticus and of the cavernous sinus which receives the ophthalmic vein, and of the carotid artery, which sends off the ophthalmic branch to accompany the nerve in its passage through the foramen. The ophthalmic vein passes through the ethmoidal fissure.)

From the chiasm to the foramen, the nerve is usually about ten millimetres long, while its orbital portion is twenty-eight to twenty-nine millimetres long. It is round, and about four millimetres in diameter. It touches the globe about four millimetres to the nasal side of the optic axis and a little below it. It passes in the midst of the ocular muscles, and is surrounded by fat and connective tissue. It is somewhat sinuous in its course, and is necessarily longer than is a straight line from the apex of the orbit to the globe, to permit free movement of the eyeball. The nerve is provided with a double sheath, the external, c, coming from the dura mater and being the thicker, the internal more delicate and closely attached to the trunk, and constituting its neurilemma (see Fig. 109). Between the sheaths is a space occupied by delicate trabeculæ of connective tissue and by lymph, b. This is continuous with the arachnoid cavity of the brain, and may be injected from it (see Fig. 10, G. & S., p. 17). A

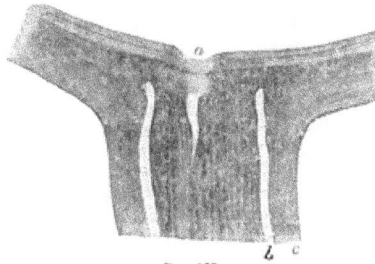

Fig. 109.

still greater anatomical refinement is given by Henle, in making a pial sheath to the nerve and two lymph-spaces. At a point varying from fifteen to twenty millimetres distant from the globe, the nerve is pierced by the arteria and vena centralis retinæ, which enter it obliquely and pass into the eye. In Fig. 109, a indicates a slight physiological excavation.

As the nerve gains the eye, its external sheath mingles with the outer part of the sclera, the internal sheath passes inward a greater distance, and mingles with the inner layers of the sclera. The nerve is subdivided into numerous fasciculi by connective tissue, but exhibits a more compact mass when seen in cross-section than do most nerves. On passing within the globe, not only are the enveloping sheaths left behind, but the septa which isolate the fasciculi also turn aside and become attached to the adjacent sclera. The fibres also lose their proper sheaths, and are reduced to naked axis-cylinders. Thus liberated from all its accompanying connective tissue, the nerve becomes transparent and of less diameter, and enters the eye under the name of optic papilla. There is a mesh-work of fibrous tissue interwoven among the nerve-fibres at the level of the sclerotic opening, which is called the lamina cribrosa. It is made up of the connective-tissue sheaths and septa above mentioned. This structure is a limit beyond which inspection of the optic nerve by the ophthalmoscope is impossible, and it is more or less visible according to circumstances. The inter-vaginal space passes into the sclera for a slight distance, viz., as far as the lamina cribrosa. The optic-nerve fibres before they gain the retina must not only pierce the sclera, but the choroid likewise ; this they do through a circular opening. The edge of the opening is sometimes in close contact with the nerve, and sometimes a small space is left, through which the sclera can be seen from within as a ring.

The facts about the appearance of the optic disc, and the distribution of its vessels, have been stated on page 32. It may be proper to say a word about the vascularity of the papilla, as shown by Leber's injections. A circlet of vessels surrounds the nerve-head, which bring it into communication with both the choroid, the sclera, and the optic sheath, as well as with the retina. A much greater vascularity is furnished at this spot than at any point outside the globe.

It will be proper to say something more in detail respecting the structure of the retina, while most of the facts have been related on page 6. It consists of ten layers, which, beginning from the inner surface, are named as follows: 1, membrana limitans interna; 2, optic-nerve fibres; 3, ganglion-cells; 4, internal granular or molecular layer; 5, internal granules; 6, external granular layer; 7, external granules; 8, membrana limitans externa; 9, rods and cones, or bacillary layer; 10, hexagonal pigment epithelium. Of these layers, the first is independent of the hyaloid membrane of the vitreous. Besides them, the retina is traversed by numerous fibres of connective tissue which run perpendicularly through it, and are known as the fibres of Müller. Of the true character of the several layers of the retina, we are as yet imperfectly informed. The adjacent diagram from Schwalbe gives his views of their relations. The layers, viz., 2, optic-nerve fibres; 3, ganglion-cells; 5, inner granules; 7, outer granules; and 9, rods and cones, are the nervous structures; all the others belong to the supporting structure of the retina, and are considered as modifications of connective tissue. The rods and cones are ranked as the most essential. At the fovea centralis certain changes occur. The optic-nerve fibres (No. 2) disappear, the ganglion-cells (No. 3) increase in number, the rods (No. 9) are not found, and the cones (No. 9) become narrower, longer, and much

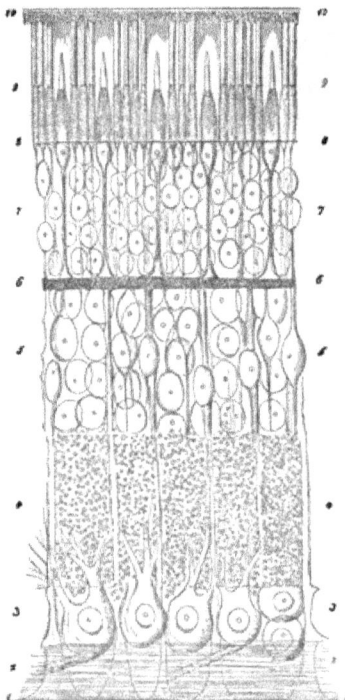

Fig. 110.

more numerous, and the external granules (No. 7) remain. All other layers are reduced to a minimum (see Fig. 5). The fovea is the most sensitive spot of the retina, the place at which direct vision occurs. As we leave this spot, visual sensibility rapidly declines, as has been mentioned on p. 10, both because the light is less correctly refracted, and because the perceptive power of the retina is inferior. The usual appearance of the optic nerve and retina through the ophthalmoscope has already been stated (see p. 32). It now remains to speak of certain peculiar congenital variations, of which the most important, and not very infrequent, is the presence of

Opaque nerve-fibres (see colored Plate I., Fig. VII.).—By this is meant that certain fibres coming from the papilla are not divested of their neurilemma, but retain it for a distance after they enter the retina. They are grouped into a brush or cluster, and exhibit considerable variety of form. They may be short and extend but a little distance, or sweep in a long, wispy plume, and usually along the principal vessels. I have seen a case in which most of the centre of the fundus was whitened by such a condition. To recognize the nature of this peculiarity, observe that the white or yellowish patch conceals the edge of the disc, has a glistening, striated surface, and the markings are in parallel lines ; the edge is marked by a hair-like fringe. The surface is intensely white or yellowish, the vessels are partially concealed, and fine glistening markings appear upon it. Surrounding parts of the fundus will be normal ; sometimes the optic nerve looks very red, but it will not be swollen. In slight cases there will be no impairment of sight, in extreme cases there will be amblyopia. While such a patch is usually continuous in surface, cases occur in which spots of opaque nerve-fibres crop out, remote from the principal locality, but very rarely does the abnormality reach beyond the macula or its vicinity.

Sometimes a pigment-spot is seen on the face of the disc, and may be congenital. It is, however, very rare. It is extremely common to have pigment-deposit on the edge of the choroid, next the papilla, and this is not ordinarily counted pathological.

DISEASES OF THE RETINA.

Hyperæmia of the retina, in the form of enlargement of the vessels, appears either as a natural peculiarity, or as a result of obstruction to the circulation. In the former case we meet great variations in the number, tortuosity, and size of the vessels. In the latter case we may have local causes, like tumors in the orbit or obstructions of the circulation in the skull ; or we may have abnormities in the heart, such as an open foramen ovale. Capillary hyperæmia of the retina cannot be recognized. Hyperæmia of the optic nerve, apart from inflammation, is almost never idiopathic. It is the sign of irritation or over-taxation of the eye. It is seen most often in hypermetropia, in troubles of accommodation, after prolonged use, excessive weeping, or irritation by foreign bodies on the cornea, etc. In short, hyperæmia of the nerve is to be accounted a symptom, and is seldom an idiopathic disease.

Anæmia in the retinal circulation is not recognized, except as the result of disease of the vessels, of the retina, or of the optic nerve, or as that condition designated as ischæmia. Mere feebleness of circulation, apart from general weakness of the systemic circulation, has no standing in ophthalmology.

ISCHÆMIA RETINÆ.

A rare occurrence, to which this name has been given, is when the retinal arteries and veins suddenly become extremely reduced, amounting almost to absence of blood-current, and with it is entire loss of sight. This is not attended with any tissue-changes in the retina and nerve, as after embolism or thrombosis. Usually, both eyes are affected. The disease was first mentioned by Alfred Graefe, who saw two cases, and six other cases are recorded. Of these, three are in American literature, two by Pooley, in "Trans. of Med. Soc. of New York," and one by Knapp, in *Arch. of Oph. and Otol.*, vol. iv., p. 448, 1875. The last case I saw in consultation. It was in a child of three years, greatly reduced by whooping-cough, who, when convalescing, suddenly became totally blind, losing perception of light. In both eyes the nerves were white; in the left, some arteries as fine as threads were discernible; in the right, no arteries could be found. In both eyes the veins were of extreme tenuity. There were no other lesions. In one of Dr. Pooley's cases both eyes were totally blind; in the other, one eye retained $V = \frac{2 \cdot 0}{2 \cdot 0 \cdot 0}$, and the other was blind. All the cases reported were in persons who had been much prostrated by previous illness. This has been thought to account for the failure of the retinal circulation, because of the feebleness of the heart's action, and something may be ascribed to the poor quality of the blood. But Albrecht **v.** Graefe made a series of observations upon patients in cholera, and examined the retinal circulation when they were in the last stages of prostration, and never failed to find some blood in the vessels, and the usual degree of sight so far as could be tested. Hence, he inclined to the opinion that some obstructive cause must be looked for in ischæmia retinæ, and thought an effusion into the optic sheath might be fairly assumed. We have never had an autopsy in such cases, and are therefore uncertain as to their real pathology. All have recovered after a few days. In all, a treatment has been adopted to invigorate health and promote the action of the heart; the patients have been kept in bed, and this alone has in some cases sufficed. In all other cases iridectomy has been done, or paracentesis corneæ, to favor the flow of blood into the retinal arteries, and with good effect. It is safe to wait in these cases for twenty-four or forty-eight hours, and if the use of food, stimulants to the heart, and rest, do not restore the circulation, paracentesis corneæ should be done. As above intimated, the prognosis is favorable.*

Hemorrhage into the Retina

occurs as a complication of other lesions, or idiopathically. It is only of the latter that it is proper now to speak. Spontaneous hemorrhage happens in persons who have the hemorrhagic diathesis, in cases of pernicious anæmia, and as the effect of disorders of menstruation; also from degeneration of the blood-vessels, from cardiac hypertrophy and valvular disease, from pulmonary emphysema after severe coughing or sneezing; also in advanced cases of atrophy of the choroid in myopic eyes, as an occasional result of iridectomy for glaucoma, or from injuries. It may or may not be possible to see the lacerated vessels. The degree of extravasation may be great or small, in a single patch or many, and upon any portion of the retina. It sometimes occurs on the optic nerve. Usually, it does not rise above the level of the membrane, but it may extend into the vitreous. Lacerations of the choroid may be attended by bleeding, which shall penetrate the retina.

In many cases hemorrhage is likely to be repeated, because the constitutional or remote causes are not removable, or only partially curable. In some cases a retinal hemorrhage is the forerunner of a cerebral extravasa-

---

* The term ischæmia retinæ is used by Allbutt (On the Use of the Ophthalmoscope) synonymously with choked optic disc, which is an error of nomenclature, for the term had been appropriated by Alfred Graefe in the above sense.

tion. It is, therefore, important in all these cases to study the remote as well as the immediate occasions for the occurrence, because not only will the possibility of recovery be often bound up with them, but the welfare, and even the life of the patient may be concerned. If the hemorrhage be not at or near the macula, the injury to sight will be moderate and be occasioned in part by slight turbidity of the vitreous, while a larger bleeding will do mischief to other regions of the retina by interfering with the conductivity of the nerve-fibres, and possibly by setting up inflammation. When it occurs in the fovea centralis, a dense central scotoma ensues, and sometimes the patient speaks of seeing things of a red hue. Abundant hemorrhage may deeply cloud the vitreous, and it is often impossible to trace the source of the effusion until the media have cleared up. The most common starting point in such severe cases is the ciliary body or choroid, while from the retina the bleeding is usually less copious, and, therefore, confined to its own tissue. Because the retinal vessels are situated in its anterior layers, it follows that recovery of sight is often possible, but if the blood should invade its posterior layers, more or less dimness of vision must be expected to remain.

The blood may be absorbed without leaving any visible alteration of tissue, while, on the other hand, a pigment-deposit may occur, or a whitish substance be left, which indicates inflammatory exudation, followed by secondary changes. When bleeding takes place into the macula, it is rare that sight is ever perfect, and even if much be gained, metamorphopsia is liable to remain. A word may have some of its letters displaced or distorted, and the letters on either side of these be natural.

It follows from the above, that prognosis of retinal hemorrhage includes questions of general health, and as to recovery of sight, improvement may usually be expected, but complete restoration is uncertain.

Treatment.—The eye must be kept at rest, glaring light avoided, colored glasses employed as needful. The indications are to prevent recurrences, and to promote absorption. Under the former head are to be included all available means of avoiding the causes which brought on the mischief, and this we leave to the treatment adapted to the special case—for instance, digitalis in heart-disease, restoration of menstruation, etc. To promote absorption, the same regard must be had to general symptoms, and while for the plethoric, leeches to the temple may be used, for the anæmic and feeble, tonics, generous diet, and sustaining measures must be employed. Sometimes mild diuretics are appropriate. For those in whom apoplexy of the brain is to be feared, special care must be given to avoid mental strain or determinations of blood to the head, either through remote obstructions or direct irritations ; hence, sufficient sleep, proper action of the bowels, and the careful observance of hygienic rules, must be insisted on, with the employment of bromides. For the cases of hemorrhagic diathesis and of pernicious anæmia, the iron mineral waters at the natural springs, as well as iron and tonics in various forms, are to be stedfastly administered. To them ergot and strychnia may be useful. The absorption of blood-clots in the retina may require from six weeks to two or more years.

RETINITIS APOPLECTICA.

(Fig. IV., colored plate.)

This disease differs from the preceding in that evidences of inflammation, both in the retina and optic nerve, are added to hemorrhage. Unlike simple hemorrhage, this disease specially affects adults and elderly persons. It is associated with degenerations of the blood-vessels, with disease of the aortic valves and hypertrophy of the heart, and with the gouty, rheumatic and syphilitic diatheses. The attack comes suddenly, with great loss of sight, and without premonitory symptoms, except sometimes headache and dizziness. The lesions are usually fully declared from the start, and in a well-marked case are as follows:

The nerve is hyperæmic and swollen, its tissue œdematous, and edges indistinct—in some cases the outline is obliterated. The veins are full and tortuous, the arteries small, and some may be thread-like. The retina has a watery look, and the fundus is spattered with small hemorrhages, many of which are in short lines running parallel to the vessels. There will also be larger and irregular patches of blood, and these usually by the side of a vessel. It will be noticed that one or more arteries are exceedingly reduced, **or** may be absolutely empty, and they will be close to or lost in one of the blood-patches. Near the hemorrhages, spots of yellowish exudations will be found, which may be extensive and numerous. If hemorrhage has **been** copious, the vitreous will be a little hazy, and further obscure the retina.

The features of various cases differ greatly. Sometimes the stress of **the** lesion is on the optic nerve, which is much swollen, infiltrated and **red,** and perhaps the seat of hemorrhage, while only a limited space in the **retina is** affected. Sometimes the retinal lesion is at the macula **and its vicinity.** Sometimes the ecchymoses run along one vessel, partly **covering it up, and** spreading around it, attended with blotches of yellow **exudation,** and not far from them will be found some empty and thread-like **vessels.** In short, the picture of these cases is exceedingly various. It was the early opinion that the starting-point of this affection was an inflammation of the retina, and that the hemorrhages were sequent to the inflammation; but there is strong reason for believing that these cases are really instances of embolic obstruction of one or more arteries or arterioles, and that the inflammatory œdema, the hemorrhages, and the exudation, follow upon the obliteration of small or larger vessels. Such effects are possible in the retina, because its vessels anastomose sparingly with any others, and if obstructed cannot obtain any aid. In fact, the branches which go to the macula and to the periphery, and also to other portions of the retina, constitute regions of so-called terminal vessels in the sense which Cohnheim has emphasized, and these localities, therefore, are liable to the peculiar changes he has described in treating of embolism. On this view the retinal process is nothing but an infarctus, with the same alterations which are found in the lungs, liver and joints under similar conditions. This explanation has not been verified, so far as I know, by microscopic investigation, but it answers better than any other to the symptoms and appearances of the disease. It has been regarded as probable by Leber, and been my own conviction for several years. The source of the embolisms might be from vascular degenerations, or detritus from valvular **or** cardiac vegetations.

Prognosis.—As regards vision, the issue **is dependent** upon the extent and locality of the lesion. Entire restitution **is** rare, partial is frequent, and loss of useful sight by no means uncommon. A breach into the macula will naturally be irreparable. Atrophy of the optic nerve may ensue, and **the** spots of hemorrhage or exudation be subsequently designated by whitish tissue; or, if the blood be fully absorbed, interstitial changes in the retina may preclude useful function. The obliterated vessels are usually not restored, and vessels which at first were pervious may slowly be reduced in size under atrophic degeneration of the optic nerve.

It must also be remembered that embolism of the cerebral vessels is possible, although these cases do not, in regard to risk to life, have an outlook so serious as do many cases of simple hemorrhage.

Treatment.—If there be much cerebral excitement—and the patients are usually much alarmed—bromides will be indicated, and to their full effect; if there be heart-trouble, with excited pulse, digitalis. But, as a rule, the pulse will be small and feeble. Local depletion is to be cautiously employed, viz., the artificial or natural leech to the temple, while enjoining rest, protection, and disuse of the eyes, general quiet, and regulation of the digestive functions. To aid absorption of the blood and exudations, the diathesis of the subject may call for alkalies or other remedies suitable for gout or rheumatism; iodide of potassium in small doses may be useful; while to persons of plethoric condition, ergot may be given. In many cases, travel and change of occupation, and measures to benefit the general health, may be the only treatment appropriate after the early period of the trouble. From three to twelve months are to be allowed for completion of the changes which will go on in recovery, and some permanent traces are generally to be found in whitish spots or glistening points, or perhaps in pigmentary deposits of greater or less extent and number.

EMBOLISM OR THROMBOSIS OF CENTRAL RETINAL VESSELS.

(Fig. V., colored plate.)

The central artery and vein are both liable to be suddenly obstructed, and the effect is sudden and usually complete loss of sight. The first case of this kind was seen by Graefe in 1858. I myself saw the patient, and the diagnosis was verified eighteen months afterward by an autopsy, and the eye was examined by Schweigger. Since that case, many other somewhat similar instances have been reported, and declared to be cases of embolism. But, in 1874, Loring published a series of cases of obstructed circulation and sudden blindness, and indicated points which, in his view, threw doubt on the hypothesis of embolism. At nearly the same time, Nettleship and Zehender expressed similar doubts as to cases of their own. In 1878, Michel published a case, with autopsy, of thrombosis of the central vein, and cited the cases which he regarded as of the same kind. In 1878, Angelucci published two other cases of venous thrombosis, with autopsies, and in 1880, pictured the case seen by Zehender in 1874, and showed by the identity of its appearance, as compared with six years previous, that it must have been one of thrombosis. In 1874, Magnus published experiments upon rabbits, by which he sought to prove that hemorrhage into the optic sheath is capable of causing lesions similar to those dependent upon direct vascular obstruction, and related two cases in which he pronounced this to be the lesion. He does not bring any post-mortem examination to support his theory.

Dependence cannot be placed upon the cases reported as embolism for a correct account of the symptoms, because not only do they differ from each other considerably, but, until Michel's paper in 1878, venous thrombosis was not taken as among the possibilities, and the symptoms which he gives are also materially different from those given by Angelucci. We must therefore, use some care in asserting what the lesion is, when we find the retinal circulation suddenly obstructed. Of embolism we have 7 autopsies, viz.: Graefe-Schweigger, 1; Schmidt, 1; Sichel, 1; Nettleship, 2; Gowers, 1; Priestley Smith, 1. Of thrombosis we have 3 autopsies, viz.: Michel, 1; Angelucci, 2. We also have one examination of an eye by Loring.

The symptoms in undoubted cases of **embolism are as follows: sudden and total loss of sight**; by the ophthalmoscope **we find the arteries** extremely shrunken **and the** smaller ones invisible, **the veins also** greatly **reduced, and more so at the** optic nerve. **Sometimes there is a** broken column **in them, and** sometimes too, within the first **few days, a to-and-fro** movement occurs in the venous contents. It also happens **in some** cases that the veins are not very small, as reported by Gowers, in **whose** case the middle cerebral and the ophthalmic arteries were filled by an embolus, and the retinal veins were swollen. Perhaps this is explained by the fact that the retinal vein may have received, by reflux, blood which was retarded in other **and** adjacent veins, because an extensive territory had been disturbed in its circulation. Pressure upon the globe, no matter how firm, will **not** evoke pulsation of the arteries or veins, a proof of the **entire arrest of circulation**. The optic nerve was in the beginning of Graefe's **case (and this** is the best type we have) decidedly pale, in other cases it has **been found** red within twelve or twenty-four hours. By the autopsy, it appeared **in some** of the cases, where **at** an early period the **nerve** was red, that the **true lesion was** one both of embolism and **thrombosis,** and the latter was naturally attended by early symptoms **of** congestion **and exudation**. Hemorrhage is not a conspicuous feature of embolism, **but it appears in** some **cases,** and these (Schmidt and Nettleship) were **a mixture of embolism** and **thrombosis**. The remarkable feature of all **cases is a gray or whitish** opacity about the region of the macula, sometimes **several nerve-diameters** in breadth, and attended often by similar opacity **in and around the** optic nerve. In the middle of the retinal haze **the** fovea centralis shines out usually as a cherry-red spot. Its brilliancy is not always the same, the red sometimes being speckled with gray; but in many cases it is a most vivid, round, or transversely oval spot. About it some small **vessels of the** macula sometimes appear, and though to some extent concealed by opacity, shine out in bright contrast to their white background. The red hue of the fovea is in most cases an effect of contrast, because the choroid is seen easily through this thin part of the retina. In most autopsies, blood was found here, and in some the choroidal pigment was increased.

The retinal opacity appears at various periods, sometimes very soon, **and** again, **as** in Graefe's case, as late as the eighteenth day. The exudation in the nerve gradually increases and invades the surrounding tissues, **and the** nerve becomes more **red (see case** by Michel: *Arch. für Oph.,* **xxiv)**. After some weeks the retina gains clearness, and the nerve takes on the white hue and density of atrophy, while some **specks and traces** of deposit remain in the retina. Usually there is **no sight at** any **time,** but, in Graefe's case, some perception of light came **on for a** short **time** in the outer part of the field. It is impossible to **excite** phosphenes.

In thrombosis of **the vein** the **arteries are** diminished, but not empty **and thready; the veins are of** usual or of increased **size**; the nerve is whitish or is early congested, **or** so covered by effusion **as** to be invisible; the opacity in the retina, when it appears, comes **soon,** is dense, and may be very extensive—in two of Angelucci's cases it was wanting; the fovea will have the bright red color, if the retina be opaque; hemorrhages may be excessive and universal, or be few and limited; pressure may not cause pulsation, yet it has been produced in the veins; vision is greatly reduced and may be destroyed, but eccentric sight sometimes remains, and improvement sometimes occurs. In some cases the eyes were enucleated, because of subsequent hemorrhages and glaucoma (Nettleship, Loring), in one irido-choroiditis occurred (Schmidt). All of these (except Loring's) proved to be of mixed character, viz., both embolism and thrombosis.

It may also be stated that the sudden blindness and the intense retinal opacity appear in other cases besides those above described, viz., after severe contusions of the nerve in the orbit (Knapp), and after section of the nerve (Pagenstecher, Berlin). The latter author has experimented on frogs and rabbits, and demonstrated that the retinal opacity invariably follows nerve-section. He calls it a cadaveric change *in vivo*, and finds it affect first the optic nerve-fibres and ganglion-cells, and finally all the retinal layers. Other writers speak of the lesion as an œdema of the retina, and one (Gowers) says the retina was filled with small, round cells, resembling its nuclear layers.

The functional and organic changes of the retina are not surprising, when it is remembered that it has few vascular relations to other parts, viz., only the few vessels of inosculation at the choroidal perforation which enter the optic nerve, and still fewer at the ora serrata. That the conspicuous opacity should occur in the region of the macula is the natural result of this being the termination of the vessels.

The cases now discussed always excite much interest, but fortunately they are few. They are not seldom associated with cerebral apoplexy and with disease of other organs. Embolism is caused by cardiac disease, aortic lesion, carotid aneurism (Knapp), and arterial atheroma, etc. Thrombosis will come from marasmus, and, if associated with atheroma of arteries, hemorrhage is likely to take place, and likewise plugging of arteries; it also results from periphlebitis or phlebitis. In these cases, effusion into the sheath of the optic nerve near the globe has been found, and there may also be acute œdema of the nerve, in which case both arteries and nerves may, at the beginning, be reduced in size, and be empty in spots (Angelucci).

As to hemorrhage into the nerve-sheath (Magnus), this suggestion was made by Dr. H. B. Sands (see "Trans. of Am. Oph. Soc.," 1866, p. 2), in a case which was certainly not embolism, and had the peculiarity that for some time eccentric vision was good, while a large scotoma occupied the centre of the field. Dr. Sands' reasoning is probable, but until positive proof is afforded by autopsy, we must hold the diagnosis in suspense. Cases of hemorrhage into the sheath, with cerebral apoplexy, are recorded (Talko) but no ophthalmoscopic symptoms are given. On this whole subject we wait for further information.

Diagnosis.—Upon the general subject of the distinction between embolism and thrombosis, it must be admitted that we must sometimes remain in doubt, especially as autopsies show that both conditions may occur. It is certain that many cases have been reported as embolism which were not of this character, and I think this may be stated of that which Liebreich figures in his "Atlas," Taf. viii., Fig. IV. This seems much more likely to have been a case of thrombosis, because the circulation was not entirely suspended, vision was not wholly abolished at any time, and it also improved. Fulness of the veins, with pulsation, reduction, but not obliteration of the arteries, is certainly more in favor of thrombosis than of embolism, or even of partial plugging by embolism. Hemorrhages are more frequent and copious in thrombosis. The milk-white opacity of the retina, and the vivid redness of the fovea are common to many kinds of **suspension** of the retinal circulation.

Prognosis, as stated, is bad.

Treatment, so far as the recovery of sight goes, has no value. Iridectomy and paracentesis, to set the blood again in motion, have been tried in vain. If glaucoma or other inflammation occur, they will be dealt with as required. There may be dizziness in some cases, and to meet complications of other organs, **such as** the brain, **heart,** lungs, etc., general treatment may be needful.

RETINITIS.

In describing **the** appearance of inflammation of the retina, we are obliged to take account of the optic nerve, because in most cases both are concerned; yet it happens that each may be inflamed without apparent participation of the other. The subjective symptoms are dimness of sight in every possible degree, whether small or great; occasional limitations of the field, and so-called micropsia, megalopsia, or metamorphopsia, and sometimes irritations of the retina in phosphenes, scintillations, or glimmerings —**there** will be no pain. Objectively the eye shows no external alterations, **either** in its blood-vessels, in the action of the pupil, or in any way, unless the retinitis be complicated by other disease. By the ophthalmoscope **we find** a variety of appearances, consisting of exudations and alterations **of** tissue and of the blood-vessels. **The** following pictures in the upright image may present themselves, and are in a certain sense typical:

First.—Capillary congestion of the optic nerve without swelling, its edges partly or wholly blurred; the fibres which pass into the retina are uncommonly distinct above and below as **fine parallel lines,** which fade into the retina gradually and **suggest** a slight **degree of infiltration.** The retinal veins are turgid and **wavy, the** arteries of normal size. There may be a whitish line around the large **vessels of** faint gray opacity. No other textural change, except possibly one or two yellowish white dots in the retina near the macula. This is **a** low grade of inflammation, with slight plastic exudation.

Second.—Deep redness of the optic nerve, the edge almost or wholly obliterated by the striation of the fibres radiating into the retina about its entire circumference, its veins full and dark, the arteries large. Both are **tortuous, and the** light streak is scarcely to be noticed—the tissue of the **retina is soaked** with transparent fluid, and is evidently swollen—the pigment-epithelium is not to be recognized (*i.e.*, the normal granular look is wanting) and the reflex **from** the fundus is reduced. This is a case of moderate serous effusion.

Third.—Slight hyperæmia of the nerve; edges not well defined; vessels **not** noticeably altered; the retinal tissue pervaded by a gray, misty infiltration, seen best at the edge of the illuminated space, or by reduced light, and this cloud occupies the middle portion of the fundus; the fovea too dark, but impossible to be defined; the small vessels of the macular-region conspicuous by their number and size, and along the large vessels a border of more positive gray; slight plastic exudations, such as are often found in syphilis (see colored Plate 1, Fig. VI.).

Fourth.—A rare condition is to find a single patch of exudation in or near the centre of the fundus, partly concealing the vessels, of considerable extent, with soft edges, and no other changes, except bright hyperæmia of the nerve. A local plastic exudation.

Fifth.—At the macula we may have a group of bright, lustrous dots, while the nerve may be a little red, a condition sometimes found in albuminuria.

Sixth.—We have the nerve swollen and hyperæmic, its edges blurred, its tissue infiltrated, the adjacent retina swollen ; a little distance from the nerve there will be white patches of irregular form ; near or at the macula lutea are similar patches, which may be rounded or arranged in radii, like an imperfect star ; along the vessels blood extravasations, and others where no vessels are visible. The white plaques or dots may be of a dull hue, like greased paper, or be intensely white and glistening. Sometimes the hemorrhages are more numerous than the white deposits ; sometimes they are few, and large surfaces of white are seen. I have a picture of a case in which the fundus reminds me of a "mackerel sky" in full sunlight, so numerous and fleecy and bright are the clouds. In some cases described under this head, pigment-deposits can be seen in small amount. In certain cases, not now under consideration, pigment-disturbance and proliferation are conspicuous, and the choroid is affected. These cases are choroido-retinitis, and have been already described (see page 261). The remarkable white infiltration of the retina found after embolism and thrombosis is not included in the present description. The appearances described are most often seen in albuminuric retinitis.

Seventh.—The picture already given of retinitis apoplectica need not be repeated.

Eighth.—There are cases in which slight infiltration of the retina, haziness and redness of the nerve, and fulness of the vessels, are associated with turbidity and floating bodies in the vitreous. This is the retinitis attendant on some cases of iritis and choroiditis.

Ninth.—In panophthalmitis or suppurative choroiditis, the retina is infiltrated with pus, but the condition belongs to pathological anatomy rather than to clinical study.

Etiology.—It is rare to find retinitis as a simple or idiopathic affection. It is rather the effect of a constitutional dyscrasia, of disease of the brain, of lesions of other parts of the eye, or of injury. Exposure to extreme light and tropical heat will cause the milder types, as seen in Arctic explorers and also among travellers in hot climates. But the chief causes are kidney disease, diabetes, syphilis, menstrual disturbances, leucocythæmia, etc. The most common is albuminuria, from whatever cause ; next may be syphilis, while association with brain disease is a frequent occurrence.

Pathological anatomy.—The changes which the retinal tissues undergo have been most studiously examined, and have excited the greatest interest. I repeat the following statement from Klein ("Lehrbuch der Augenheilk.," p. 393) : "1st. We find a homogeneous, hazy exudation containing a few round cells (lymph-corpuscles), especially along the vessels. The exudation becomes gradually more turbid as it is permeated with fat-granules. 2d. Serous infiltration of the tissues, with formation of spaces and vacuoles, especially in the nerve. 3d. Proliferation and thickening of the connective tissue and supporting framework, particularly of the fibres of Müller, whereby the surface of the retina is made uneven, and its thickness increased. 4th. The external granule-layers are hypertrophied, and consequently the retina is thickened and thrown into waves, whose heights and depths the rods and cones must follow, and they are in no small degree destroyed. 5th. Fatty degeneration, both by deposit of fat-granules—for the most part in granular layers—and also by fatty infiltration of the sup-

19

porting fibrous tissue. **6th. Sclerosis of the** nerve-fibres. They seem to be transformed into a glistening opalescent substance, and present remarkable swellings, which are fusiform, retort-shaped, club-shaped, and of various forms. The rods and cones are very seldom sclerosed. 7th. The walls of the fine vessels become sclerosed, and the adventitia of the larger ones become thickened. The veins and capillaries exhibit distortions and ruptures, and newly formed vessels appear. 8th. About the optic disc we sometimes have hypertrophy of the interstitial connective tissue; it becomes infiltrated with lymph-corpuscles, shows gray and other degenerative processes. **9th.** Alterations of the pigment, and of the vessels of the choroid, belong **to late stages of the** process, and we **may** have detachment of the retina, and also molecular haziness of the vitreous, with increase of its cellelements."

The above-mentioned lesions explain the various ophthalmoscopic appearances. **The** brilliant white and yellowish spots and patches consist mainly of fatty infiltrations of the granular layers, of the connective tissues, or of sclerosis of the nerve-fibres. Which of these is present, can to some degree be determined by their relation to the vessels which course **only in** the anterior layers of the retina.

It was once thought that certain ophthalmoscopic appearances could be considered typical of the constitutional cause of the disease. Hence, the terms albuminuric retinitis, syphilitic retinitis, have grown into use as signifying certain intra-ocular pictures; **whereas** we are entitled to use these **terms, not** by what we find in **the** eye, but by our discovery upon proper **investigation** that the patient has either Bright's disease or syphilis or diabetes, **etc.** It is, nevertheless, useful to call attention to some of these forms of retinitis under such special types.

Nephritic retinitis (see colored Plate 2, Figs. XII., XIII.).—The efficient cause of albuminuric retinitis seems to be the uræmic state of the blood, and hence we find this lesion in all forms of kidney disease, and under all the conditions which call them forth. It occurs in the large white kidney, in the waxy and in the fibrous kidney. It appears during pregnancy and after scarlatina, as well as in the spontaneous cases. While hypertrophy of the heart often exists, this is not essential **either** to the inflammation **or to** the hemorrhages. In many cases the renal disease has been recognized before the eye-complication set in. On the other hand, it has happened many times that the examination of the eye has given the first announcement to the patient that his kidneys were seriously diseased; he had had no disturbance of function to excite his attention until his sight began to fail. It will generally be found in these **cases** that there have been other symptoms, viz., headache, nausea before **breakfast,** and frequent micturition, especially at night. Sometimes the only token of trouble in the eye is a group of minute, glistening, white dots at the macula, and some hyperæmia of the nerve (see colored Plate 2, Fig. XII.). While one eye may contain from six to twenty, the other may have very few. In other cases we may find the situation such as is sketched in colored Plate 2, Fig. XIII., and the variety is extreme. The factors consist of white or gray or greasy-looking spots, small or large, which are grouped at the macula or near the nerve, and seldom at the periphery of the fundus; swelling of the retina, and possibly of the nerve—the vessels are tortuous **and** full, or little affected; the nerve is sometimes swollen and infiltrated, or again flat, and either hyperæmic or dark and slaty, or dead white and vessels not turgid; in some cases hemorrhages constitute the chief feature, and the fatty sclerosed plaques are very few; again, the proportion is reversed. Detachment of the retina has been seen, but is rare. The above-mentioned pictures are what appear to the ophthalmic surgeon, while to the general practitioner, who scrupulously looks at the inside of the eyes in every **case** of Bright's disease, it may happen that he will find a stage earlier than the above, where, as it is said, the **signs** are those of simple diffused haze, **very** faint and gray, with possibly some fulness of nerve-capillaries **and of** retinal vessels.

It also happens that the stress of **the** disease may not fall on the retina, but **on the optic** nerve, causing in it **great** swelling and infiltration, **hem-orrhages,** new-formed vessels, turgid veins, and reduced arteries. Such a **case is** pictured in Liebreich's "Atlas," and I can confirm the correctness **of the** representation (colored Plate 1, Fig. IX.). The period at **which the** retinal complication appears is thought by Earles, the most recent writer **on** this subject, to be when the kidney disease is well established, and that it signifies a grave deterioration. He (see *Birmingham Med. Review,* Febru-ary, 1880) examined 100 cases of kidney disease to ascertain the nature and frequency of the eye-lesions. Out of the 100, 94 had albuminuria when examined—almost **all were inspected but once ; in** 28 **there** were retinal lesions ; **in 3 there were lesions of the optic nerve.** Of **the 31,** about one-third were **unaware of any visual defect ; in** 12 of **the 28, both eyes** were affected ; **in 16, but** one eye was concerned—a result **unlike** what previous observers have recorded, probably because **the cases** studied by others have been those in which conscious visual defect occurred. In **a large** number the lesions were simply small specks of fatty deposit **or of hemorrhage,** which required close scrutiny to be discovered. By **other writers, the** proportion of **cases** of eye-lesion in albuminuria are thus given : **by Gale-**zowski, in 104 cases, as high as thirty-three per cent.; by Lecorché, in **206** cases, at twenty-one per cent.; and by Lebert, at twenty per cent. It **seems** to be most common in the granular kidney, but it belongs to all **varieties, as** already **stated.** In pregnancy, it has been seen as early as the **third** month, and occurs up to any subsequent time. I have several times seen it after miscarriages. After scarlatina it is much more frequent than after measles. A scarlatinal case reported by Dickinson ("Diseases of the Kid-ney," p. 230) had detached retina in both eyes, besides the retinitis. An-other form of eye-trouble in Bright's disease, **is** from urænic convulsions, or from urænic poisoning without convulsions. But few **cases** are re-ported. I had the opportunity of examining such a case in **the** person of a friend who was under the care **of** Dr. Dudley and of Dr. **A. L.** Loomis. I saw him after an attack of stupor, followed by **an** urænic convulsion ; his sight had been very dim and had somewhat improved. For thirty-six hours he had suppression of urine. There **was** decided infiltration and œdema of the optic nerve and retina, giving it a foggy **look.** The nerve looked waxy, the arteries were very thin, the veins wriggly. The patient was exceedingly blanched, having suffered from what was declared to be waxy disease of kidneys, liver, spleen, and intestines, **and** been an invalid for many years. He had so little blood that congestion of any tissue seemed impossible. As the urænic attack passed, his sight returned.

The course of the disease is various in duration, but it rarely results in recovery. The event of the kidney-lesion carries with it the issue of the lesser malady. In one case the patient was carried off, twenty-four hours after my examination, by a large cerebral hemorrhage, while sitting in con-sultation with two physicians. Very often the growing complications of the general disease make the patient indifferent to the state of his sight. But if the kidney-lesion be **of** long duration, or take on a favorable character, it is not unusual to find the retina regain much of its normal structure. The hemorrhages may absorb, the fatty degenerations grow faint and disappear, and vision improve. In the cases succeeding pregnancy or scarlatina, there may be expectation of great improvement, and, perhaps, of restoration. If, however, as appears by the presence of pigment-deposit, the deeper layers of the retina have been seriously involved ; and if, too, the nerves show infiltration, and the vessels have grown small, indicating atrophy, the future becomes much less auspicious.

Treatment.—The principal point is to protect the eyes from strain and irritation. Local bloodlettings are inappropriate, and nothing is to be done which will reduce the health. Confinement in dark rooms is not to be thought of ; colored glasses may be used if the light be offensive. The whole subject of treatment is contained in what is most suitable for the renal affection. All the rules of hygiene, climate, clothing, food, and exercise, as well as the proper medication, are to be adopted which would be enforced if the eyes were not at fault.

Retinitis from diabetes mellitus is much less often seen than the preceding. With it, its features are to a great degree identical. There can be no doubt as to the competency of diabetes alone to cause retinitis. While it happens that in some cases sugar and albumen may coexist in the urine ; the alteration in the blood is the occasion of the lesion. It would seem that hemorrhages are more frequent and abundant than in albuminuric retinitis. They break into the vitreous, and therefore opacities in this structure are common. A case which I described in 1869, was one of diabetes without Bright's disease, and occurred in a lady sixty years of age. When first seen, in 1867, the retinal lesions were slight, but they soon increased, and hemorrhage was abundant. Improvement occurred to a marked degree, but a little more than two years afterward the lesions returned, and sight was worse than before ; the general health had also materially fallen off. This patient had iritis before the retinal trouble. Galezowski reports a case in which iritis serosa and afterward acute glaucoma followed the retinitis ; iridectomy proved of no value and enucleation was done. The essential lesion seemed to be copious hemorrhage and its transformations. Nothing need be said as to treatment, because it is included in that of the general disorder.

Retinitis syphilitica (see colored Plate 1, Fig. VI.).—This may take place at any time after the early stages of syphilis, but is more common with the secondary symptoms. It may or may not accompany the lesions of the eye. Under the form of neuroretinitis it may result from syphilitic brain disease. We find little tendency to hemorrhage, and very rarely fatty degenerations or sclerosis. On the contrary, we have exudation of lymph-corpuscles, diffuse infiltration of serum and fibrin, and formation of connective tissue. The nerve-fibres suffer atrophy and so do the ganglion-cells, while the outer layers undergo degenerations by hyperplasia of connective tissue and other alterations. Complications with choroiditis is extremely common. The optic nerve is seldom swollen, is moderately congested, edge blurred, the veins enlarged, the arteries normal, or even reduced ; the retina at the nerve-border is faintly striated, sometimes is gray or dark, and about the centre of the fundus it has a faint haze, most easily discovered by weak illumination ; along the larger vessels, the whitish exudation is more intense ; bright white or yellow spots are uncommon, and a considerable patch of yellow exudation can occur, but is still more rare. If the vitreous be hazy, the veins will, by their distortion, best indicate the fact of retinitis ; but, under these conditions, the diagnosis of retinitis requires great caution. One must see retinal exudation to make the diagnosis certain.

The lesions have been thought to belong to the somewhat advanced periods of syphilis, as when the glands are affected, and to adults chiefly. I have now a youth, seventeen years of age, under treatment, who, with syphilitic gummy iritis had, ten days afterward, optic neuritis in the same eye, indicated by plastic exudation, pale color, and reduced vessels (a rather unusual picture) followed in a month by œdematous swelling of the nerve, and diffused faint infiltration of a large extent of the retina; meanwhile the iritis had disappeared and the media had become clear. The other eye took on iritis, and through the moderately hazy media its optic disc was seen to be œdematous and slightly swollen; the veins were turgid, and the capillaries congested. The special tendency of syphilitic retinitis is toward the central part of the fundus, and it is exceedingly prone to repetition. An attack may last from three to six weeks. The macula alone may be the seat of lesions, and only the faintest indications in a finely granular or speckled appearance may be discernible.

The disturbance of sight is variable, and in no proportion to the visible alterations. With little tissue-change the vision may be very bad, and vice versa. Metamorphopsia takes place from exudation into the macula, by which the cones are thrown into disorderly arrangement—either spread out, or packed together more closely. In some cases objects appear too small, because fewer cones receive the image (micropsia), and the contrary may happen, at least theoretically, if too many cones be clustered into a given area (megalopsia). This last condition is seldom, if ever, observed, because if the cones become condensed they usually undergo atrophy. No kind of glasses, either spherical or cylindric, can relieve this condition. Other disturbances are scotomata, which may be central or eccentric, and sometimes are ring-shaped, with central vision unimpaired, and the ring may be complete or incomplete. Förster has noted a distinction in these cases of positive and negative scotomata. The former appears to the patient as a dark spot in the visual field, and is most emphatic in feeble illumination. Consequently, a patient so affected seeks a strong light, and then reads fairly well. The lesion may be due to retinitis circumscripta (at the macula), to choroidal hemorrhage, or other choroidal disease. A negative scotoma is not recognized by the patient under any degree of illumination, whether strong or feeble; it is a color-scotoma, and chiefly for red. The negative scotoma occurs in disease of the optic nerve, and in the amblyopia from tobacco and alcohol. Color-perception in syphilitic retinitis is good, except in the event of atrophy. Scintillations and phosphenes are frequently complained of—one eye alone, or both eyes, or each in succession, may be affected. Recurrences are frequent.

Treatment is constitutional, and consists of mercurials and iodide of potassium. Regard must be had to the patient's general condition, and to the stage of the disease. If tertiary, the proportion of iodide will be larger; but mercurials, to the degree of toleration without ptyalism, are essential. Should the disease be hereditary, the same remedies are to be used, while cod-liver oil and means of promoting nutrition are to be especially regarded. In many cases serious complications occur in other tissues of the eye, or there may be tokens of the disease in the brain, as I have seen.

Prognosis depends on the extent to which the system has been tainted, and its responsiveness to treatment. Affection of the deep layers of the retina is more unfavorable than of the anterior layers; therefore, cases attended by pigment-changes, **or** by choroidal exudation and atrophy, are unfavorable. Atrophy of the optic nerve, and of the retina, sometimes ensues. Of the uncomplicated cases, very many recover, with no injury **to** sight. While the duration **of many cases** is from six to eight weeks, **others are** more obstinate. **When the macula** alone is concerned, there **is great reason for anxiety, because serious** impairment of sight is **prone to remain, and we may find choroidal changes, or** permanent **exudative deposits, or development of nerve-atrophy.**

Retinitis from leucocythemia **has been described. It resembles the forms** seen **with albuminuria,** except that **a** peculiar **yellow hue is cast over the** fundus and the exudations. It is **a** rare disease.

Other causes of retinitis are found in septic processes, such as abscess in the soft parts and in joints, putrid bronchitis, purulent pericarditis, pleurisy, peritonitis, etc., as proved by pathological examinations (see Dr. M. Roth : *Virchow's Archiv.*, Bd. 55), as quoted by Schweigger.

Retinitis proliferans is a name recently applied to a condition in which connective tissue is developed in the vitreous. I have observed a case of this kind for more than three years in a young girl, who had no other malady than menstrual irregularity. At the beginning, the symptoms were those **of** mild neuro-retinitis, and in a few months it passed away. It re-**turned,** and then membranes pushed out into the vitreous from the optic disc and the adjacent retina. They were perforated by holes and not vascular ; the retinal haziness reappeared. At a still later time, rounded spots of choroidal atrophy, with pigmentation, were found not far from the nerve. At **this time** sight was quite dim, and not likely to be recovered. The **case** was referred to in speaking of choroido-retinitis. Treatment seemed **of** small value.

PIGMENT-DEGENERATIONS OF THE RETINA.

Retinitis pigmentosa.—The affection which is thus designated is a chronic disease, consisting of proliferation of connective tissue in all the layers of the retina, atrophy of the nervous elements, and of intrusion of pigment into its tissue, with a tendency to follow the blood-vessels. The same features belong to choroido-retinal affections already considered, but with the difference that, unlike them, the alterations are not confined to the **outer** and middle retinal layers, but concern them all. Such is the anatom-**ical** distinction which Leber points out. The chief feature in the disease **is the** pigmentary deposit, which comes, we are told, from the hexagonal epithelium, and so far from showing any of the usual signs of inflammation in enlarged vessels, exudation, and hazy vitreous, we have simply the tokens of atrophy. The walls of the blood-vessels are thickened, and their calibre reduced ; the pigment-epithelium is in part atrophied, and in other regions its cells are multiplied greatly and penetrate the whole thickness of the retina. These lesions appear without **the** occurrence of any other disease, and without any outward sign. On **the other** hand, they also arise **in a** secondary way after other maladies, such **as total** corneal staphyloma, irido-choroiditis with closed pupil, exudative choroiditis, etc.

In regard **to the** histology of this disease, much has been written, and the process is interesting, but **we** must pass by most of the details. The periphery of the retina is the usual seat and beginning of the disease, and it advances centripetally toward the macula lutea. The rods and cones are destroyed at an early stage, and sometimes the retina is in spots converted wholly into **connective tissue.** Where the pigment is most thick, the retina and choroid adhere **together.** The sub-epithelial or vitreous layer of the choroid is studded with masses of colloid-thickening, and the optic nerve is atrophic, even up to and beyond the chiasm. Sclerosis of the choroidal vessels occurs to some degree.

The subjective symptoms are peculiar there is a loss of peripheral **sight, which** gradually advances toward the centre, and for a long time the **central vision** may, in good light, remain correct. But the peripheral **blindness** impairs the patient's freedom of movement, because it **compels him to** constantly turn his head to acquaint himself with surrounding objects. In addition, these patients when in dull light experience a grievous reduction of central vision. At night they become almost helpless, and their malady has hence been called hemeralopia. When walking at night they keep their eyes on the sky to help guide their steps, and grope in much uncertainty. After **a time** central vision, **even** with good light, becomes affected, and in the **end** total blindness **ensues.** The symptoms in **most cases** begin in early **life,** while in **a few no trouble** was noted until **fifteen or** twenty **years** of age. The **consummation of the** disease usually comes after twenty to thirty years of age. It is found in families, and intermarriage of kindred has been considered greatly instrumental in its production. **Leber finds about** twenty-five per **cent. of the cases** within this category.

Special peculiarities sometimes appear in the symptoms. I have notes of three cases in which the central vision was good, while exterior to this was a zone of blindness, outside was another zone of good vision, and at the periphery again was blindness (ring-scotoma). Such cases prove that for a certain zone in the periphery the rods and cones remain intact, while across the interior blind zone, where the bacilli are destroyed, the optic fibres continue uninjured. Again, Leber, who has studied this subject with care, puts under the same general head certain cases of congenital night-blindness, or amblyopia, without concentric field-limitation, and which exhibit no ophthalmoscopic changes. They may remain *in statu quo* for a lifetime, and certain remarkable examples of heredity are given, extending so far as through six generations (see G. and S., Bd. v., Th. v., p. 650). In the congenital cases nystagmus is frequent. The pupils act slowly, but respond to light.

To the ophthalmoscope the appearance is striking. Dots and network of pigment are scattered over the periphery of the fundus (see Fig. 111). Often the choroidal vessels are strongly brought to view by atrophy of the epithelium. They may also show yellowish or buff outlines, which indicates their sclerosis. The retinal vessels—both arteries and veins—are small ; their walls thickened, possibly bordered with a whitish line, and upon them pigment will lie in greater or less quantity. The optic nerve I have, in certain instances, seen to be red, although not swollen ; in other cases it will be gray, and ultimately a dirty white. In all cases it is opaque, by interpolation of connective tissue. In the last stage the nerve loses its capillary vessels, and the emergent arteries are reduced to threads. Opacity at the posterior pole of the lens is of frequent occurrence , it gradually intensifies, and finally may become complete cataract. The rate of progress is extremely slow. Vitreous opacities are not common, but I have observed them. I have also seen irido-choroiditis ensue. Some cases which strongly resemble retinitis pigmentosa are syphilitic, either acquired or hereditary. In them the disease may be confined to one eye, but in the typical cases both eyes are involved. A certain proportion of deaf-mutes have this lesion, and among idiots, as Liebreich showed, it is not rare.

Diagnosis.—Very little difficulty can arise in ordinary cases. Choroiditis disseminata is either isolated in spots, or most extensive, and presents exudations and atrophy, with white blotches and pigment-deposit in greater

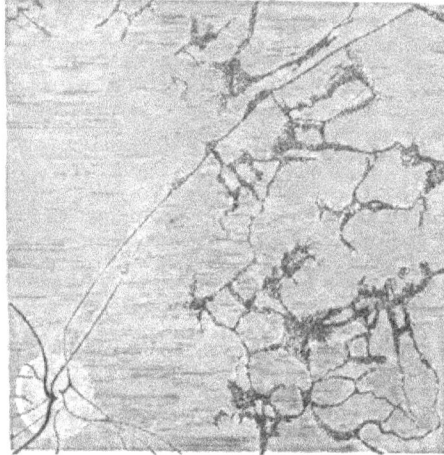

Fig. 111.

quantities. Sometimes choroidal lesions will strongly simulate the retinal lesions, but neither is the difficulty frequent, nor would an error be important. It cannot be admitted that syphilis is capable of causing the real retinitis pigmentosa.

Donders has shown that the essence of the disease is not the pigment-deposit, but the atrophy of the retina. The insensitive region of the retina he found to be in advance of the pigment-district, by throwing upon it, with the mirror, a very small flame, which was not perceived, although not resting upon the pigmentary region.

Hemeralopia, without pigmentation, is less prone to increased loss of sight, and more frequently comes to a stand.

Prognosis is unfavorable, although the rate of advance is slow.

Treatment is unavailing. Alteratives, derivatives, strychnia, etc., have no effect. Attention need only be given to proper care of what sight remains, and its economical employment.

DETACHMENT OF THE RETINA.

(Colored Plate 2, Figs. X. and XI.)

Displacement of the retina by fluid effusion between it and the choroid, when very pronounced, and the retina opaque, can be seen by the naked eye, but as ordinarily found, is to be recognized only by the ophthalmoscope. It presents various appearances, according to the nature and quantity of the fluid. It may be partial or total, and may occur at any part of the fundus. Its more frequent seat is near the equator. It often develops within a few hours, but it may take place during one or two weeks. The subject is sometimes unconscious when it occurs, and usually there is no pain. In some cases there are premonitory flashes of light or of color. Sight is always impaired. The visual field is very defective in sensibility, and may be entirely blind. A degree of light-perception is often present over the damaged part. When the central part of the retina is not visibly disturbed, it may have good sight, but frequently this is not the case. Secondary changes may take place in the separated retina, and we may infer that similar effects extend to other regions. It is very common, both before and after the occurrence, to hear complaint of muscæ, and examination reveals floating vitreous opacities, sometimes very numerous. It is almost the rule to have diminished tension. Examination by the inverted image gives an idea of the extent and relations of the fluid, while the upright image tells the depth of the effusion and the details. What arrests attention is, that the retina bulges over a greater or less extent toward the middle of the eye. If the membrane itself is not clearly noticed, its disturbance is seen by the position of the retinal vessels. They suddenly bend toward the centre of the eye, become tortuous, tremulous, and dark in color, with no light streak. They may take an abrupt change of course, follow long curves, or wriggle in short bends. They will generally undulate. The retina may be clearly visible, because of a drab, gray, speckled, or glistening appearance ; on the other hand, it may be transparent and difficult to recognize. If the amount of fluid is great, little trouble will be had in diagnosis, but if small and transparent it may take close inspection to discover the disease. Nothing but a rippled surface may be seen, and this will sometimes be the condition upon a region where effusion has taken place and afterward disappeared. In some cases the sac may look like mahogany, and then the effusion will be bloody. I have seen zones of different color, in different parts, passing from mahogany to yellow and then to transparency. It is very common to find whitish spots on the retina, or bright streaks, and sometimes cholesterine. It does not generally happen that the region of the yellow spot is involved, but when this does occur, and the amount of fluid is moderate, the fovea appears bright red, and contrasts vividly with the adjacent retina. This hue is in consequence of the thinness of the membrane at this point, which permits the choroid to shine through it. This is essentially the same explanation of a similar contrast in embolism. In one case of this kind, which I studied many years ago, I was convinced that the fovea had remained upon the choroid and been torn from the surrounding retina. Precisely the same look I have not since met with, although I have seen other instances of a detached fovea, and I therefore regard a central laceration as possible. It is not uncommon to find a rent in the retina at its peripheral part ; there may be as many as three, near together and parallel. Sometimes a loose tongue is torn up, and through the opening the choroidal vessels are visible. Liebreich pictures such a condition most truthfully.

The fluid beneath the retina is albuminous ; even when it is not bloody it contains blood- and lymph-corpuscles, fat-cells and crystals, pigment and epithelium, etc. In a large number of cases the vitreous next the retina is liquefied, as is proved by the tremulousness of the sac. After a time, but not at an early date, the rods and cones become macerated and destroyed, or by interstitial inflammation fibrous tissue may be developed, and atrophy of nerve-elements ensue. In the vitreous, fibrous-tissue is almost always found, and floating bodies are almost invariable.

The above description belongs to the cases which may be called simple or idiopathic in character. But we find separation of the retina as an outcome of a large number of morbid conditions of most various origin, in which cyclitis or choroiditis bears a part. Almost every case of chronic irido-cyclo-choroiditis has detached retina. It is a pathological condition very common in museum preparations. It appears often as a total detachment, and in its last results presents simply a cord running from the optic nerve to the ora serrata, within which no recognizable traces of the vitreous remains (see Fig. 82, p. 212). In condemned eyes with closed pupil the total want of perception of light often warrants ante-mortem diagnosis of the above condition. As a part of the suppurative process in purulent choroiditis or retinitis, or perhaps in panophthalmitis, separation of the retina occurs. I have seen in such a case flecks of blood scattered thickly over the whole retina; I opened the eye at the equator, found yellow serum between choroid and retina, and in the posterior surface and substance of the latter innumerable small extravasations. Cystic degeneration of the retina is described by Iwanoff and Leber; both in the late and earlier stages of the retinal disease, it is only seen by the microscope.

A choroidal tumor is usually attended at an early stage by subretinal effusion, and may by it be masked from view, but such occurrences are rare. So too the cysticercus cellulosus has been seen behind the retina.

The causes of subretinal effusion of the simple variety will be understood by the following statistics from my note-books. I find 58 cases in fifty-four patients. The table of causes is as follows:

Myopia........................		30
Inflammation, viz.:		
Iritis	1	
Choroiditis	5	9
Neuro-retinitis	3	
Cerebral symptoms.................		2
Injuries...........................		6
Unknown		11 Total.... 58

Four patients had the lesion in both eyes.

The ominous frequency of the lesion in myopia, amounting to fifty per cent., is startling. In almost all of these, the vitreous was hazy, but this must be regarded both as a result as well as a precedent condition of the effusion. It is a prevalent belief that the vitreous opacities, i.e., fibrous tissue, by their shrinking, can draw off the retina, but, as Schweigger remarks, this cannot be by direct traction when the floating fibres which we see are perfectly free. I found the lesion in subjects of whom the youngest was seven years of age, with M—12 D, and the oldest was sixty-eight years of age, with M—O, 75 D. The greater number had M greater than —3 D. Some were feeble, others robust. Some had, others had not used their eyes to excess. It is evident that an effusion of serum from the choroidal vessels is more likely to occur in eyes whose distention has produced great tenuity of the choroidal vessels.

A very plausible theory has been advanced by Raehlman (*Archiv. für Oph.*, xxii., iv., 233), and in substance is supported by Priestley Smith (loc. cit., p. 138) to the following effect: the vitreous near the retina undergoes structural degeneration and chemical change; it liquefies, and we find shreds floating in it. The change in its density alters entirely the balance of osmosis between it and the blood in the choroid ; for some cause which we cannot explain, the fluid portions of the vitreous drain away by absorption ; a serous fluid from the choroid is necessarily effused, and, being albuminous, is unable to make its way rapidly through the retina. It therefore pushes it off, and perhaps actually bursts through it. This theory supposes an absorption of liquefied vitreous as the beginning of the morbid movement, and if the process were always gradual, it could be easily accepted as an explanation. At present we fail to understand how the frequent suddenness of the attack may be accounted for. The loss of ocular tension favors the theory, and as Priestley Smith contends that the ciliary body has most to do with vitreous nutrition, his remark is pertinent, that the cases of total retinal detachment in spoiled eyes always show extensive disease of the ciliary body. It is true, in a few exceptional cases, that subretinal effusions have occurred in glaucomatous eyes. It must then be assumed, as Raehlman says, that the vitreous has been converted into a substance of abnormal diffusibility, and this must have special relations to an albuminous fluid. He has produced retinal detachment in animals by injecting into the vitreous, solutions of common salt, the result appearing within less than an hour. An increase in the saline constituents of the vitreous (which naturally are less than one per cent.) greatly increases the tendency of albuminous solutions to transude, and their passage through the retina is much more difficult than for the saline fluid ; hence the effusion accumulates outside the retina. The explanation of hemorrhagic effusion is not so difficult, but we must admit that even by Raehlman's theory, all difficulty is not removed. As to the frequency of its occurrence, Hirschberg reports 113 cases among 22,500 patients, while Cohn gives 191 cases among 20,000 patients.

Prognosis is, as a rule, unfavorable. An advantage attaches to the descent of the fluid to the lower part of the eye, which is a common occurrence, because blindness above the horizon is less injurious than blindness below the horizon. The replaced retina generally recovers part of its function ; it happens, too, that some absorption is quite common, and it is of the greatest consequence to have the macula escape. But, even when it does not seem to be involved, we may find metamorphopsia. I have already mentioned the dulness of sight due to torpor of the retina, and degenerative changes are quite probable ; moreover, a detachment is prone in time to become greater. It is a frequent thing to have cataract occur ; I found it twelve times among 58 eyes. Spontaneous recovery takes place in a few cases ; I have observed it four times—all were myopic persons—two were not treated at all ; one was treated antiphlogistically, and he had irrecoverably lost one eye already by the same lesion ; the fourth was operated on by scleral puncture, and with no good result for three months, but after that time a great portion of the fluid disappeared.

Treatment.—Two opposite modes have been resorted to. One, recently commended by Samelsohn, and considerably adopted, consists in confinement to bed, and constant use of a pressure-bandage for four weeks. The other method is to use such means as will best and most rapidly promote the general nutrition, withholding the eyes from all use and taxation. This means abundant food, exercise, and recreation, and avoiding everything likely to cause congestion of the head, and is not inconsistent with light purgatives and such medication as circumstances may indicate.

A choice between these opposing methods may be determined by the patient's general condition. If the health should be such as to contraindicate the treatment by bandage and close confinement, it would certainly be unwise to enforce it. If not injurious to health, it is most suitable to resort to it at the outset. The supine posture need not be rigidly insisted on, because the descent of the fluid to the bottom of the eye is best promoted by the upright position. On the other hand, there can be little doubt that improvement in the quality of the blood is a **most** important aid to absorption, and hence out-door life, travelling, horseback exercise, and all hygienic advantages, will sometimes be the manifestly proper course. Much is to be left to individual judgment on this topic. It is certain that customary antiphlogistics are out of place, except when the lesion is traumatic or the effusion is hemorrhagic ; even in the latter case it may not be advisable. Hypodermic injections of muriate of pilocarpine, gr. ¼ to gr. ½, have been lately commended—in some cases with good result, but in others no benefit ensued. Some experiences are given by French and German writers of an indecisive character.

Operative treatment has been resorted to for many years. Von **Graefe** and Bowman punctured the sac to let its contents diffuse into the vitre**ous.** Some gratifying results ensued, but rarely were they permanent. Some cases were rendered worse ; I have had two such unpleasant experiences. At present this treatment has been abandoned. Wecker intro**duced** a gold-wire suture into the sclera. This has not found general approval, and some cases of destructive irido-choroiditis have been caused. Puncture of the sac through the sclera was done in 1859 by A. Sichel, and has been repeated since by Arlt, Alfred Graefe, Hirschberg, and others. Wolfe has lately called attention to the value of laying open the conjunctiva so as to clearly expose the sclera before puncture, but the suggestion has a minor value. A paper by Hirschberg, reporting ten cases of scleral **puncture, authorizes** the conclusion that no harm is done, that in most cases improvement is gained, but that relapses are to be expected. Wolfe gives a case in which the relief was complete, and had remained fifteen months. Higgins reported double optic neuritis with subretinal effusion in both eyes, and in one tapping was done once, and in the other eye was done three times, and in neither case with any benefit.

The conclusion would then seem to be that, for the first four or six weeks, there should be no operative interference. That, during this period, such treatment should be employed as the general condition of the patient indicates, and this may comprise both the pressure-bandage and muriate of pilocarpine, and the resort to health-promotion. Should nothing be gained by such measures, and the retinal fluid have descended, scleral **puncture** may be performed, and by the delay no harm is incurred, be**cause the** separated membrane generally preserves its normal character **for** months. The puncture is done by pushing into the eye obliquely, in the meridianal direction, a broad needle or narrow iris-knife upon the site of the effusion, and turning it so that the fluid shall flow out with the aid of light pressure, made upon the globe with the finger of the other hand. The patient may sit up or lie down, and turn the globe far up. Care must be taken to locate the effusion correctly before puncture. Mr. Wolfe first opens the conjunctiva and holds the wound apart by strabismus-hooks before piercing the sclera. By the first method the fluid spreads beneath and distends the conjunctiva, and, if needful, may be let out by another puncture, but rarely is this necessary.

There is no objection to an immediate and hasty glance at the interior of the eye by the ophthalmoscope. There should be no bleeding, and the retina should not have been wounded. A padded bandage should be applied, the patient put **to bed for** a few days, and be kept for three or four weeks under observation. Usually, no important reaction follows. The definitive result is uncertain, and will be determined by **the** precise nature of the pathological conditions which caused the lesion. As **to** the treatment of cataract, consecutive to detached retina, **it** offers extremely little encouragement. I have consented to operate, and have found needling result in almost no absorption. I have extracted and gained **the** object of **removing an** objectionable blemish from the pupil with no gain **or** loss to **sight. It is to** be expected that **a partial** detachment would by **the** necessary **wound** become more **extensive**; but I did not **find** this **true, in at least one instance.**

Leber reports a case, shown to him by Stilling, of a patient who had had five attacks of detachment, followed each time by complete recovery.

GLIOMA RETINÆ.

Fungus hæmatodes of the eye.—The subjects of **the** above disease are often infants, and none have yet been recorded whose age was above sixteen years—the great majority **are** below ten. **The** cases are usually discovered by a bright reflex from the interior of the eye, which arrests attention. Both eyes may be implicated simultaneously, or in succession. There will have been no preliminary symptoms, and the anterior part of the eye may be normal.

The appearance varies with the extent of the growth. When observed early, the ophthalmoscope or oblique illumination will detect a white or yellow, or reddish yellow surface, with blood-vessels whose arrangement is more or less plexiform and unlike the retinal circulation, and they will be **minute,** or possibly very scanty. The surface may be smooth or **nodulated.** Inspection is easy, because the mass approaches the nodal **point. The** retina may be in position, or, more usually, it is detached by **effusion. The** vitreous and lens **are clear.** As the tumor grows, the lens is pushed forward; the pupil **becomes** sluggish and dilated—may become adherent; the anterior chamber **is shallow**; the anterior ciliary vessels are distended. If a longer duration and greater growth be permitted, the coats of the eye give way **and** bulgings appear. After **a** time the growth crops out, becomes ulcerated, bleeds, gives forth **sanious** and fetid discharge, and **a** projecting, hideous tumor may be **formed.** Pain begins with distention and external growth, **and** cachectic **symptoms soon** arise. Surrounding parts, such **as** the bones of **the head or the brain,** become involved, and metastasis may **take** place to other **parts of** the body. Death, either by exhaustion or by brain-disease, **is the final event.**

While fatal if left to itself, **or if not cut short at an** early period, experience has shown that a **very early removal of the eye** may save life, because in the beginning the **disease is local. We** have many authentic cases **to** verify this statement. I have knowledge of two in my own experience. **When** relapse occurs, it **is** within a few weeks, perhaps within less than a month, **or** within three months. Several children in the same family have been **known** to be affected. Wilson reports eight children attacked, of whom **three** died by this affection; the others survived ("Oph. Hosp. Rep."). The most usual mode of death is by extension of the disease through the optic nerve to the brain. Should a relapse occur, a fatal issue is almost certain. A single case is given in which the patient survived after removal of the secondary tumor.

The pathology of the growth has been carefully studied by Knapp, Hirschberg, Delafield, and others. It consists of small round cells, similar to, but not, as once supposed, identical with the granular layers of the retina, with a scanty amount of fibrous tissue and numerous blood-vessels. The starting-point is most frequently from the external granular layer, but it can begin from any other layers. Delafield remarks that the disease is anatomically identical with small round-cell sarcoma, and might be so called. Hirschberg and Bull **concur in** this view. It may grow outward or inward, more **frequently the latter.** Its prevalent mode of extension **is** along the optic nerve.

Diagnosis.—There are several possibilities **of error.** **1.** Glioma should not be mistaken for choroiditis metastatica. In the latter **case** there will be a history of a preceding sickness, either of the brain or of the spinal cord, and redness of the eye may have been noticed. The eye itself will usually show a shallow anterior chamber, the pupil be adherent **and irreg**ular, the periphery of the iris retracted, the lens pressing **against it, the** globe a little reduced, tension minus ; the yellow mass may be **vascular, and** the retina may or may not be detached. The symptoms and history **of an** inflammation are the clue. It is conceivable that an eye having glioma **may** be seen during the period of glaucomatous outbreak and the existence **of** deep disease never have been recognized, and at the time be masked by turbidity of the media. Wadsworth reports one such case, where the **cornea was** suppurating. An incision to relieve the suppuration discovered **the** tumor.

2. Raab (*Arch. f. Oph.*, xxiv., iii., 163) gives the anatomical description of three eyes enucleated **as** gliomatous, which had a peculiar deposit of fibrous tissue behind the lens, resulting from cyclitis or from choroiditis. In one case the lens and iris were pushed forward, the pupil dilated to an extreme degree ; there was a greenish white reflex, upon which red streaks could be seen ; tension a little plus ; slight punctate opacity of the cornea ; the peri-corneal and ciliary vessels were injected. In the other two cases the facts were somewhat different, but need not be repeated. In all, there was increased tension, absence of inflammatory tokens in the anterior part of the globe, and the presence of a light-colored mass in the depth of the eye. It must be admitted that a mistake in diagnosis would, under such circumstances, be pardonable.

3. I have seen two cases, in young subjects, in which there **were** intra-ocular deposits, resembling glioma in all macroscopic appearances, which ultimately terminated in the disappearance of the mass. There were no signs **of** inflammation, and, for want of better knowledge, they were called **cases** of "strumous" deposit.

4. White sarcoma of the choroid **is not found usually at** the early age when glioma occurs, yet I have **one** such **case recorded, in** a child eight years of age.

Treatment.—Medication is useless, and the only safety lies in the earliest possible removal of the globe, with as much of the optic nerve as can be excised. There is encouragement that life can be saved if the disease be extirpated at an early date—the operation would be proper at a later period to relieve pain, **but at** the stage of fungus growth the undertaking may be too formidable. **The** surgeon will act according to his discretion, in view of all the circumstances. Should the painful condition present itself of glioma in each eye, I should not hesitate to enucleate both, if, by so doing, a fair prospect of life could be secured.

To remove an eye which is the seat of some disease that resembles, but is not glioma, **is a needless mutilation** ; but it may be justly argued that **it** is better to **err on this side than to fail** to remove an eye which is truly **at-**tacked by this **formidable** malady.

There is one case **on** record of preservation of life when **the** disease **re-**appeared in the orbit and **was** again removed. All its contents must **be** taken away, and the walls **of the** orbit treated with chloride of zinc. But the expectation is against **recovery.**

CHAPTER XVI.

DISEASES OF THE OPTIC NERVE.

Hyperæmia.—Apart from inflammations, turgescence of the capillaries and of the larger vessels of the nerve is found under a great variety of conditions, and the greatest care is required to distinguish from each other the symptomatic and the idiopathic. By far the larger number of cases are symptomatic. Of these, the majority are the effect of fatigue of the accommodation, as found in hypermetropic astigmatism, spasm of the ciliary muscle, or other causes of asthenopia. Another cause is cerebral hyperæmia, more particularly of the meninges. It is not to be inferred that every case of cerebral hyperæmia will reveal itself in congestion of the optic nerve, but the concomitance occurs. In apoplexy, the nerve may be deeply red, and it may be pallid; no uniform sequence can be asserted, as was once supposed. In cases of fracture of the skull (or other injuries) hyperæmia is often seen, and, in general, morbid processes at the base of the skull are apt to cause optic congestion, if not inflammation. It is not safe to argue from this symptom alone, but in a given case it will take its place among other phenomena, and often have decided value in diagnosis. With patency of the foramen ovale of the heart and general cyanosis, venous stasis in the optic nerve is also seen. In those who use alcohol to excess the nerves are red. In plethoric persons there is always fulness of the optic circulation, and the greatest scope must be allowed for anatomical and physiological varieties. Hyperæmia of the optic nerve is, therefore, of very uncertain value as a pathological symptom.

Anæmia of the optic nerve appears simply as a part of general feebleness of circulation. A temporary arrest of circulation occurs in some cases at the beginning of an epileptic attack, but the contrary has also been seen. Upon an attack of syncope the nerves become pallid, as I have witnessed. In cholera, Graefe found the current of blood still in motion through the nerves during the last stages of the disease, and remarks that the flow in the veins was intermittent or jerky. Both arteries and veins were all extremely small.

NEURITIS OPTICA.

In describing inflammation of the optic nerve, writers are accustomed to lay stress on the presence or absence of swelling. We find cases in which this attains a remarkable degree, and, because in other respects the nerve may seem not much altered in structure or function, a special name has been devised to describe it, namely, "choked disc." So, too, the term papillitis is used to express inflammation of the head of the nerve. The term "choked disc" will continue in use, doubtless, having come to us from Albrecht v. Graefe, and been founded upon his theory of its causation; but the theory has been given up, and the term is of no strict pathological value, but useful as a colloquialism. We shall confine ourselves to a statement of the appearances of the optic nerve when inflamed, and later discuss the subdivisions and etiology of the lesions.

We have capillary hyperæmia, loss of transparency, change of texture, blur or obliteration of outline, swelling and tortuosity of veins; arteries sometimes enlarged, sometimes reduced; there may be hemorrhages. According to the acuteness of the process, the nature of the effusion, or the stage and duration of the disease, we may have very dissimilar pictures. Take a few illustrations. A boy with syphilitic iritis, with gummy exudation, has, after a little time, neuritis optica, the media being unusually transparent. The nerve is slightly swollen, its tissue not clear, but of a pasty look, the edges not blurred, the arteries a little full, the veins distended and tortuous. Here there is plastic infiltration quite abundant, and smothering, so to speak, the capillary hyperæmia which would have been expected. In the other eye, at a later time, neuritis also occurred, but the nerve was red, not so opaque as in the other case, more swollen, and both arteries and veins turgid; the nerve-edge not distinct, although definable. In this case the infiltration was less intensely plastic and more mixed with serum.

In certain cases of œdematous infiltration (choked disc), we have swelling measured by 2, 3, or even 6 D; the mass is almost globular, and shows an extreme degree of parallax when the observer shifts his point of view from side to side. The substance is bright red, the outlines are almost or quite obliterated, and fine lines radiate into the retina; the arteries seem small, and the veins are strangulated (see colored plate, Fig. VIII). Pulsation may exist in the veins, and possibly in the arteries. In some cases the surrounding retina is much infiltrated, and again it is not. If it be not, the border of the swollen nerve will be well marked, and its increase in size greatly emphasized.

In some cases the small vessels are extremely abundant, while the circulation is evidently impeded, because they wriggle in and out in a manner which suggests the head of Medusa (see colored plate, Fig. IX.). Sometimes minute, or even considerable hemorrhages appear.

Another class of cases show little swelling; the nerve will be deeply red, while its texture will no longer be transparent, its edges will have a corona of fine lines, and the arteries and veins will be turgid. There may in this and other cases be infiltration along the retinal vessels. If the nerves have a normal excavation, this may be recognized; but it is not likely that the lamina cribrosa can be seen, because of the infiltration. (For the opaque lines along the vessels, see colored plate, Fig. XVI.)

Again, more extensive changes can arise in which both optic nerves and retinæ participate, and are swollen, infiltrated, and hyperæmic, while the notable fact is, that white or buff-colored specks or patches appear in the retina, both in the neighborhood of the optic nerve and of the yellow spot. They may be clustered in radiating streaks in the latter locality, such as are found in albuminuric retinitis. This is not a frequent occurrence, but one notable case is on record by Schmidt (*Archiv. für Ophth.*, xv., iii., 253), and I have seen the same condition. The lesion now referred to is rare, it is true, but its occurrence without albuminuria compels caution in assigning the etiology of an ophthalmoscopic picture.

While writing this chapter a case in point occurs, in an unmarried and delicate looking woman, twenty-two years of age, who was brought to my notice by Dr. Shorter. On March 17th she was taken with severe headache and vomiting. The headache did not keep her from sleep, but

20

she was obliged to go to bed. About a fortnight later, Dr. Shorter found choked discs and vision $\frac{20}{20}$ in each eye. For six weeks there has been more or less headache with some vomiting, and vision has declined to $\frac{20}{40}$. A week ago the left rectus externus became paretic, causing distinct convergent squint. Now, in each eye the optic nerve is a mass of **reddish** gray exudation, with glistening patches of white ; its diameter doubled ; its height equal to 4 D; the macula being emmetropic ; its border fringed with striæ, fine vessels and hemorrhages mingle together upon **it,** and the veins are enormously swollen and serpentine, their hollow **curves dip** under the tissue, and their surface everywhere is blurred by **effusions ;** the arteries are small, and also tortuous. Exudation reaches into a **wide,** surrounding zone of retina, and on the horizon stretches half way to the macula. Around the yellow spot a series of bright yellowish lines are arranged in equidistant radii for about one-third the area of the circle. In the right eye they are distinct as ever seen in Bright's disease, and each one is broken into dots. In the left they are more delicate, continuous, and fewer. The patient's urine has no traces of albumen, but the intraocular appearances are such as would correspond to the most flagrant form of albuminuric neuro-retinitis. At the same time the weight of evidence is strongly in favor of cerebral tumor, especially as impairment of the left sixth nerve is added to the other usual symptoms.

Figs. XIV. and XV., colored plate 2, are representations of a case which I observed of neuro-retinitis, which exhibit noteworthy exudations. **There was** distinct evidence of brain disease, and none of kidney lesions. Fig. XIV. was during the early **stage, Fig.** XV. after several weeks had passed.

It thus appears that inflammations of **the nerve** and retina merge into each other, and may exist in various proportions. It has been noted in some cases that swelling of the nerve has not been greatest at the centre, but at the margin, and has left the centre comparatively depressed. Such a condition has been called perineuritis (Galezowski), but we have no good evidence of the truth of the assumption that the sheath of the nerve is the starting-point of the disease.

Such are some **of** the pictures of optic neuritis, but no language can fully delineate what only the pencil can depict.

When acute processes are abating, other features come out. Some portions of the nerve may have a gray opacity, from formation of connective tissue, and another part be red. The cases of most acute infiltration may by slow gradations pass into a gray or buff, into a bluish or white color, and the result be connective-tissue development and atrophy of nerve-structure. On the other hand, the tissue may regain transparency and its normal hue.

The length of time required for the culmination of acute neuritis, and for its entire retrogression, is impossible to be determined. Months may usually be counted on, and Matthewson ("Trans. Fifth Internat. Oph. Congress," p. 63, 1876) reports a case of choked discs in which the optic nerves remained *in statu quo* for three years. As above stated, the termination may be in complete recovery, or in partial or total atrophy of the nervefibres. A state of swelling, with a white **and** apparently flocculent texture, sometimes remains a long time, and this has suggested the term "woolly" as a fit description of it (see colored plate 2, Fig. XV.), (Hulke). **Gradually** the swelling subsides, and may eventuate in concavity. The borders for some time remain fuzzy and obscure, but at length the choroidal margin comes out black and distinct. White lines bordering the vessels continue for a period, and at length may disappear (see colored plate 2, Fig. XVI.). If a case be noted for a sufficient time, the atrophic appearances succeeding severe optic neuritis will in nowise differ from those which are seen in cases of primary atrophy of the nerve. It has been thought that a distinction could be made in this regard, but if sufficient time be allowed, both processes will bring about the same ophthalmoscopic result. During a considerable epoch we may with some confidence say that the atrophy in progress has been preceded by inflammatory conditions. This fact bears upon the question of the concurrence of brain disease.

Subjective symptoms are often vague. Phosphenes and limitation of the field and scotoma may be present or absent; hemianopsia may exist as a complication in special cases. Knapp has called attention to a measurable enlargement of the blind spot. Impairment usually occurs, but what is most remarkable, there may be very little or no interference with vision. Emphasis must be laid upon this notable fact. For some time after the first cases of distinct optic neuritis without injury to sight were reported, the statement was received with incredulity. The simple reason was, that inspection of the optic nerve was not made until the patients complained of damaged vision. Inasmuch as the larger proportion of cases of optic neuritis arise in consequence of brain disease—especially tumors, whose symptoms are often obscure—it need not be said how ready we should be to seize upon a symptom so important. In presence of intra-cranial disease, it follows that an ophthalmoscopic examination is indispensable. Complaint of loss of sight is not required to justify it any more than in cases of albuminuria; for both the one and the other, ophthalmoscopy should be routine practice. The examination should be repeated at intervals, notwithstanding it may have been negative, because the time at which the optic neuritis may appear is undetermined. When this shall be adopted at hospitals as a prevailing rule, we may hope to arrive at a solution of some of the obscurities which yet surround the pathology and true significance of optic neuritis.

In the great majority of instances both optic nerves are inflamed, and this is almost the rule where intra-cranial affections are the cause. But cases of one-sided neuritis occur in which the symptoms clearly indicate intra-cranial lesions. Such cases are noted by Magnus, Pagenstecher, Fieuzal, etc. I have a woman now under observation with this condition. We also have six cases of cerebral tumor recorded, in which only one optic nerve was affected, viz., one by Reich * (details not given), two by Hughlings Jackson,† one by Pooley,‡ one by Parinaud,§ and one by Bouchut.‖ But for monocular neuritis optica we are usually to seek the cause in the orbit, or at least below the optic chiasm. It is common in such instances to have other nerves involved. I have seen the third and fourth combined with the optic, and the lesion was doubtless at the sphenoidal fissure. I have also seen neuritis optica associated with paresis of the rectus internus muscle, and the attendant exophthalmus and pain indicated the situation of the disease to be in the orbit.

While inflammation of the optic nerve does not in itself produce pain, it is in a great number of cases dependent upon causes which give rise to

* Klin. Monatsblätter (Zehender), vol. xii.
† Ophth. Hosp. Reports. vii.
‡ Arch. für Oph. und Otol., Bd. vi., p. 27.
§ Annales d'Oculistique, lxxxii., 19, 1879.
‖ Bouchut: Ophthalmoscopie Medicale, p. 144.

severe suffering, and these may be either in the orbit, in the head, or in the general system.

Etiology of neuritis optica.—As local causes, we have in the orbit cellulitis, either primary, or as the effect of erysipelas, periostitis, tumors, or injuries. As general causes, we have disease of the brain, of the kidneys, diabetes, syphilis, uterine disease, disturbances of menstruation, typhoid fever, lead-poisoning, cerebro-spinal meningitis, great loss of blood, etc.

But intra-cranial disease is the most frequent cause, and is said to cover four-fifths of the cases (Mauthner). We find it in meningitis, both tubercular, acute and chronic; in abscess, in pachymeningitis, in meningitis basilaris, but not usually in hydrocephalus, either external or internal, and pre-eminently do we find it in intra-cranial tumors. The way in which the nerve becomes inflamed in cases of intra-cranial lesion has been the subject of much discussion and of diverse theories. Two ways are recognized as the direct physical route, one by way of the nerve-sheath, and the other by the nerve-trunk. The latter seems obvious, and is called neuritis descendens; the former seems to be more frequent, and its explanation has excited the chief discussion. It was observed by Stelwag ("Ophthalm.," Bd. II., p. 618, 1856: Hydrops nervi optici), and later by Von Ammon (1866) that the sheath may be distended into a conical form just behind the bulb. Schwalbe (1869) showed that the intra-vaginal space was continuous with the arachnoid cavity, and that lymphatics run into the head of the nerve from the sheath. Schmidt and Manz and others quickly took the ground that here was the channel by which the morbid process advanced from the brain to the papilla. Their conclusion was that papillitis results from effusion which has trickled into the optic-sheath from the arachnoid cavity. As a fact, the sheath seldom shows signs of inflammation. An exceptional case of peri-neuritis is given by Alt, caused by meningitis, the cavity of the sheath being obliterated by exudation of fibrin and round cells. But generally there is simply dropsy, and that near the globe. It has been assumed that the serous effusion caused the papillitis, but it is not proven that the papillitis may not cause the effusion. That the lymph may have a degenerating effect on the fibres in the papilla has been rendered probable (see Kuhnt: *Von Graefe's Archives*, xxv., iii., 256), but that does not concern the question as to which comes first in order of causation, the papillitis or the dropsy. To present all the facts and considerations which belong to this topic would involve a long discussion, and I shall try to state in condensed form the latest views and researches.

The first theory was that of Von Graefe, to wit: that intra-cranial pressure acting upon the cavernous sinus, caused stasis in the ophthalmic vein, and hence the swelling. This theory has, for various good reasons, been laid aside. The second, which still has supporters, is that the papillitis comes from percolation of fluid from the arachnoid space through the sheath-space, and causes choking of the lymph-vessels. The third originated with Benedikt, in 1868, who has restated it with additional argument in 1876, and it is favored by Mauthner ("Vorträge, Gehirn und Auge," p. 566, 1881). It assumes that from different parts of the brain an irritation is propagated to the vaso motor nerves which govern the vessels of the papilla, and cause the effusions in it and also within the sheath. This theory makes the sheath-dropsy the consequence of the papillitis, while the second (Schmidt) makes it the cause of the papillitis. For an admirable review of the question up to 1875, see a paper by Loring (*Am. Journal Med. Science*, p. 361, Oct., 1875.) Hughlings Jackson (*Med. Times and Gaz.*, p. 311, March 19, 1881) refuses to accept Benedikt's theory, but offers an explanation which seems to differ from it more in terms than in **ideas**. He says that a tumor or other irritating cause first excites an instability in the gray matter surrounding it, and that this leads to discharges upon the vaso-motor nerves, which produce contraction and dilations of the vessels of the optic nerve, *i.e.*, inflammation. He also quotes Pagenstecher's words (" Oph. Hosp. Reports," p. 162, 1873), which really adopt Benedikt's theory. Fourthly, Parinaud (*Annales d'Oculistique*, p. 26, 1870) takes the ground that in every case of papillitis there will be found œdema of the brain-substance—that is, interference with the lymph-circulation—and that this is propagated to the head of the nerve either by its trunk or through its sheath, and occasions the swelling. He says that in most cases there will be intra-ventricular, or subarachnoid effusion, or both.

As to Schmidt's theory of what Mauthner calls transportation of fluid, cases are recorded of papillitis without any dropsy of the sheath, and a notable one is by Treitel (*Arch. für Oph.*, xxvi., p. 105, 1880), and the same is asserted by Parinaud (loc. cit.), 1879. He (Parinaud) says that the lymphatics of the lamina-cribrosa are not easily filled by fluid injected from the cavity of the sheath, but that the injection must be made under the internal sheath, where we come into the lymphatic system of the nerve-trunk, and this view he supports by experiments. A communication exists between the lymphatics of the lamina cribrosa and of the sheath-cavity, but the vessels are few. That the quantity of fluid usually found in the sheath is small, that it seldom fills the whole distance between the eye and the apex of the orbit, and that it concentrates in a conical shape at the globe—all these facts militate against the theory of transportation. On the other hand, Manz has shown that in acute meningitis it is common to find dropsy of the sheath, but seldom does papillitis occur.

The assumption that œdema of the brain is the immediate occasion of papillitis, is not opposed by any known facts. It is favored by the cases which do not easily find explanation by other theories. For instance, there are cases of tumor of the brain without optic neuritis. Reich gives one without details ; Hughlings Jackson cites one of tumors in each cerebellar lobe, and also one in a cerebral hemisphere (*Med. Times and Gaz.*) ; Parinaud details one in the left spheno-frontal lobe of large size (*Annal. d'Ocul.*, vide supra, p. 19). These cases are hard to explain on the theory of vaso-motor irritation of Benedikt, because tumors in these sites have, in other cases, caused the neuritis. It has happened that only one eye has been the seat of papillitis, when there was cerebral tumor. Hughlings Jackson has two cases, each in the cerebral hemisphere, and the opposite nerve affected ; Pooley has one case, and the nerve on the same side was affected ; Bouchut has one case, where the left pedunculus cerebri was the seat of a large fibro-plastic induration (" Ophthalmoscopie Médicale," p. 142, 1876). The size of the tumor is of no account in the production of papillitis, neither is its situation. Both large and small tumors, and wherever located, will cause the lesion. Acute meningitis ranks next to tumors in frequency of production of papillitis. Tubercular meningitis, in the minority of cases, has this effect. Garlick reports swelling of the disc in fourteen out of twenty-four cases of autopsies (" Med. Chirurg. Trans.," lxii., 447, 1879), but the degree was generally small, and other authors do not lay stress upon it. He, however, finds changes in the optic disc in eighty per cent. of the cases of tubercular meningitis—that is, hyperæmia and effusion. In hydrocephalus there is generally atrophy of the nerve, but in some cases acute œdema has been found (Parinaud, loc. cit., p. 14). The readiness with which the bones of the skull yield to the internal pressure, accounts for the ordinary absence of optic œdema, while the pressure of the intra-ventricular fluid on the chiasm and optic centres favors the production of atrophy of the nerves.

Reich (*Zehender's Klin. Mon.*, xii., 274, 1874), who collected forty-five **cases** of cerebral tumor with ophthalmoscopic appearances, and autopsy, says that three cases did not exhibit neuritis or any optic-nerve alteration. It should be added that this symptom often occurs late in the progress of cerebral tumors. Jackson gives the autopsy of a man who had symptoms of tumor for nine **years,** and it was only within six weeks of death that papillitis appeared.

It has also **been known to run its course and** to get well before death occurred ; **but the rule is to the** contrary. **Gowers** says that four-fifths of the cases of cerebral tumor have neuritis. **It** occurs too in abscess of the brain, but very seldom in abscess of the cerebellum. Among twenty-three cases of choked disc, with autopsy, Jackson found tumors seventeen times, abscess three times, apoplexy twice, softening once. Cerebral hemorrhage rarely causes papillitis. It has been seen in some cases of lead-poisoning in which, according to Parinaud, the brain is affected by œdema.

With all the above facts before us, it would appear that cerebral œdema may fairly be regarded as a link in the chain of causation, and this theory certainly is more intelligible than the vaso-motor theory. It does not exclude the concurrence of direct transmission of fluid by the sheath-cavity, for this does occur ; but it connects the lymphatics of the optic trunk with those of the brain, and founds an explanation upon recognized anatomical data. Whatever can excite cerebral œdema, general or local, may cause papillitis ; and if this should not occur, the **optic** lesion would be wanting. It must be remembered, in this **connection, what** influence effusions into the ventricles may have in producing obstruction in the lymph-circulation, and this is present, as a rule, in cases of papillitis.

Jackson says he has seen optic neuritis in atrophy of the brain. This bears hard upon the theory of Parinaud. In the existing state of our knowledge we may hold the theories *sub judice* and wait for further facts ; but those above stated prove that the examination of the nerve is of importance ; and also show that we are not authorized by the state of the nerve to attempt to decide where the disease is located, nor what may be its nature. We are compelled to put all the symptoms together and from the sum total to form our diagnosis. Mauthner ventures the assertion that swelling of the disc, amounting to $+5$ D. $(+\frac{1}{8})$, is indicative of tumor.

The form of inflammation called neuritis descendens, is more common **than** the above, and presents only the usual tokens of inflammation, viz., hyperæmia and infiltration to greater or less degree. It needs no special description, and offers no peculiarities or difficulties in explanation. It may or may not be concurrent with brain-disease, and its existence is perfectly authenticated.

The *pathological anatomy* of choked disc consists, as would be expected, in serous effusion which separates the fibres from each other, leaving wide spaces between them, in an increase of blood-vessels, especially veins, sometimes in hemorrhages, and if the disease be long in duration we have increase of connective tissue and atrophy of nerve-fibres. The most signal alteration is a peculiar varicose swelling of the nerve-fibres : they may become club-shaped, beaded, or like ganglion-cells with granular contents. The changes are the same as those which are found in albuminuric retinitis.

Another form of neuritis is the interstitial, which is by far **the** most frequent, and may be primary, or succeed the œdematous condition just described. There is enormous cell-infiltration and **increase** of connective tissue, affecting both the neuroglia of the nerve-fibres and their intervening trabeculæ. Round cells surround the fibres and the blood-vessels, and the latter are often greatly increased in numbers and cause extreme swelling of the papilla. As the round cells subsequently develop into connective tissue, which undergoes shrinking, the blood-vessels and nerve-fibres in turn diminish **or** disappear, and atrophy ensues. According **to** Alt, whose description **is** what, in the main, is being followed, two kinds of atrophy occur. **In** one of them the nerve-fibres become simply thinner, and we find lying **between** them fatty cells, probably neuroglia-cells undergoing regressive metamorphosis. In the other form, the nervous element is **represented** by a grumous substance, formed of molecular fat-drops, **that is detritus.** This change may involve **much or** little of the nerve-structure, and it may at any stage become stationary, or **be** continuously progressive.

The optic layer in the retina also becomes thin, while perivasculitis and interstitial retinitis are observed. The sheaths **of the** nerve may be inflamed and hypertrophied.

In some cases **of** extreme hemorrhage **at** the base **of** the skull, whether spontaneous or **after injury,** blood has penetrated **the** optic-sheath.

Acute myelitis **has, within a** few years, been found to be accompanied by affections of **sight.** The **first** case was published **by Steffan** and Erb, another by **Dr. Seguin, of this** city, and I have **joined Dr.** Seguin in contributing another. **The** eye-symptoms are those **of acute,** but moderate, neuritis optica, with remarkable impairment **of** the **visual** field and of central vision. There may be entire loss of **direct** sight; there may **be any** kind of irregularity in the fields, including total abolition on both **sides** or affection of one only; there may be repeated recoveries of sight **and** relapses. The singular peculiarity of the cases has been that vision, **both** direct and indirect, should undergo such great and unexpected **variations.** The lesion of the cord was in its lower and middle portions, **as** was fully manifested by symptoms of the bladder and the lower limbs. No explanation of the optic neuritis has been offered, although we may bear in mind that a root of **the** tractus has been traced by Stilling into the crus cerebri at the red nucleus of the "hood." The agency of **the** sympathetic nerve **has been invoked to** explain the optic neuritis, but **this** is nothing better than surmise. **All** the **cases** have gotten well, both **in** respect to sight and **to** the functions of **the** cord. In my own case, large doses of iodide **of** potassium **were** employed, gradually reaching three hundred grains daily, **and were** well **borne.** The case occupied about four months in its evolution.

Neuritis optica may occur at any age, and even congenital cases have been seen. When children are attacked, which **is** not uncommon, the cause is usually intra-cranial, and most frequently it is **due** to some type of chronic meningitis. The cases may be of considerable duration, and the results may be little or severely damaging to sight. The disease in adults frequently has a chronic course, although the acute type is most common. Persons may be attacked more than once, and there may be good recovery. Such **cases** are generally syphilitic, and at present I have one under observation. **The** man has had two attacks in each eye. From **the** first he fully recovered, and now is getting well from the second.

The prognosis in regard to sight is in the large proportion of cases unfavorable. But it is necessary to carefully weigh the general symptoms, and to form, if possible, a proper notion of their connection with the eye. If no progressive, or permanent, or essentially destructive cause can be discovered, it is fair to believe that, in many cases, a valuable degree of sight may be retained.

Treatment.—In many cases of intra-cranial disease the lesion of the optic nerve receives no special attention, and does not demand it. Being symptomatic, its fate is bound up with that of the dominant disorder. It is, however, true that full doses of iodide of potassium are very frequently appropriate both for the eye- and the brain-trouble, especially if the latter be chronic. In other cases bromides are clearly indicated by the cerebral phenomena, and they will not be less suitable because the optic nerve may be implicated. On the other hand, should some other cause be active, such as uterine or menstrual disarrangement, or there be great debility, iron and tonics may be called for, but no special suggestions are demanded because of the nerve-lesion. Local depletion by leeches, artificial or natural, is of small value. There is generally no photophobia, nevertheless the force of custom usually directs the wearing of shaded glasses. Cases of syphilitic lesion will require mercurials and iodides, according to the phase of the disease ; both are usually needed, and the dose of the latter may in some cases have to be very large. We may have the direct effects of syphilis upon the nerve, or its remote effects through lesion of the brain. In the latter case, the visible alterations of the nerve may seem quite out of proportion to the brain-symptoms, and even to the tokens of impaired sight, both in visual acuity and in the field. The nerve-disc may be scarcely swollen, be but little infiltrated, and the hyperæmia moderate.

Neuritis in albuminuria or glycosuria simply calls for greater efficiency in the use of such remedies as may combat the general disorder, and the same remark applies to all the conditions of a constitutional or remote kind under which this ailment occurs. While the lesions sometimes come with precipitous haste, and at other times are slow in reaching their climax, in all cases recovery is deliberate, and may, when it does take place, be protracted through months. If, on the other hand, atrophy ensue to a greater or less extent, as is most probable, the indication will be to stimulate to their best activity such of the nerve-fibres as remain intact, by strychnia, phosphorus, and cod-liver oil. Sometimes iron and arsenic are useful. This matter will be again referred to under the head of atrophy of the nerve.

Orbital neuritis optica.—Certain cases of loss of sight occur whose symptoms lead, by exclusion, to the hypothesis that the orbital portion of the optic nerve is the only part involved ; Graefe first suggested this explanation. Usually but one eye is affected. The damage to sight may be either complete or partial ; it may be entirely lost in a few hours, or within a few days. There will be no pain, although headache is a possibility. By the ophthalmoscope the nerve looks nearly normal. The pupil is active. There are no brain symptoms, and no albuminuria and no hysteria. With this absence of marked opthalmoscopic signs and of brain-symptoms, the orbital part of the nerve is supposed to be the seat of inflammation. We have never yet had an opportunity to verify the diagnosis by autopsy. The following case will illustrate the features and history of the trouble :

Mrs. M., thirty-four years of age, wife of a Methodist clergyman in New Jersey, came to me on December 12, 1878. Has been married thirteen years, has one child, is in good health, except slight indigestion and occasional rheumatic pains. The last menstruation was four days too soon. In right eye V.$=\frac{20}{20}$; in the left merely sees movement of the hands on the outer side of the field. Six days ago, on awaking in the morning, she found that the left eye only had ability to discern the situation of the window. By noon this had been lost for the central region, and remained only in the extreme temporal part of the field, as found on my first examination. The pupil is normal, no headache and no head symptoms, but had "a feeling of deadness about the brow and the opening of the orbit." Menstruates about every three weeks, and flows copiously. By opthalmoscope find the optic nerve injected, and a little swollen and indistinct on all sides, except on the outer part, the veins a little enlarged, the arteries rather small. All the retina rather hazy. Lesions greatly out of proportion to the loss of visual function; patient put upon iodide potass. After a week she began to gain a little better perception—slowly, the improvement continued. She took the iodide for about four weeks, but no notes were taken until six months passed, when in right eye V $=\frac{1}{10}$, left eye V. $=\frac{2}{10}$. The nerve was decidedly pale, veins large, arteries unchanged, and near the macula were many whitish dots. Color-perception very deficient: leaves on the trees appear black, and the only color which she readily recognizes is yellow. In November, 1879, viz., eleven months from the beginning, find O. D. with + 48 c. 90° V. $=\frac{2}{10}$, O. S. + 60 c. 90° V. $=\frac{2}{10}$ visual field, O. S. normal, color-perception bad, nerve pale, the outer half most decidedly.

The above case presents recognizable ophthalmoscopic features, but these are in some cases almost or entirely wanting. The symptoms may have a sudden onset, or be days, or even weeks, in reaching a consummation. The loss of sight may consist in central amblyopia or in central scotoma. The latter effect is common. One eye or both may be implicated. The attack may follow upon febrile disorders, measles, menstrual derangements, rheumatism, syphilis, etc. A great dilation of the pupil is often observed—sometimes decided reduction in the retinal vessels may occur. It is thought by Graefe that there may sometimes be thrombosis of the vessels. The seat of the disease may be anywhere between the globe and the optic chiasm. The central scotoma and loss of color-perception, which may or may not go together, is an interesting symptom. It has been thought by Leber (G. and S., b. v., 834) to be produced in the following way: he assumes that the optic fibres which go to the macula, and to the region between it and the nerve, are found in the temporal half of the disc, and lie next the sheath of the nerve-trunk. Moreover, that the peripheral retinal fibres lie in the axis of the trunk. From this arrangement it follows that inflammation on the surface of the stem will cause central retinal defect. Such superficial lesions of the optic nerve have been anatomically discovered, although not in any case where central scotoma had previously been observed. It must be admitted that great obscurity hangs about these cases from the want of exact pathological data.

Treatment must to a great degree be deduced from the assumed cause of the affection, and while no decided indications can thus be obtained, iodide of potassium or saline diuretics, the Turkish bath, in some cases the artificial leech, careful general régime, constitute the proper proceedings.

ATROPHY OF THE OPTIC NERVE.

(Colored plate, Figs., XV., XVI., XVII., XVIII.)—The appearances of this condition are somewhat various. The nerve-disc is always opaque, and in the greater number of cases is white; but we also find it gray, leaden, bluish, or "dirty." Very often the lamina cribrosa is conspicuous, appearing as a mixture of white and dark dots, or intersecting fibres. The nerve is flat, or more often concave, and is especially apt to be saucer-like; the degree and kind of concavity will be modified by the original form of its surface, whether or not it may have had a physiological excavation. The outline is always in advanced cases sharply defined, and is often deeply pigmented. There may be a time, if atrophy follows inflammation, when the border is ragged, or striated, or ill-defined. According to the nature of its surface, both as to color and form, the nerve may be uncommonly bright and luminous, or of a dull hue. There are **cases** of partial atrophy where the temporal half is white and pallid, while the nasal side is red. Care must be taken not to hastily pronounce on such a condition, because such an arrangement is often normal. The vital point in diagnosis is want of transparency in the atrophic nerve-tissue, while the healthy substance always is transparent. A want of capillary vessels, and the development of connective tissue, is necessarily implied in the above description. As to the larger vessels, the arteries will be small, sometimes thready, and the veins, although larger, will also be of reduced size. Sometimes the vessels are not much changed in calibre, and in other cases they are almost entirely wanting. It is not rare to find the vessels bordered with gray or whitish lines, **so-called** peri vasculitis. Often there will be traces in the retina of a concomitant or pre-existing lesion.

We have primary and secondary atrophy, i.e., the process begins in the **nerve,** or it may have been preceded by inflammation. Again, the lesion may originate in the nerve, or be propagated from the brain or the spinal cord. In primary atrophy a distinction is sometimes founded on the color, **viz., into** white and gray, while there are mixed forms. It must also be **remarked that** in old age the **nerve** often loses its clearness, and may become gray or leaden, with **no evidence of** impaired function. For proper recognition of changes **of color, the direct** examination is much to be preferred, and the illumination **should be** moderate. The mirror should be a feeble reflector, plain or unsilvered glass, or **the flame** must **be** turned **low,** or a slip of smoked glass may be interposed. The size of **the** disc is never smaller, because this is determined by the scleral opening, but the nerve-trunk is diminished. Reduction **of** bulk finds expression in the disc by the recession of the **surface,** i.e., the excavation. Sometimes much care must be taken not **to mistake a** simply pallid nerve for atrophy, and occasionally the distinction **is** difficult to make.

Causes.—According **to a** table of 166 cases, gathered by Galezowski, about 50 per **cent. are** due to diseases of the brain and spinal cord, 13 per cent. are **traumatic,** 9 per cent. are due to alcoholism, and 8 per cent. to syphilis. **The** remaining causes are of the most varied kind. Choroidal and retinal diseases will entail nerve-atrophy, and retro-ocular neuritis has the same result. Lead and tobacco are among the toxic causes, **while** erysipelas, congenital and dentition troubles, menstrual disturbance, sexual excesses, fevers, diabetes, etc., are other causes. Often **we are left in** entire uncertainty, and even the assumed causes may rest **upon slight evidence.** It is proper to remark the frequent occurrence **of optic** atrophy in locomotor ataxy. The disc is usually gray, and the **vessels** not reduced in size. The lesion is ordinarily a precursor **of** the spinal cord disease, and sometimes by a long interval. So too it sometimes precedes the general paralysis of the **insane,** and occasionally disseminated sclerosis. I have seen a congested and turbid condition of the optic nerves in general paralysis of the insane a number of times, and this stage is succeeded in time by the appearances of atrophy. Recently I made an autopsy of a case of general sclerosis and atrophy of the brain, with enormous increase in the lateral ventricles and their fluid contents, in a man, forty-seven years of age, in whose case the initial symptom was white atrophy of the optic nerves. Afterward mental disturbance appeared, but there were no symptoms of lesion of any other nerves.

Primary or idiopathic atrophy usually affects both eyes, and is much more frequent in men than in women. Secondary atrophy may follow lesion of the chiasm, of the cerebral centres, or of the cortex about the supra-marginal or angular gyrus (Ferrier). Pressure on the nerve by exostosis, tumor, or aneurism, will cause atrophy; the narrowing of the optic foramen by hyperostosis **causes** the same effect. Internal hydrocephalus more frequently gives rise to atrophy than to neuritis, although the former may succeed to the latter. Pressure upon the chiasm by dis-**tention of** the third ventricle is the mechanism of the lesion. Meningitis, chronic **or** acute, may cause atrophy without recognized neuritis, the probability being that the retro-ocular part of the nerve has been implicated in inflammation, far more than the papilla. Diseases of the choroid and retina are often accompanied or followed by nerve-atrophy, as in retinitis pigmentosa, or hemorrhagica, and atrophic choroiditis; blows and injuries of the globe, fractures and fissures of the orbit, may cause atrophy. Berlin has shown that fissure of the orbit is far **more common** than has been imagined, and is the **true** explanation of many **cases of** amaurosis which at first presented **no** ophthalmoscopic changes, and subsequently exhibited optic nerve atrophy. Glaucoma simplex consists largely in nerve-atrophy, while the chronic inflammatory form leads to the same result. Rheumatic **or** syphilitic **thickening** of the sheath, and various orbital processes can cause atrophy.

Morbid anatomy.—The most correct division of atrophy **is into** the parenchymatous and the interstitial. The former begins in the nerve fibres, the latter begins in the connective tissue. Abadie and Charcot have insisted on this division rather than on the visible appearance **of** the nerve, whether gray or white. The parenchymatous occurs in locomotor ataxy, and the interstitial in sclerosis and as the result of inflammation. In the **former the nerve** looks a grayish white, like the posterior columns of the cord in ataxy. The fibres are softened and granular, and easily break into fragments. They do not entirely disappear, but are transformed into indifferent fibres. While they **keep** their arrangement into fasciculi, with proper surrounding **connective** tissue, they become lessened in bulk and their sheaths more loose. Generally the change begins at the periphery of the nerve, and this gives a clue to certain symptoms in tabetic atrophy. The lesion may ascend **to the chiasm** of the tracts, sometimes to the external corpora geniculata. It may be limited to certain parts, or be irregularly diffused.

Interstitial atrophy displays **increase of** interfascicular connective tissue, and consequent choking of the nerve-fibres. The latter lose their myelin, and are reduced to axis cylinders. Sometimes only a certain sector is concerned, and this **may** be either central or superficial. In the retina the layers which suffer are the optic fibres and the subjacent ganglion-cells; other parts, **as** a rule, remain intact.

Symptoms.—The impairment of function affects both the acuity of sight, **the extent of the** field, and the color-sense. In many cases the pupils are not affected, while in others they exhibit suggestive peculiarities. Sometimes there is a degree of hypersensitiveness to light, and vision is apt to be better on a cloudy than on a bright day. Going suddenly from bright daylight into an obscure place causes much embarrassment. At a more advanced period an intense light is sought for. Sometimes there are phosphenes. At first all objects seem overspread by mist, and central acuity declines more and more. Peripheral vision soon suffers, and the study of the field is important. In essential (parenchymatous) atrophy, the limitation is usually concentric. In the atrophy which precedes or attends locomotor ataxy, the outer part of the field is most invaded, according to Förster ; gradually the outline becomes more irregular, angles are cut out which point toward the optic nerve, and the last remaining part of the field will include the macula and the region between it and the optic nerve. Impairment of color-perception is very common, and begins with inability to recognize green, then red, and finally blue and yellow. The proper limits of the fields of the respective colors will first be reduced in area, and afterward color will not be recognized at all. Sometimes the first color which is lost is red. A central scotoma for red is significant in cases of tobacco-poisoning, which is sometimes evidenced by atrophy ; a similar remark applies to alcohol-poisoning. Abadie ventures to make a distinction between parenchymatous and interstitial atrophy, founded upon personal observation, to this effect. If a patient has vision $\frac{1}{2}$, cannot recognize green, and knows red and yellow without difficulty, he has parenchymatous atrophy. If, however, vision be less than $\frac{1}{15}$, and color-perception is fair, it is more than probable that the atrophy is interstitial (Annal. d'Ocul., lxxx., p. 196, 1878).

Atrophy usually affects both eyes. Being in so many cases associated with other nervous disease, its presence demands careful inquiry into accessory symptoms. Especially are signs of ataxy to be looked for. Charcot thinks that, in the very large proportion of cases of primary atrophy, this lesion is precedent at a longer or shorter interval to spinal-cord disease, i.e., ataxy, although additional symptoms may be long in appearing or in assuming a marked character. The knee-reflex must not in these cases be forgotten. Again, we have to look for attendant cerebral symptoms, both mental and organic. Yet a large number of cases yield no information as to other relations, and we may merely speculate on what has been the reason of the nerve-trouble. It must be stated that it is not very rare to have only one nerve involved.

Prognosis.—Seldom is it otherwise than bad. Proper weight must be given to surrounding and causative circumstances, because the nerve-lesion is so often concomitant, and its character largely dependent on the chief affection. Certain cases retain remarkable sight, and they are more likely to be the interstitial. Such an instance is the following :

Miss M., aged twenty five, in good health, when ten years old had acute meningitis, became totally blind in both eyes, and remained so for two years. Improvement then began. When sixteen years old, she had scarlatina, which again greatly reduced her sight and left her with some nervous disorder, for which she was treated by several physicians, taking nux vomica and zinc, being cauterized on the neck with hot irons, etc. She consulted many oculists, and all gave an unfavorable prognosis. Three years ago the left externus muscle was cut to correct divergent strabismus. I saw her, at the request of Dr. Shorter, November 11, 1880. She is subject to headaches and to supraorbital neuralgia ; for the last year she has had pain and constrict on about the waist. Pupils natural. Find O. D. V. $= \frac{4}{8} \frac{8}{8}$, O. S. perception of light. Visual fields at twelve inches ; O. D. fixation is on a point 5° to temporal side of nerve, and reaches to 15° above and 15° below the horizon. The field is oval ; its long axis equals 18°, and its short axis 10°. O. S. fixation is central. She says the fields vary in extent at times, and vision is also variable. She has been taking ergot, by advice of Dr. Shorter, and with benefit. I suggested salicin, because of malarial symptoms, and in lieu of quinia, which disagreed with her, and to be followed by full doses of strychnia. I was told in May, 1881, by Dr. Shorter, that the fields had permanently enlarged, and vision in the right eye had reached $\frac{4}{8} \frac{8}{8}$. Ophthalmoscopic appearances remained unchanged, being those of pronounced white atrophy. Such recovery of sight, and its retention for so many years, are uncommon, and give a glimmer of hope upon the calamitous future which we are in most cases obliged to predict. I regard the case as remarkable, while I freely admit the probability of exaggeration in the statements as to the degree of blindness when the first trouble occurred.

Mere pallor of nerve must not be mistaken for atrophy. The state of the field and the quality of the color-sense are to be duly considered, as above remarked. The rule of Abadie may be taken for what it is worth, the interstitial being a more favorable condition than parenchymatous atrophy. **One** consolation is often possible, viz., that the rate of progress to the bad is slow.

It is affirmed that regular concentric limitation is more **unfavorable than** fields of irregular outline caused by deep entering **angles of darkness.** Cases which show a similarity in the irregularities **of the two fields** are to be traced to brain-lesion, and they may or may not be curable. **Our** ignorance of the true pathology of many cases should restrain us from **dogmatism,** and, while we utter only the truth so far as we know it, **we may** hesitate to pronounce a doom which many patients regard as worse **than** death.

Treatment.—The first indication is to correct any and every departure from normal function which we can discover or control in nutrition, sexual organism, lungs, etc. ; also to counteract the syphilitic, rheumatic, or gouty diathesis, and scrofulous tendencies ; to discriminate, so far as may be possible, diseases of the brain and spinal cord, and, even when there may be no token of syphilis, iodide of potassium in high doses is justly esteemed. Under these heads are included many possibilities of treatment suited to the peculiarities of each case. In former times setons and blisters and moxas were much employed ; now the actual cautery is applied to the scalp and to the skin over the vertebræ, when the brain or the spinal cord are thought to be congested. When, however, these proceedings have been tried or do not seem to have any claims for trial, we are reduced to the use of a few remedies. The most important, and the one favorably regarded for a long time, is strychnia. Nagel has the credit of having taken it up systematically and with energy, and claims special benefit from its hypodermic administration, giving of the sulphate of strychnia $\frac{1}{15}$ to $\frac{1}{10}$ grain once daily in the temple. His method has been largely adopted, but it is to many patients inapplicable, because of the constant attendance on the physician which it necessitates. I have therefore substituted the internal administration in granules, which may be $\frac{1}{60}$ or $\frac{1}{50}$ grain each, and give a quantity sufficient to bring about manifest constitutional symptoms. There will be a tendency to cramps in the legs and in the lower jaw, sometimes colic, and sometimes exalted nervous excitability. The amount requisite to obtain this result will be different according to the subject, but $\frac{1}{4}$ grain daily is not a large amount for an adult male, and I have given $\frac{1}{2}$ grain with impunity. This amount is to be gradually reached by adding to the number of granules every third day, until the proper symptoms appear. When the full effect of the drug has been reached, the dose is to be kept up for three or four weeks, or so long as continuous gain is observed, and this may go on during three months. If, after three weeks of strychnia-symptoms, no improvement in sight is discoverable, the remedy may be considered unavailing. Another remedy, which is less positively effective, **but is** to some extent useful, is phosphorus. It is sometimes given with strychnia. It is not advisable to push it to the production of constitutional symptoms, and it is not usually carried higher than $\frac{1}{12}$ grain daily. A combination of the following kind is a tonic of high value, both to these cases and in many debilitated subjects. It has the virtues of both remedies, and came to my knowledge from Dr. Gray, of the State Asylum for Insane at Utica, N. Y.

℞. Acid. phosphoric. dil. .... ℨ iij.
    Strychniæ .... gr. j.
Take thirty drops in water three times daily.

The proportions may be varied as desired.

As to zinc and nitrate of silver, I can say nothing from experience.
Cod-liver oil is often helpful, but not as a specific like strychnia. Quinia
and iron are many times indicated. Electricity has failed to vindicate its
pretensions to any real value, although, by its capacity for exciting phosphenes, it fosters the hopes of a credulous incurable.

Amblyopia and Amaurosis.

Under these heads are designated certain cases of dimness or loss of
sight, which are not attended by any perceptible intra-ocular lesions.
To this category hemianopsia does not belong; neither is retro-ocular
neuritis thus included, because, in most instances, faint though dispropor-
tionate lesions can be found. By amblyopia we mean partial, and by amaurosis total loss of sight. There may or may not be limitation or irregularity
of fields, or scotomata. The cases may be ranged under the heads of .
1st, traumatic; 2d, hemorrhagic; 3d, toxic, such as by tobacco, alcohol, or
osmic acid; 4th, by uræmia; 5th, by diabetes; 6th, hysterical; 7th,
migraine.

Traumatic cases have been denominated concussions of the retina.
They may be caused by blows upon the globe or upon the orbit, upon the
head or by shocks to the vertebral column. A blow upon the eye by a
blunt instrument may not cause any visible lesion of tissue, either without or within the globe, yet be followed by decided amblyopia. One remarkable feature which sometimes appears is an immediate rigid contraction of the pupil. This has been noted by Berlin, and probably by others.
I have seen it a number of times. The contraction is to the smallest size,
and resists for hours the most vigorous use of atropin. So small have I
found the pupil, that ophthalmoscopic examination has been impossible.
Berlin (*Klin. Monatsbl.*, xi., p. 42, 1873), writing upon concussion of the
retina, divides these cases into two groups. First, those in which cen-
tral vision is moderately impaired (say to $\frac{1}{60}$ or to $\frac{1}{40}$), while peripheral sight is intact. In a few days sight is fully restored, and he thinks
irregular astigmatism of the lens is the proper explanation. Second, cases
in which the optic nerve, either in the orbit or in the brain, has suffered
lesion. An explanation is to be given of not a few cases, which was brought
forward by Berlin in the Heidelberg Ophthalmic Congress, 1878, to the
effect that fissure of the roof of the orbit is far more frequent than is generally supposed, and it extends often through the optic foramen or through
the sphenoidal fissure. To discover it the dura mater must be stripped
from the bone, which is rarely done at an autopsy, and it was found by
Dr. von Holder to occur in ninety per cent. of the cases of fracture of the
base of the skull. Naturally, by hemorrhage into the optic sheath, or by
laceration of the fibres of the nerve, sight would be injured, and atrophy
might ensue, while for a long time no signs might be seen within the eye.
Berlin, l. c., describes cases of direct injury to the globe, in which he has
seen a spot of whitish opacity of the retina at a point opposite the place of
injury, and sometimes also on the site of the blow, which began to appear
within a few hours, and vanished after two days, with restoration of vision.

In experiments upon rabbits he always found a subchoroidal hemorrhage at the situation of the retinal opacity. Aub (*Archives for Oph. and Otol.*, vol. ii., 173) reports a case of metamorphopsia after a blow, which implies disturbance of the retinal elements. There may be innumerable complications of such injuries in other lesions of the globe, but they are not now under consideration. Continual pressure on the eyeball will cause blindness. Testelin reports it in a man who, when drunk, laid for many hours with his eye pressing upon his hand. Graefe cured a case of severe blepharospasm, which had lasted eleven months, by section of the supraorbital nerves, and the child had become almost blind. In the course of a month sight returned, simply, as Graefe believed, because the pressure was removed.

Traumatic amblyopia, or amaurosis, may occur through a great variety of injuries of the skull or brain, to which no clue can be found should the patient survive, but in which some disorganization of tissue undoubtedly has occurred to structures concerned in vision.

Concussions of the spinal cord may cause loss of sight. I saw a man who, by a railway collision, received a sudden and severe blow upon the lower end of the spine, which had an effect such as would have been produced by a blow from above. He suffered extreme pain at the base of the skull, along the spine, while his sight was, as I remember, about $\frac{20}{100}$, and the visual fields were contracted to an angle of about thirty degrees. In both eyes there was extreme hyperæmia of the optic discs, both in the large and small vessels. For a number of weeks the condition remained unchanged, and I do not know how it finally turned out. In this case a paralysis of the fibres of the sympathetic might well be assumed as the cause of the vascular dilation. What caused the extreme limitation of the fields is purely conjectural.

I saw a case where a very large man fell into a hole in the street, and struck the outer edge of one orbit on the pavement. He lost sight in the eye of the injured side immediately, and after a few days the opposite eye became very amblyopic. In neither was any lesion to be seen by the ophthalmoscope. A fissure of the orbit might have extended to both sides. It is important to inquire for bleeding of the nose, and to search for subconjunctival ecchymosis, which may come to view several days after the injury. Both these signs would be strongly indicative of fissure, but would not be indispensable as symptoms.

Prognosis in traumatic amblyopia and amaurosis is good, for the mild cases, while for severe injuries of the orbit, of the skull, of the brain or the spinal cord, it must be guarded.

Treatment at first will be such as the special conditions of injury call for, and when the primary symptoms have passed, resort for restoration of sight may be had to strychnia, either by injection or by the stomach. The mild cases of traumatic amblyopia will get well spontaneously within a few days (see case in Hirschberg: *Centralblatt*, April, 1881, p. 100; by Reich, neuroretinitis partialis after injury of skull—recovery; also see case in Gowers, p. 318, fracture of orbit, etc.).

Amblyopia, or amaurosis from hemorrhage.—This subject has been elaborately presented by Fries in an inaugural dissertation ("Beilageheft zu Zehender," 1876), based upon 106 cases recorded from 1641 to 1876. A number of instances have been published by other observers since Fries. The sum of the matter is that, after severe hemorrhage, loss of sight sometimes takes place, and may be immediate, may be within a few days, or be deferred as late as the eighteenth day. In a number of cases signs of neuritis, or retinal hemorrhage, or of retinitis, or of atrophy of the nerve, **were** found, while in other cases no visible lesions appeared. The pathological connection between cause and effect is not understood. The source of the bleeding may be most various: most often it is from the stomach **and** intestines; next, in frequency, **it comes from** the uterus either in childbirth, from abortion, or during menstruation; it may be from the lungs, the bladder, the urethra, or by venesection, or by injury. The last is the least common **cause.** It **is** sometimes attended with peculiarities in the visual field, such as irregular defects or scotomata. In 90 per cent. (Fries) both **eyes** are affected; in 47 per cent. the loss of sight is permanent, the pupils being dilated and perception of light wanting; in 31 per cent. there **was** improvement, and **in some** of these cases this occurred in only **one eye; in** 21 per cent. entire recovery was obtained. The time when recovery set in was variable, that is, from **a** few hours to three or four months, and in one case to nine months. The pathological appearances would naturally have great influence on prognosis, and in the bad cases inflammatory signs are most pronounced.

Treatment would be modified by the general condition of the patient and by the source of the bleeding. The most efficient means of aiding recovery **is:** first, the vigorous use of strychnia by injection or by the stomach; second, by galvanism; third, by dry-cupping about the temple; **and** fourth, by general invigoration, to improve the action of the heart **and the** quality of the blood. That the sight does not more frequently suffer by large loss of blood is in part due to the comparative independence of the intra-ocular circulation as compared with the systemic; but this, of course, is only true within certain limits. Gowers ("Medical Ophthalmoscopy") has condensed many of the observations on this subject (**see** pp. 184–188).

Amblyopia from alcohol and **tobacco.**—**No little** controversy has been held on this subject, based, perhaps, not so much on scientific data as upon ethical opinions and upon personal prejudices. In view of the extreme prevalence of indulgence in both stimulants, and the comparative rarity of eye-lesions among such as practise it, some have denied the causative connection. But the evidence is quite as strong as for many etiological conclusions about which medical observers have no scruple. The difficulty often is to **separate** the two causes from each other, and, while many **are** willing **to admit** the pernicious effect of chronic alcoholism on sight, **they** are averse to accept that tobacco-poisoning could cause the same re**sult.** In point of fact the two very commonly combine their evil influence **in** the same individual. In both cases the ophthalmoscopic lesions **are faint.** In alcoholic amblyopia we usually find a dull red nerve **with** swollen veins, rather hazy borders, and torpid circulation; atrophy may subsequently ensue. In tobacco amblyopia the nerve is brighter, and more nearly normal, or it may show tokens of atrophy or of interstitial inflammatory exudation. Generally there is very little **lesion to be** recognized. The acuteness of sight may be reduced to $\frac{20}{30}$ or to $\frac{20}{100}$, **or** be sometimes less; **there** is rarely **any** limitation of the **field.** The conspicuous symptom, which is pathognomonic, **is** a central **scotoma** for red. Förster first called attention to this. The outline of the scotoma is usually a horizontal oval. It extends beyond the optic nerve **often, but not** always. To detect it a very small test-card must be used, viz., one not more than one-half inch square and on the end of a stiff wire. It is also needful to bear in mind the tendency to speedy fatigue in looking at bright colors. It is well to use a red background and pass the test in front of it, and thus avoid this liability to error.   Or by using two test-cards simultaneously, one for direct looking and the other coming in from the periphery, the central scotoma will more certainly be developed.   Only a few minutes' continuous looking at the cards is to be permitted.   I have abundantly satisfied myself of this symptom, and could give many diagrams of fields.   Nettleship gives one in which there was a central scotoma for red, and beyond it a ring of un-impaired color-sense, and outside of this an annular scotoma, while the periphery was normal.   It is also alleged that perception of green **is de-**fective as well as of red, but this is far less common.   In advanced cases **there will** be marked color-blindness because of the considerable nerve-atrophy.   It is impossible to say what degree of narcotic indulgence is **at-**tended by this diseased state, but the various susceptibility of individuals is extreme.   I have seen a youth of nineteen, who had smoked cigarettes in large quantities for five years, exhibit a classical picture of nerve-atro-phy and red scotoma, with $V = \frac{3}{8}$?   But Förster has found more subjects above the age of thirty-five than below it.   Mr. Hutchinson gives many cases in the "Ophthalmic Hospital Reports," and some are under thirty. The approach of the lesion is very insidious, and often is not attended with any other signs of poisoning—neither headache nor other symptoms. Both eyes are alike affected.

Prognosis is relatively good.   Marked atrophic appearances, of course, have a bad outlook.

Treatment.—Abstinence is the *sine qua non*—total and unequivocal. The next remedy is strychnia, and its potency is undeniable.   Not all apparently moderate cases get well, but their chance is very good.   Mr. Hutchinson found three-fourths of his cases either get well or greatly im-prove.   My own experience has been similar, except that, with habitual drunkards who have become very amblyopic, perfect restitution is not to **be** hoped for.

Other poisons which cause amblyopia are lead, osmic acid fumes (see case in "Trans. Am. Oph. Soc.," seen by myself), nitro-benzol containing aniline (see case by Litter, in Hirschberg's *Centralblatt*, p. 118, April, 1881); the patient was in coma, and the surface of the body, as well as the eye-grounds, were intensely blue.   Silver and mercury are said to cause amblyopia; quinine, in large doses, has had the same result, as reported by Roosa and Grüning.   Grüning concludes his article on quinine amaurosis (see *Arch. of Oph.*, p. 81, March, 1881, by this statement): "The patient, after the ingestion of a single dose or of repeated doses of quinine in various quantities, sud-denly becomes totally blind and deaf.   While the deafness disappears within twenty-four hours, the blindness remains permanent as regards peripheric vision, central vision gradually returning to the normal after some days, weeks, or months.   The ophthalmoscope reveals ischæmia of the retinal arteries and veins, without any inflammatory changes."   Salicylic acid is, by Riess, reported to have had the same effect.   Full doses of santonine do not impair sight, but make all objects look yellow.   I have no knowledge of the appearance of the fundus in these cases.   In regard to lead, we find inflammation of the nerve and retina, atrophy, and also amblyopia without visible **lesion**.   There are also cases of brain-lesion.   For lead amblyopia, treatment by iodide of potassium gives good results ; for inflammation and atrophy, the prospect is unpromising.

Uræmia from kidney disease, and as found in diabetes, causes amblyopia, and even amaurosis.   It is not needful to enlarge on the subject, which has been mentioned on page 292.   In pregnant women, a sudden and some-times complete loss of sight has been known to occur, without any evidence

21

of urænia or of albuminuria. In the eye-grounds only slight lesions could be found, but the optic nerves showed tendencies to atrophy. This occurrence has been known to happen twice to the same person. It is very grave, and has raised the question of premature delivery as the only means of preserving sight. I can offer no advice concerning it. Sudden amaurosis from suppression of menstruation is reported by Samelsohn. After typhus and typhoid fever, sight is sometimes impaired or lost, **and** in due time there will usually be atrophy of the nerve.

Hysterical amblyopia, or, as it is sometimes called, retinal anæsthesia, is a recognized condition, and has been studied by Charcot, Landolt, and others. It is temporary, irregular, and attended by other hysterical symptoms. Hemi-anæsthesia is sometimes a characteristic of the cases. There may be only one eye affected, and but one-half of the field. Other eye diseases have been simulated in these cases, viz., glaucoma, and for this iridectomy was done by Cuignet. Prognosis is good, and treatment must not be too serious. The main question is diagnosis.

Migraine, or megrim—Scotoma scintillans—Epilepsy of the retina.—Certain persons are subject to attacks of transient loss of sight, which may or may not be attended by other symptoms, such as headache, disturbance of stomach, and less often by impairment of mind, of touch, etc. What is called a blind headache, or sick headache, is the most frequent and mild form of the disorder. The visual **disturbance** may be the most conspicuous of the phenomena, and should **not** be regarded as of serious moment. It sometimes occasions total blindness in both eyes, **and** then naturally causes no little alarm. An illustration is the following:

Mrs. W., a widow, forty-three years of age, was brought to me by Dr. Burchard. She seems well nourished, but is said to have lately lost flesh. Has an anxious expression, been greatly worried by business cares, and been under much excitement. Three months ago she began to have partial obscurations of sight in the right eye. In a little time the sight would "go out," as she said, every twenty-four hours, and everything be dark for fifteen to thirty minutes. This condition affected the right eye only until a few evenings since. Then, while at the opera, she became totally blind for about one-half hour, and, as she says, was in total darkness. Since then the same thing has happened every evening, and lasted from a few moments to half an hour. At my interview with her in the office she exclaimed that sight was leaving the left eye. Her vision had been previously examined and found to be $\frac{20}{20}$ in each. While the dimness existed I inspected the left eye with the ophthalmoscope. The arteries were reduced in size, the veins were normal, no other notable appearances. The attack passed in a few minutes, and then the arteries grew larger and were like those of the other eye. There were no signs of effusions or inflammation. **My advice** was to use bromide of ammonium, to take nourishing food, and try to secure exemption from anxiety and care.

Another case came to my notice within a year, in the person of Dr. **B.**, Asst. Surgeon U.S.A., thirty-four years of age; in vigorous health, has refractive error as follows: O. D. −24, −10, 10°, V.=$\frac{2.0}{2.0}$. O. S. −5¼, −12, 160°, V.=$\frac{2.0}{2.0}$. Has diverging strabismus for distance and converging strabismus for near objects. Both optic nerves are hyperæmic. Has had megrim for many years, consisting in severe headache and central scotoma of that kind described by English writers as being bounded by the fortification line, that is, its outline is zig-zag, with angles like those of a fort, and the edge is brightly chromatic. Dr. B. has used many remedies, but with little advantage, and finds that some disturbance of the stomach is most likely to cause the attack.

Many writers have put on record their experiences in this way. As indicated in the above cases, the blindness may be total or partial, may affect both eyes or but one, and sometimes it has taken the form of hemianopsia. A partial ischæmia of the retina, due probably to spasmodic contraction of the arteries, seems to be the pathology, and the treatment must be directed to those general indications which may be found in the digestive, or uterine, or other functions. The attack, so far as sight is concerned, involves no danger.

One caution should be observed in these cases: not to confound attacks of megrim with the temporary obscurations which occur in glaucoma. For this reason, the tension of the globe, the state of the optic nerve as to excavation, and the limits of the field, must be exactly determined.

Temporary loss of sight may occur as the aura of an epileptic attack, and also in fainting. Hallucinations, visions, and other visual disturbances, need not be considered, because they belong to the domain of mental disorders. It may be mentioned that cases are recorded where patients who had power of speech and ability to write correctly could not read what they had written, although sight was perfect. There has been found at the autopsy a lesion of the brain not far from the gyrus angularis (see Robert, " Troubles Oculaires dans les Maladies de l' Encéphale," p. 440, 1880). This condition is called (not very aptly) blindness for words. An instance, which was very carefully studied, is given by Dr. A. B. Ball, in the *Archives of Medicine*, April, 1881 (New York : Putnam's Sons), with autopsy.

Prolonged exposure to strong light causes disturbance of sight, which in some cases consists in reduced sensibility (torpor) of the retina, and in other cases in exalted sensibility (hyperæsthesia) of the retina. We have, therefore, the words *hemeralopia* and *nyctalopia*, whose significance has been confounded by disagreement as to the Greek words from which the terms were derived, but to which the meaning now attached is that the first means good sight by day and poor sight at night, while the latter means the reverse. The first is synonymous with night-blindness, and may occur to persons travelling in the tropics, to those who work in furnaces, to soldiers who drill on very sunny parade-grounds, etc. It is noticed that insufficient or poor food is a factor in the disease. The persons see well by day, but when the light is reduced below a certain degree, whether by nightfall or artificially, their vision is far worse that it ought to be. There is no impairment of the field and no appearances to the ophthalmoscope. Sailors designate this condition as " moon-blindness," and attribute it to lying on deck in the moonlight.

Snow-blindness, as it is called, is a mixed condition, consisting of intense photophobia, spasm of the eyelids, conjunctival irritation, and sometimes chemosis. The bright glare and the severe cold combine to cause the condition, and the treatment consists in soothing applications to the lids, warm water, with a little tinct. opii, protection by shades, goggles, veils, etc., or, if possible, shelter indoors. Some of the cases exhibit anæsthesia (torpor) of the retina, instead of undue sensibility. For the first-mentioned cases, prolonged exclusion from light, with good diet, and in many cases antiscorbutics or cod-liver oil, will almost surely afford relief. Cases of night-blindness now mentioned must not be confounded with retinitis pigmentosa, of which some cases are congenital and exhibit no pigment in the retina. These latter always have a remarkable limitation of field, which may be attended with reduced central vision, and distinguishes them from the cases now considered.

Another category are what may be called cases of *hysterical hyperæsthesia* of the retina. It is found both in men and in women. It is often associated with some error of refraction, and is brought on generally during an attack of illness. Light becomes unpleasant, and the patient desires its partial exclusion, until at length nothing but absolute darkness may be tolerated. One such instance is as follows : a man of active mind and good health was submitted to an operation for varicocele. During confinement to bed he amused himself by studies in mathematics. His eyes after a time gave him trouble, and he caused the windows to be shaded. Anxiety about the condition of his genital organs made him morbidly sensitive, and this aggravated the growing irritability of his eyes. After he was cured of the varicocele he remained in a perfectly dark room for several weeks, and was unable to bear the least light without great distress. When he came to me he was encased in wrappings about his eyes, which he was very loth to remove. By urgency and insistance, and assurances that his fears of blindness were needless, I finally succeeded in examining and testing his eyes, and found a high degree of hypermetropia. Suitable glasses were given, and he soon gathered courage to face the light and use his glasses, and was perfectly restored. I relate this case from memory, and cannot adequately convey the extremely distressing state to which a highly educated and capable man had been brought.

An instance in a young woman, about twenty-three years of age, was so intense in its character as to be absurd. To shut out all light when going out-doors, she had constructed a visor of pasteboard, cotton wadding, and green cloth, which covered her head and face to the end of the nose, like a huge mask, and was tightly tied behind. Over this she wore a thick veil. These things were removed in a dark room, and then no persuasion could induce her to unclose her eyes. She finally consented to be chloroformed in a dark room, and while she was unconscious the blinds were thrown open, and, on waking, it was some time before she observed the increased light, and thenceforward improvement took place. She had muscular asthenopia, and there was a relapse afterward, but the hyperæsthesia ultimately disappeared.

In another case, of which I have full notes, the lady had been for eight months in darkness. When she had laid aside the wraps about her face, I succeeded, by the help of the atomizer playing on the lids and by hopeful talk, in getting her to open her eyes, and in two hours she was able to bear the ordinary light of a room. Afterward atropia was used. There was marked conjunctival irritation and spasm of accommodation. Encouraging assurances to the patient greatly aided her recovery, and by abductive prisms, to correct muscular insufficiency, she was able to get moderate use of her eyes.

More need not be said on this subject, except to remark that attention must be given to the general condition of the patient, both physical and moral, and not only will medical skill, but personal tact, be required to win confidence and impart courage. There will often be found refractive or muscular errors to be corrected, as well as irritation of the conjunctiva, and perhaps nasal catarrh.

HEMIOPIA, OR HEMIANOPIA (MAUTHNER).

The first word denotes half-sight; the last denotes half-blindness, and is the term to be preferred, because the loss, and not the preservation of function, is the topic of consideration. Theoretically, we might have the fields divided into an upper and lower half; but such cases are so rare as not to merit discussion, provided we leave out of view all in which there are no manifest intra-ocular lesions. The fields are divided by a vertical line passing through the centre, and both eyes are involved. The frequent cases are those in which homonymous sides are wanting, *i.e.*, both right or both left sides; the less frequent are those in which the external or temporal halves are blind. Least frequent, and very rare, are cases in which the median, or nasal sides of the fields, are absent. In both the temporal and nasal cases the line of division is apt to be irregular, and not on the vertical meridian, and the blind parts of the field are not wholly deficient in light-perception. But for homonymous hemianopia the **line** of division is sharp, is on the vertical meridian, and the blind sides **of the** fields have no sensation of light. These are **cases** of intra-cranial, and often of cerebral disease. As a rule, the interior **of the** eye presents no lesions. The optic nerves come into relations with **a large** number of ganglia at the base of the brain, and until recently they **were not** supposed to penetrate any farther; but we now know that they pass onward to the cortex of the posterior occipital lobe, at its hinder part. Ferrier fixed upon the gyrus angularis as the terminus; later investigators, especially Munk, include adjacent parts. Munk has shown that this region in the ape has a most remarkable correspondence in function with the retina. If certain parts of the cortex be destroyed, a part, and always the same part of the opposite retina, becomes blind.

Experience in experimental physiology, in anatomy, and in pathology, has shown that hemianopia may come from a lesion, 1st, of the cortex of the posterior occipital lobe; 2d, of the intermediate ganglia, viz., the thalamus opticus, the corpora quadrigemina, corpora geniculata, etc.; and, 3d, of the tractus or of the chiasm. In the case of the chiasm, the symmetrical binocular hemianopia is not so certain to occur **as** in the other lesions.

Monocular hemianopia can take place only by lesion of one nerve in front of the chiasm; one such case without autopsy is given by Mauthner. Within a few years much discussion has taken place relative to the struc**ture** of the chiasm, especially whether the tractus cross in whole **or in part.** The partial crossing was the old doctrine, and, although **recently denied,** it has come to be fully substantiated again by a great variety **of** evidence. We are still in the dark as to how the crossing and direct fibres are related to each other. We know that their arrangement is complex, that it does not correspond to any scheme yet imagined, and that probably it is not the same in all individuals. It has been held that in the optic nerves, up to the optic foramina, the direct fibres lie to the outer or temporal side; but this is now in doubt. It is certain that each nerve is composed of more fibres which have crossed than of direct fibres, and the ratio is about as three to two. In the chiasm it has been shown that there are commissural fibres, known as the commissura inferior, or of Gudden. They do not enter the optic nerves, but go to the peduncles of each side respectively. There is also a commissure of Meynert, which does not belong to the chiasm, but lies upon it. The crossing of a part only of the fibres of each tractus in the chiasm was absolutely proven by Nicate in cats. He cut it from before backward on the median line, and the animals continued to see with each eye; that is, on the temporal half of each field, which would have been impossible did the fibres all cross in the chiasm. The same has been proven in man by examinations of the brain in persons who have had total atrophy of one optic nerve; in them, both tractus are partially atrophied. On the other hand, in a case of destruction of only one corpus geniculatum, both optic nerves were partially atrophied (Gudden: *Arch. für Ophth.*, xxv., i., Taf. iii., Fig. 14). On this point there is now no doubt. As to the cerebral centre for vision, Charcot assumed that lesion on one side caused total blindness of the opposite eye, and not binocular hemianopia. This has been rendered unlikely by Munk's experiments on dogs and apes, while a case of softening of the occipital lobe seen by Curschman, and another by Westphal (Mauthner: "Gehirn und Auge," 480, 1880), are conclusive that lesion of this region, without influence of other parts, causes hemianopia of both eyes.

I find records in my note-book of 13 cases of hemianopia. Of these, 11 were homonymous and 2 were temporal. Of the 11, two were partial—that is, not the entire half, but corresponding parts of the half of each field were lost. Of the 11, six were blind on the right and five on the left side. Of these 11, three had no other affection and no ophthalmoscopic changes; eight had complications which were as follows: slight temporary dilation of one pupil in one; loss of sensation in three; aphasia to slight degree in three; hemiplegia in one; lesions of nerve or retina in five cases. Eight patients were men, five were women; their ages were from thirty to eighty-two. In all, the attack was sudden. In some, central vision was much injured; in others it was good; in some, it was at first much worse than it subsequently became. In all there was from the onset severe headache, lasting for days or weeks. In some there was impaired memory. It thus appears that the eye-symptoms may exist with no other concomitant than severe headache, while it may be attended by other signs of brain-lesion. The persons are, in most cases, beyond middle age. If the right side of the field is lost, the inconvenience is much greater than when the left side is destroyed, because we read and write from left to right, and unconsciously let the eye run forward to the next word, which, when there is right-side hemianopia, is impossible. It happens that less than one-half of each field may be lost, as in the following case:

Miss S., aged thirty, came to me in April, 1879. She was forewoman in a large establishment, and, being extremely busy, she had not slept well for many months. Four weeks ago she had an attack of severe pain, running from the left clavicle and shoulder to the head, which drew it to one side, and prevented sleep for two nights. Two weeks afterward she found herself totally blind on the right side. There were no other symptoms. Examination found no disease of the kidneys, nor of the heart; she had not had rheumatism. Vision in each eye ⅖. By the ophthalmoscope: right disc not swollen, vessels full; near it the retina a little œdematous, and has some dots, while a bluish zone surrounds almost all the disc. Left eye about the same. Visual fields have this peculiarity: that while there is symmetrical hemianopia, perception remains on the extreme periphery of the blind sides. Treatment was iod. potas., gr. x., ter. in die. After a month her condition, which had been one of great excitement, because of her impaired sight, was more calm, and she slept well, and felt very tired. Had some numbness of the left leg. Visual fields then showed a clearing up of the lower half of the previously blind portion.

In August following, the fields were about the same, and in the retinæ there were fewer specks than at first. The right optic nerve was more hyperæmic than the left. The patient then passed out of observation.

It is almost certain that the lesion must have been in the occipital lobe of the left side and was, probably, a small hemorrhage, which was absorbed, and left much of the tissue uninjured. That the periphery of the field was not injured at the first, points to the limited character of the brain-lesion. I could cite another case of partial hemianopia still different in the arrangement of the fields. I will relate one of the only two cases of temporal hemianopia which I have seen.

A sea-captain, aged forty-eight, in good health and of good habits, after exposure to a tropical sun on a voyage in the Gulf of Mexico, was attacked with severe headache, nausea, and diminished sight, which in five days amounted to total blindness.

After a few days, sight began to return, and he observed that he saw badly on the outer side of each field. After three months I saw him, viz., in November, 1880, and found that the condition of the fields was what he represented, but that there was not total, but partial blindness, *i.e.*, amblyopia in each temporal half. The direct vision of each eye was good. The interior of the eye showed no special lesion.

For explanation of such a case, we must look to the region of the chiasm and on the median line. The symptoms lack entirely the sharp and definite quality of cerebral hemianopia, and the initiatory total blindness precludes such a supposition. In fact this kind of half-blindness is both rare in occurrence and vague in symptoms.

Nasal bilateral hemianopia has been very seldom recorded, and is to be accounted for only by the supposition of a double lesion acting upon each side of the chiasm, or on the outer side of each optic nerve. While this is possible, it is excessively unlikely to occur. Dr. Knapp's supposition that atheroma of both carotids in the foramina can cause it, is not considered probable.

Diagnosis scarcely needs to be alluded to, yet for want of attention, or from neglect to examine the visual field, this condition is often unnoted in general practice. Physicians at the present day seem almost as reluctant to investigate eye-troubles, as in former times they were eager to assert their full competency to deal with them. It certainly is not too much to expect a general practitioner to ascertain that a patient has hemianopia.

Prognosis.—For most cases, recovery of the lost fields is not to be looked for. Where central vision is bad that will often improve. This will be found in many cases to be connected with intra-ocular lesions. A larger question is as to the patient's general welfare. We have to deal with a brain or an intra-cranial disease. It may be an apoplexy, an embolism, softening, a new growth, a periostitis, or an injury, etc. We must put together all the symptoms and thence deduce the best judgment we may. In most cases we have to do with apoplexy.

Treatment is to be regulated by the general symptoms. Most frequently iodide and bromide of potassium will be suitable. Syphilis may underlie the local lesions. For the eyes nothing is specially indicated.

Simulated blindness.—If a person falsely pretends to be blind in both eyes, we examine the behavior of the pupils to light and upon the effort of accommodation, we note the correct fixation of both eyes to the same point, we inspect the interior of the eyes by the ophthalmoscope. Of course, if all the above conditions correspond to the normal state, we are not authorized to declare the person a malingerer, however much we may suspect it. We must simply leave him to take care of himself, as if he were not blind, and wait for something to turn up which may surprise him into betraying himself. Goldsmith found that a mad dog on the street quickly cured a blind beggar, as he tells us in verse.

With pretended blindness in only one eye it is easier to cope. A stereoscope can be arranged with a part of a diagram in each half of the slide, which, when combined, make the proper whole. Wafers of different colors may be pasted on opposite halves of the slide and expose the cheat. A prism of 6° to 12°, with base up or down, may be put in a spectacle-frame, and if the individual sees a dot or a line or a distant candle double, he has betrayed himself. Rotating the prism will often deceive the malingerer. A bit of ground glass may be put over the seeing eye and vision tested in the one which is alleged blind; a red glass can be employed to bring out contradictory statements. The above proceedings are required in cases of draft for military service, **or** in prisons.

I have known a person to be impressed with a belief **that** one eye was blind because it had been hurt, and was declared by **an** oculist to be seriously damaged. The ophthalmoscope discovered opaque optic-nerve fibres, and the oculist being deceived, deceived the boy. With much difficulty I persuaded him to let me bandage the eye which he believed to be **the** only good one, and by the time when the next meal was served he found his sight improving.

CHAPTER XVII.

THE ORBIT.

Anatomy.—The cavity which **contains** the eyeball is a quadrangular pyramid, in which we distinguish **the base** or opening, the four walls, **and** the apex. The angles of the base are rounded, and the infero-temporal angle is at a lower level **than the** infero-nasal angle. In other words, the upper and lower sides have a slope downward toward the temple. The edge is somewhat **sharp on** three sides, because it overhangs the interior surface. On the **fourth, viz.**, the internal side, the edge rounds off and slopes toward the median line. The orbital margin **is capable** of great resistance, because the bone **is** dense, and is buttressed **by the** zygomatic arch, which expands into the malar bone, and by **the arch of** the frontal bone. On **the medial** side, where strength **is not needed**, the bones are very thin, **but** their celluloid arrangement gives great **capacity** for dispersing the force of shocks. On the wall of the inner margin we have the groove in which is lodged the lachrymal sac, in front of which is the insertion of the orbicularis, and behind which is the insertion of the tensor tarsi. This edge is formed by the ascending process of the superior maxillary, which joins the nasal process of the os frontis. The groove for the lachrymal sac is mostly channelled out of the lachrymal bone. At the supero-nasal angle the supra-trochlear arteries and veins are found, and **a** little behind the edge is the loop through which passes the tendon of the superior oblique (trochlearis) muscle. At the distance of from six to ten millimetres from the supero-nasal angle we have a foramen, or, it may be, a notch which gives passage to the supra-orbital nerve, and to a small arterial twig. Beyond this the edge overhangs the cavity more decidedly, until at the supero-temporal angle we find behind it a decided fossa, in which is lodged the lachrymal gland. Passing around the outer to the lower border, we find, a little inside the middle of the latter, the region where the infra-orbital nerve emerges. It comes out of the bone about eight millimetres below the edge. The orbital margin is composed of the superior maxillary, the frontal, and the malar bones. The prominence of the frontal sinuses make this the most elevated part of the region of the base, while at the temporal side the bone is thickest and most dense. The inner wall of the orbit is smooth and polished, presenting the surface of **the** os planum of the ethmoid, and here we have two foramina in the suture between the os frontis and the orbital lamina of the ethmoid ; the anterior foramen gives passage to the nasal branch of the ophthalmic nerve into the skull, and the latter to an artery into the nose. The superior wall, or roof, of the orbit is slightly concave and smooth. Immediately next to it is the levator palpebræ superioris ; it is very thin, and at its anterior and inner part it separates into two layers, between which is the frontal sinus. The surface of the outer wall is nearly flat, inclines outward from the median plane, and is composed of the greater wing of the sphenoid, and of the malar bone. The inferior wall is thin, and furrowed by the groove for transmission of the infra-orbital nerve ; below it lies the antrum. Between the outer and the inferior walls, at their place of junction, is the spheno-maxillary fissure, which sweeps from without and below inward and upward in an imperfect right-angled bend, bounding the posterior part of the body of the superior maxillary bone, and opening into the muscular mass and the vessels which lie about the pterygoid process. It is through this fissure that a cut is made in excision of the upper jaw. The fissure is sometimes called the inferior orbital. At the angle of junction between the superior and outer wall is another fissure, shorter than the preceding, which separates the lesser and greater wings of the sphenoid, and is called the superior orbital, or sphenoidal fissure. It gives passage to all the motor nerves of the eye, to the ophthalmic nerve, and to the ophthalmic vein, while it has upon its cranial side the cavernous sinus, and is occupied by dense connective tissue, which shuts up the aperture firmly. At the apex of the orbit, above the inner end of the sphenoid fissure, is the optic canal or foramen, which perforates the sphenoid at the junction of its wings with its body. The canal is cylindrical, is rather larger in front than behind, i.e., funnel-shaped, is from eight to nine millimetres long, and about six millimetres in diameter on the average. Its course is from below and outward, upward and to the median line, and the canals of opposite sides converge to each other. Posteriorly they open into the middle cranial fossæ. The canal contains the optic nerve and the ophthalmic artery. The sphenoidal fissure running outward, and the spheno-maxillary fissure running downward and outward, meet and become continuous with each other just below the optic canal. These fissures are of variable length and breadth ; they may differ on the opposite sides of the same person, and usually grow larger in later life. The spheno-maxillary fissure contains fat and connective tissue, and some vessels which communicate with the internal maxillary.

The dura mater adheres to the sphenoidal fissure and to the optic canal very firmly. It clothes the surface of the optic canal, and, curving forward, is continuous with the outer sheath of the optic nerve and with the periosteum of the orbit. There is also an inner sheath of the optic nerve which is continuous with the arachnoid, and which at the optic canal is firmly attached to the outer sheath by meshes of connective tissue. In many cases this meshwork is sufficiently open to permit fluid to be injected from the cavity of the skull into the space between the two sheaths of the nerve, as was proven by Schwalbe. But this is not invariably possible, and in all cases the optic nerve is so closely attached to the wall of the bony canal that it cannot be pulled away from it. Schwalbe says this adhesion is most intimate at the upper part of the canal. The ophthalmic artery coming from the internal carotid and about two millimetres in diameter, lies in the canal below and to the outer side of the nerve. The recti muscles and the superior oblique originate around the opening of the optic canal, the rectus externus having two roots, between which passes the third nerve.

The orbits stand to each other in such relation, that their axes form an angle opening forward of about forty degrees, whose apex would be at the middle of the anterior clinoid process of the sphenoid. The floor of the orbit slopes downward, forward, and outward ; the shape of the cavity thus inclines the eyeballs to assume a position looking outward and downward. The globe is placed so as to lie nearer to the outer wall than to the inner, while the optic canal is above the level of the middle of the eye.

CELLULITIS ORBITÆ.

This appears under the form of inflammatory œdema and of phlegmon. It may be occasioned by a variety of lesions, both traumatic and idiopathic. The mildest form of œdematous cellulitis is as follows: a delicate boy, of pale skin, about nine years old, complained of dull pain about the right eye; there were no signs of ocular inflammation. After two or three days the globe began to advance from the orbit, and I saw him. There was a little swelling, but no redness of the lid, no chemosis nor conjunctival redness; the eye stood forward several millimetres, and turned outward. There was difficulty in movement, and occasional diplopia. Pressure on the globe when firm was unpleasant; no hardness or tumor could be felt on pushing the finger between the globe and the orbital margin. Vision was perfect, pupil and fundus natural. The symptoms continued the same for several days and the eye finally returned to its place. Such a mild attack of cellulitis is most likely to occur, as I have found, in young and not robust children. It is not dangerous, and needs only mild external applications, such as warm infusions of opium or of chamomile-flowers, or the liquor ammonii acetatis, 1 part to 5 of water.

Phlegmonous inflammation of the orbit is either idiopathic or symptomatic. It may be ushered in with a chill and rise of temperature. There will be pain, swelling, and duskiness of the lids, especially of the upper, and the eye will advance. Chemosis of a yellowish red color, with conjunctival vascularity, will appear; the eye will move with difficulty, and in the height of a severe attack it will be absolutely rigid; pressure on the globe may or may not make it recede, and will cause deep pain; exploration by the finger in the circumocular sulcus will find the tissues firm, tense, solid and painful, some parts being more tumid and tender than others. This last symptom of resistant and painful infiltration of the orbital tissue is the important feature. In bad cases the eyeball becomes involved by infiltration and opacity of the cornea, and perhaps even to its suppuration. The optic nerve, the sclera, and the interior textures may in turn participate, and the end be panophthalmitis. The cases vary in severity and general features, but the above facts are the chief symptoms. Such a lesion may be metastatic from puerperal fever, remote phlebitis, septicæmia, carbuncles, or typhus fever; it is more frequently coincident with facial erysipelas, with disease of the neighboring bony walls, or with acute inflammation of the lachrymal gland. It has been observed as a complication of purulent meningitis; but in such cases there is strong reason for regarding thrombosis of the cavernous sinus or neighboring veins as the middle factor of the process.

Prognosis is of necessity serious, and much turns on the age, health, and habits of the patient, and on the cause of the trouble.

Treatment will vary according to the period at which the case is seen, and with its cause and complications. For a case seen early, cold or warm applications, as the feelings of the patient dictate, and free use of leeches on the brow and temple—six or eight of them, will be judicious. When the swelling rises high, and where exploration beneath the rim of the orbit detects deep infiltration and resistance, and if the swelling of the lid be of a hard and brawny type, and there be much pain, an incision should be made at the point of greatest tension and tenderness with a straight, narrow bistoury, close and parallel to the wall of the orbit and generally above the eye. The knife should enter for one-half inch or more,

and while the point may not cut widely, the opening through the skin and
fascia must be one-fourth to three-fourths of an inch in length. By this
incision the tension of the oculo-orbital fascia is relieved, the vessels
are unloaded, serum finds vent, and the tissues are relaxed. It is not
necessary to find pus, but in case such a focus has formed, the knife must
aim for it and go to any depth to reach it. My convictions are strong in
favor of an early incision with sufficient external opening, as a means of
arresting the phlegmonous inflammation and the formation of pus. I
quite disagree with a tendency to long tarrying until pus begins to show
at the surface, because meanwhile grave mischief can befall the optic
nerve and likewise the cornea, as the result of tension, pressure, and contiguous inflammatory action. The wound is to be kept open by a tent of
borated lint and washed freely with warm borated or carbolated water to
promote bleeding and subsequent discharge, and the external lotion of
warm water or acetate of ammonia must be continued; or hot poultices
of ground slippery-elm bark or of spongio-piline wrung out of hot water,
should be kept up. The patient's general condition will modify the local
treatment, because if laboring under a grave general disease, fever, pyæmia, or meningitis, there will be, perhaps, some hesitation about inflicting
pain or incurring much loss of blood. The principles of general surgery
must be our guide. When, however, orbital cellulitis and blepharitis
complicate erysipelas of the face or head, incisions are to be made early,
when the skin assumes the tense and dusky hue and hard feel of phlegmonous infiltration. The circulation being strangulated by the effusions,
deep incisions offer the best chance of preventing the sloughing of tissue
and injury to the eye. In these cases there is less danger from undue
bleeding, because the vessels are choked by the infiltration.

It need not be said that in many of the complicated cases, the general
disease demands the chief attention, in stimulants, supporting food at
intervals, quinine, mineral acids, etc.

It may happen that the eyeball passes into general suppuration or
suppurative keratitis. For the latter, and for early stages of the former,
warm fomentations are to be sedulously used. Paracentesis of the cornea,
or its free division, may be required. For suppuration of the globe, where
it has become tense, and is giving great pain, an incision may be made into
its anterior half to evacuate the vitreous, at least in part. In these cases
enucleation may be done, as I do not hesitate to do when its suppuration
is primary and the affection of the orbit secondary. It is fair to say that
there are, in literature, a number of cases of fatal results following abscess
of the orbit, but, naturally, discrimination must be made as to those which
are associated with other and grave disorders. An extensive abscess of
the orbit, when purely local, threatens risk by extension backward to the
brain, and it may also, in some cases which terminate in recovery, cause
so much contraction of the connective tissue as to interfere with the motility of the eye; this has been seen to take place, especially between the
levator palpebræ and rectus superior. Such mishaps are, however, uncommon, and recovery usually occurs with entire restoration of function. Impairment of sight is unfortunately not so rare, consisting in lesions of the
optic nerve and retina, inflammatory and atrophic, also in intra-ocular
hemorrhages and in detachment of the retina. Some cases of amaurosis
or amblyopia do not show any visible lesions, and in them the cause may
lie in the stretching of the nerve, or in pressure on the nerve as it goes
through the optic canal, by the collecting of exudation in the sheath. It
is said, too, that by pressure, the axis of the eyeball may be altered, giving

rise to hyperopia or to myopia according to the direction in which its force is chiefly exerted. But even from visual dangers, most cases are ultimately safely delivered.

INFLAMMATION OF THE OCULO-ORBITAL FASCIA.

Tenonitis.—It is not intended to refine needlessly upon the varieties of inflammation in the orbit, but we meet cases sometimes whose distinctive features justify us in designating the above tissue as the seat of the lesion. For example, a girl, fourteen years of age, came to me, within a year, in whose left eye was to be seen a yellowish bleb over the insertion of the rectus externus. The circumocular conjunctiva at the equator was moderately injected; the globe was slightly prominent, its movements a little uncomfortable; pressure upon it caused pain. Such an attack had occurred twice within six weeks, **and** disappeared each time in less than two weeks. The girl seemed in good health, and **no** syphilitic or strumous taint **was apparent.** Such a case would seem like episcleritis, but the prominence **of the eye, and** its tenderness on being pressed into the orbit, located **the disease farther** back, and pointed out its **seat** to be in the ocular part of the capsule of Tenon.

It happens that the symptoms of the above case may become a little more pronounced, so that instead of a local and well-defined bleb, chemosis may begin at and surround the whole equator and **reach** the cornea, attended by no distinct symptoms of scleral, conjunctival, or other disease of the front of the eye, but attended by swelling of the upper lid, slight proptosis, and slight restraint of motion. Such is the picture **of** well-marked tenouitis. Its recognition is practicable only when the effusion is serous; its origin is usually rheumatic. Treatment should be mild, in soothing lotions and choice of suitable rheumatic remedies, according to the indications: alkalies, iod. potass., salicylate of sodium, etc. Mild cases will get well in a few days. Dr. Bull has recorded cases of the disease following operations for strabismus, and I have mentioned one of the same kind Other operators have seen the same.

PERIOSTITIS ORBITÆ.

This appears as an acute and a chronic condition. The causes are traumatic, rheumatic, syphilitic, and so-called **scrofulous.** The most common are syphilitic, and the disease is most frequent in children. The favorite locality is the margin of the orbit. There will be pain; œdema of the lids; some chemosis, beginning equatorially; and the distinctive symptom is that the edge of the orbit is very tender when pressed by the finger. I have seen this symptom in exquisite degree in cases of syphilitic acute periostitis, in which the diagnosis was perfectly palpable *(sit venia verbo!)*. The chronic form is diagnosticated with more difficulty, except when nodular, bony swellings appear within reach of the finger, and they may interfere with the function of the nerves or muscles, or even with the position and movements of the eye.

The consequences of periostitis are extremely various. In severe cases acute phlegmon of the orbit ensues, with all the features previously described. In other cases only a limited and small amount of pus is found; it may be at any point of the orbital wall. Such a case befell a friend of mine some years ago. She was a lady, over fifty years of age, who had constant and troublesome pain over the forehead, aggravated at night, and lasting for some weeks. It became localized over the supra-orbital notch, and was attended by swelling of the lid in that region. I ventured, after some delay, to pass in a narrow knife, and on probing reached a spot of rough bone. Pus in small quantity escaped, a fistula was established, and after many months, and persevering treatment by injections, it was closed up. It may happen, that when not evacuated in front, such an abscess, breaking down the bony tissue, shall find its way into surrounding parts, viz., into the ethmoid, the antrum, the frontal sinus, or into the cavity of the skull and brain. The consequences of these several events are readily understood. Again, caries or necrosis may take place, and when the dead tissue reaches the surface, and comes out through the formation of an abscess, a fistula ensues, and, when it heals, deformity of the skin and lids will ensue, viz., ectropion, etc., of various degrees. A very notable case of this kind came to me some years ago, and was published in the "Trans. Am. Oph. Soc.," p. 129, 1870. The disease began at one year of age, and when I saw the young lady she was sixteen. Pieces of bone were discharged during many years. When I saw her, there was protrusion of the globe from the orbit, the cornea was opaque and turned downward ; the upper lid adherent to the upper margin of the orbit, and fully everted ; the conjunctival surface covered by thick and coarse granulations ; the globe constantly exposed, and requiring to be covered by a pad or the hand. In the orbital edge a deep sulcus, from which bone had been exfoliated. The other eye intolerant of light, and in a state of chronic irritation. The treatment adopted, and which was the only resource, was enucleation of the globe, removal of all of the conjunctiva, bringing down the lids to proper position, and closing up the orbit by flaps of skin so as to cover it in completely. The deep part of the orbit was filled by growth of connective tissue, its cavity was narrowed by hyperostosis of the walls, and there was no possibility of wearing an artificial eye.

Treatment.—During the acute period usual antiphlogistic measures will be employed locally, but iodide of potassium will be vigorously used ; in syphilitic cases, gr. xx., ter. in die, or perhaps in larger amount, while in rheumatic cases, gr. v., ter. in die, may suffice, but the suitable dose and remedy will depend on experience of the case. Should an abscess threaten, the same treatment will be proper which has been before stated, and my judgment is in favor of an early incision, always using a long, narrow knife, or bistoury, for a sufficiently deep puncture. If a fistula has formed, one must be very prudent in its exploration by a probe. Often the deep parts are readily excited to inflammation. The proper way is to secure a full external opening by dilating it, at first with laminaria probes, and later by small sponge-tents, until access is gained to the deeper parts of the sinus. Syringing with a fine tube should be practised daily, so long as any secretion is pent up, and when the parts have become callous, or indisposed to heal, stimulating fluids may be introduced, but always under strict limitations of prudence as to possible over-effect. An attack of acute cellulitis is not hard to awaken, and will be liable to be disastrous.

If rough or dead bone is felt by the probe, it is not wise to attempt its removal, nor its solution by injections, but simply keep the outlet patent, and use warm antiseptic injections of carbolic acid, 1 to 100, or of aqua chlorinat., 1 to 20, or acid. boracic, 1 to 25. The general condition of the patient is meanwhile to be properly cared for in administration of good food, ol. morrhuæ, iron, especially syr. ferri iodid.; giving mild mercurials, viz., biniod. hydrarg., gr. $\frac{1}{15}$ to $\frac{1}{10}$ ter. in die, with small doses of iod. pot., the object being to bring the nutrition up to the state in which healthy tissue shall replace the dead bone.

The possibilities of serious complications involving life are not to be **forgotten in these cases,** because of the near vicinity of the **brain.**

THROMBOSIS OF THE ORBITAL VEINS

occurs necessarily in cases of phlegmonous inflammation of **the orbit, and** cannot be distinguished from that condition. The process may extend **to** the cavernous sinus and thence to other sinuses, causing cerebral symptoms of a recognized character, according to the parts involved. When the lesion extends to sinuses in both sides of the skull, then we may have obstruction to the venous circulation in both orbits simultaneously, producing exophthalmus, œdema of the lids, with severe brain-symptoms. When, however, the cavernous sinus becomes plugged by a clot, as a result of previous thrombosis of cerebral sinuses, and in connection with cerebral disease, the order of events will be reversed. For example, phlebitis of the sinuses may come from otitis interna, and if pyæmic symptoms occur we may get orbital symptoms on both sides. They will be exophthalmus, œdema of the lids, immobility of the globe, hyperæmia of the retinal veins, papillitis, paralysis of muscles, mydriasis. In these cases the combination and order of symptoms is of special importance in order to diagnosis. Purulent meningitis without thrombosis may cause the orbital symptoms, but if some starting-point for thrombosis can be found in a local caries of bone or phlebitis, the true nature of the process **becomes** explained.

Prognosis and treatment of these cases need no special remark. For further discussion of this subject see Berlin in G. and S., vi., p. 540.

EXOPHTHALMIC GOITRE, GRAVES' DISEASE, BASEDOW'S DISEASE.

Under these names are described a condition of palpitation of the heart, hypertrophy of the thyroid gland, and protrusion of the eyeballs. While the fully developed disease includes these three items, any one of them may be wanting. Moreover, while both eyeballs are usually extruded, and can be nearly pressed back into their proper place by the fingers, one may be more advanced than the other, and sometimes only one is affected, as is noted by Stellwag, and as I have once observed. Other symptoms are : extreme excitability of the patient ; he is readily startled, and has flashes of heat ; pallor, and flushing of the face quickly alternate ; the action of the heart is very irregular and thumping, its pulsations may be habitually one hundred, or mount to one hundred and sixty ; there may be some consecutive hypertrophy and systolic bruit, and also a bellows murmur over the large vessels. The thyroid presents variable and sometimes unsymmetrical enlargement. A choking in the throat (globus hystericus) is common ; the patient may be unable, for an instant, to catch the breath or swallow. The eyes stand forward in a peculiar stare, and show the sclera above the cornea, **and,** as Graefe noted, this look of surprise or fear is aggravated by actual retraction of the upper lid, which exposes the globe more than the pushing forward of the eye by a tumor is observed to do. Stellwag also noted the infrequency and slowness of the action of the lid in winking. By exposure the cornea and conjunctiva are irritated and congested, and ulceration of the cornea has been observed. Sometimes it is never fully covered by the lids, and especially is exposed in sleep. The pupil usually is natural in its action, although mydriasis, has been noted ; movements of the eye are unimpaired, diplopia is rare and transient, vision is not involved, the circulation in the fundus is ordinarily not peculiar, yet Becker has seen the retinal arteries pulsate. The disease is complicated with anæmia, and in women often with amenorrhœa and chlorosis, and the patients are hysterical. We may also see digestive disturbances : nausea, vomiting, diarrhœa, bloody stools ; also cough, profuse sweating, and Bulkley has cited two cases with urticaria. There may be nodules of inflammatory exudation, or patches of transient redness, and increased heat of the skin, varicose dilations of vessels may take place on the nose or cheeks, and there may be ephemeral tumors on the eyebrows and lids, attended sometimes by dilated vessels. I have seen one case in which undoubted exophthalmic goitre was attended by firm tumors about the opening of one orbit at its lower border, and which, under the microscope, seemed to be composed of enlarged lymphatic gland tissue ; only one eyeball was protruded. A case of Heymann's had repeated paroxysmal attacks of conjunctivitis with membranous exudation.

The disease occurs with greater frequency among women, according to Emmert (see *Arch. für Oph.*, xvii., p. 203), in the ratio of nine to one. The cause is not definitely ascertained ; most of the lesions are traceable to disturbances of the sympathetic nerve, but where their origin may be is not determined. Some consider the cervical sympathetic, and others the cervical portion of the spinal cord, and Hammond ("Diseases of Brain," p. 819, 1881) considers both the brain and medulla oblongata as the starting-point of the disease. In all these structures, autopsies have found lesions, but not with uniformity. The heart is often a little, seldom greatly enlarged, and there may be insufficiency of the mitral valves ; the thyroid gland shows in old cases some increase of connective tissue and colloid cystic degeneration, but at death it usually collapses ; in the orbit rarely is anything abnormal discovered, and the ocular protrusion disappears. It is, therefore, justifiable to assign the cause of exophthalmus to vascular enlargements, and Snellen has corroborated this opinion by showing that, with a stethoscope, a distinct vascular murmur can he heard during life. Recklinghausen has found fatty degeneration of the ocular muscles.

The disease is slow in progress in most cases, while a few have the good fortune to gain an early recovery. There is great emaciation and prostration, and in fatal cases the end is brought about by asthenia or by phthisis.

Treatment has naturally been very diverse. Some have given remedies to act on the heart, especially digitalis ; others, simple general tonics, quinine, iron, strychnia, etc.; but the best results are gained by galvanization of the sympathetic nerve in the neck for five to ten minutes daily, with a current as strong as the patient can bear ; one pole on the nape of the neck and the other passed up and down over the region of the sympathetic in front. Bartholow in three cases observed improvement from the outset of treatment by putting the negative pole of the primary current on the epigastrium, and the positive pole so as to include the cilio-spinal region, the cervical sympathetic and the pneumogastric within the circuit. Hammond has injected the thyroid with fluid extract of ergot, using twenty minims. I have used fluid extract of ergot internally, and bromides with benefit, and find in Hammond the following formula, which is well aimed : R. Ferri pyrophosphatis, zinci bromidi, ãã ʒ j.; digitalis tinct., ʒ v.; ergotæ fl. extract, ℥ iv. Dose, a teaspoonful three times daily. Generous diet,

moderate alcholic stimulation, cheerful surroundings, avoidance of excitement and of overwork, and the suitable adaptation of occasional remedies
to symptoms, are important considerations. For the exophthalmus no particular treatment is to be commended, except, if the cornea and conjunctiva
become dry through exposure, a little purified (white) vaseline or cosmoline may be put between the lids two or three times daily. A pressure-
bandage is sometimes comforting, and pushing back the globes into the
orbits gives some relief; it may be necessary to hold the lids in approximation by a strip of rubber plaster, or even to do an operation for partial
closure of the lids at the outer angle. Partial division of the levator, as
suggested by Graefe, is not practised.

PULSATING EXOPHTHALMUS.

Under this term are included affections of diverse nature, whose common features are protrusion of the globe and pulsation, which may be felt
by pressure, and whose sound may be recognized by the ear. It is impossible, in most cases, accurately to distinguish between vascular growths,
or angiomata, arterio-venous aneurisms, varicose dilation of veins, true arterial aneurisms, and thrombosis of the cavernous and adjacent sinuses. The
literature of the subject is large; exact knowledge about it is small. In
1869 I tabulated the cases known to that date, where the common carotid
had been tied for pulsating exophthalmus. Since then the subject has
been summed up by Rivington, Harlan, Nieden, and Schlaefke.

Schlaefke has catalogued ninety-three cases of pulsating exophthalmus
(*Archiv. für Oph.* xxv., iv., pp. 112–162), and to this I add one more, published this year, making a total of ninety-four. The cases may be traumatic or spontaneous; they are usually rapid in development, vary in degree of protrusion, are liable to attacks of transient inflammation of the
conjunctiva, and, in some cases, show tokens of retarded circulation in
chemosis or in swollen vessels. Seldom is there pain; the patient is often
conscious of a pulsating bruit, and may have dizziness. Sometimes an
enlarged vessel will be found, projecting at the upper and inner or at the
lower and inner angle of the orbit, and it will pulsate; vision generally
is unaffected; motions of the eye are natural in extent and co-ordination; there is no diplopia; in the fundus the vessels are enlarged, and
sometimes pulsate. In some cases there are dilated vessels of the skin
in the neighborhood of the lids. (Aneurismal dilation of the capillaries
and small vessels of the adjacent skin, which may extend into the orbit
and cause protrusion of the eye and pulsation, are excluded from consideration.) Pressure on the common carotid stops the palpebral pulsation
and also the bruit, both to the patient and to the examiner, while the
globe retires a little into the orbit. The eye can be pressed into the orbit
a certain distance, and the firmer the pressure the harder the pulsation.
Stooping forward increases the protrusion and the pulsation, and is unpleasant to the patient, because of the sensation of weight.

Formerly, cases having most or some of the above symptoms, were
styled aneurisms of the ophthalmic artery. Guthrie published a case of
double exophthalmus, in 1803, in which, at the autopsy, he declared that
there was an aneurism of each ophthalmic artery. Later examinations
have failed to find a single case of this sort, and that of Guthrie is now
discredited, because of the meagreness of his description. In truth,
diagnosis in these cases is approximate, and seldom exact. There have
22

been, **as I** make out by combining the catalogues of Harlan ("Trans. Am. Oph. Soc.," 1875) and of Schlaefke (*Arch. für Oph.*, xxv., iv., 112–162), fifteen autopsies, including the untrustworthy case of Guthrie.   Harlan includes three cases of malignant orbital tumors, which had been diagnosticated by competent surgeons as diseases of the blood-vessels, and for which the common carotid was tied.   If these be added, there are eighteen autopsies.   Leaving them out, we find in the fifteen that the lesions were, thrombosis of the cavernous or other sinus by clot, dilation of the sinuses and veins, communication between the carotid artery and the cavernous sinus, and aneurism of the carotid as it passes into the skull.   It is evident that the displacement of the globe has been simply by direct or by indirect obstruction of the orbital veins, and the pulsation has been communicated through the column of blood from the carotid artery.   The pressure on the large vessels, or communication between the carotid and the cavernous sinus, have occasioned the murmur.   In a case of Nelaton's, injury **on** the opposite temple caused the ocular protrusion—a splinter of **bone** penetrated the **walls of the carotid** and **of the cavernous sinus.** Generally, in traumatic cases, the lesion of bone is on the same side.   **Some** cases have healed spontaneously.   They are much less likely to be published than those in which treatment has been adopted.   In most cases the patient, because of deformity, discomfort, or repeated inflammations, desires relief.   The important point to be decided is, as to the presence of a malignant growth.   I may be permitted to reproduce the summary of symptoms which Schlaefke gives as distinctive of pulsating exophthalmus, due to the ordinary causes, viz., obstructed or dilated sinus, arterio-venous communication, etc.   There is protrusion of the globe ; at the upper and inner angle of the orbit is a soft, subcutaneous tumor ; there is visible and palpable pulsation of the globe and of the tumor ; there is a continuous bruit, with rhythmical increase—by pressure on the carotid the pulsation ceases—the patient hears its murmur ; the cutaneous veins of the upper lid and forehead are distended.   I may amend this description by saying that a pulsating or distended vein may appear either at the lower and inner angle of the orbit, or at the upper and inner angle, and that the angular artery may be greatly enlarged.

In malignant tumors, as against these symptoms, while there will be protrusion of the globe and pulsation and bruit, there will be the following features : the protrusion has come less rapidly, there will not be an outcropping pulsating **vessel** at the upper or lower and inner angle of the orbit, the murmur will be less vehement (but a strong murmur is not necessary **to** the other condition, as my case proves), a resisting tumor will soon be detected in the orbit, and, if time enough be allowed, the case will soon declare its true character, if it be a tumor.

The case which has been referred to as coming under my own observation, is as follows :

M. **M.**, aged twenty-two, native **of** Ireland, single, was attacked with an illness **four** years ago, which kept her in bed **for** five weeks.   She had fever, great pain in the head, nausea and vomiting ; constant **noise** in the left **ear**, with some deafness, came on during the last two weeks.   As she was recovering, **she found** on waking one morning from sleep, that the left eye was swollen, red, and **protuberant.**   There was no pain nor loss of sight.   The exophthalmus soon attained **its** maximum and the eye seldom gave her trouble.   She had a few mild attacks of inflammation in it.   It did not annoy her, except that if she stooped it would come farther out and feel very heavy.   For nine months previous to the eye-trouble her menstruation had been very scanty.   When on her way from Ireland to this country she stopped at Limerick, and there an attack of inflammation began, which continued until her arrival.   She came to the New York

Eye and Ear Infirmary on May 9, 1881; the left eye projected half an inch beyond the other. There is some chemosis and anterior ciliary injection. Media clear, pupil normal; nothing wrong in the fundus, except that the veins on the papilla are enlarged; V. = 5/8. Below the eye, along the border of the orbit, is a pulsating swelling; the angular artery is much enlarged, pressure on the eye makes it recede into the orbit. By auscultation, no thrill is heard but a low pulsating murmur. Pressure on the common carotid stops pulsation in the vessel below the globe. **This** vessel comes **from the inner** side of the orbit, and pushes out under the skin **of the lower** lid like a large varicose trunk.

On May 18th, patient etherized; the angular artery tied, and the protruding vessel exposed, and tied at the inner and lower side of the orbit; it was then cautiously dissected up and traced into the cavity of the orbit, until it reached the groove for the infraorbital nerve, where it dipped down. A ligature was put about this end **and the** vessel excised. It proved to be a vein, and was larger than a crow's quill. No severe reaction occurred. In eighteen days both ligatures came away. In fifty days the patient was discharged. Pulsation had ceased, and the eye had retired one-fourth of an inch.

November 18, 1881.—The **eye back to its proper place.** The **optic nerve normal**; vessels of correct size, V. = 5/8.

Treatment.—Sufficient **time** should be allowed for development of symptoms to enable one **to** form a fair judgment of the probable nature of the case. For a vascular, and perhaps for a cancerous tumor, if any operation were proper, it would be excision, while for a lesion of blood-vessels an attempt at such a proceeding would be most likely attended by dangerous hemorrhage and disastrous results. Pressure on the globe is unavailable and ineffective; injections into the orbit of astringent fluids have been practised successfully, but doubtless these were cases of vascular tumor; injections of iodine have been made, but with fatal results. Pressure on the carotid by the fingers, or an instrument, has in some cases given happy results. Harlan relates a case in which digital pressure was kept up for some time in the hospital without apparent benefit, and the man was discharged with instructions to practise it himself as much as possible. He was taught to find the common carotid in the neck, and when not busy in his duties as brakeman on a railroad train, he pressed the artery by the rounded end **of a stick,** which was about sixteen inches long, and kept it up for many hours **daily.** After several months he showed himself at the hospital entirely **cured.** In my case, ligature of the dilated and pulsating vein, which protruded at the lower edge of the orbit, and simultaneous ligation of the angular artery, procured in six weeks the complete retrocession of the globe, and entire cessation of the faint murmur which had previously existed. In the event of failure of these methods, or in case there be good reason for not attempting a trial of them, ligation of the common carotid is the remedy. Among eighty-eight cases commented on by Nieden, in which, of course, an exact diagnosis of the disease is not vouched for, forty-seven resulted from injury, the rest were spontaneous. Of the idiopathic cases, **the** greater number were **women.** In forty-nine cases the common carotid was tied, in **one of them** both were tied. Of the forty-nine the results were as follows:

Good in .... 33, or 67.4 per cent.
Partial in .... 6, or 12.2 per cent.
Indifferent in .... 3, or 6.1 per cent.
Fatal in .... 7, or 14.3 per cent.

Out of 12 cases where digital compression had been made, five were completely cured and seven derived more or less benefit ; in 9 cases injections were made into the orbit of ergotin, sesquichloride of iron, etc., and in five success followed ; in 3 cases there was spontaneous cure, in 6 death took place, and in 9 no treatment was adopted. Total, 88 cases.

Exophthalmus may also be caused by disease in neighboring cavities. We have to consider how the orbit and the eye may be affected by disease in the frontal sinus, in the ethmoid cells, in the antrum, in the sphenoid sinus, and from the spheno-maxillary fissure, and in the cavity of the skull. It is not my purpose to attempt to compass the extended field of pathology which is thus mapped out, but rather to call attention to the possible influence of various diseases in these several parts upon the functions of the orbit and of the eye. We may have collections of mucus, or abscesses or tumors of various kinds, either filling or springing from these localities— the tumors may be fleshy or bony, may be malignant or otherwise. Disturbance in the position and movement of the eye will in many cases be the only symptom, but in certain cases special symptoms arise which deserve notice.

Distention *of the frontal sinus by mucus or by pus.*—This condition may be consecutive to severe nasal catarrh, syphilis, or periostitis within its cavity. Polypus has been found in the cavity, and also small exostosis. Without going into detail of the various phases of the above morbid states, it is pertinent to call attention to the variable size and extent of this cavity. In young subjects it has no existence, but becomes of notable size after thirty years of age, and beyond that period of life it may present the most remarkable variations in extent. Mackenzie has written the best chapter on this whole subject (see "Diseases of the Eye," pp. 93–121, Am. ed., 1855), in which he collects cases from the older writers. Wells relates a case of mucus accumulation in the frontal sinus ("Diseases of the Eye," p. 787, 3d Am. ed., 1880), and I have seen one or two myself. The locality will always excite suspicion as to where a collection of fluid may reside, and if an exploratory incision demonstrate that it does not belong to the soft tissues of the orbit, the wall of the sinus may be carefully opened by a strong knife, or, if needful, by a small drill, to decide whether the disease be there concealed. Afterward, long treatment by antiseptic and astringent injections will be required.

Distention of the ethmoid cells of a similar kind is described by Dr. Knapp (see "Report of Fifth International Oph. Congress," 1877, p. 55). The patient was a girl, fourteen years of age, who had a tumor at the inner and upper corner of the orbit, resembling in all respects an exostosis. The surface of the bone was exposed by a free incision, and as a chisel was applied to its base for its removal, its walls promptly gave way and disclosed a cavity filled with stringy mucus. The opening was freely enlarged, the contents were fully evacuated, and it was found that some of the fluid used in syringing escaped from the nostril. In about a year the case was cured.

Tumors in the antrum press on the floor of the orbit, and may displace the eye, and I have met with a case in which a fibro-plastic tumor came up from the spheno-maxillary fissure and pressed the globe forward.

There are cases on record of congenital malformation, chiefly in the neighborhood of the lachrymal bone, and often in both orbits, by which the brain comes into direct relation with the orbit ; its cerebro-spinal fluid pushes down the dura mater as a cyst through an aperture in the bony walls.

TUMORS OF THE ORBIT.

There is an enormous literature on this subject, and the obscurity which attends an exact diagnosis of the nature, extent and relations of the disease, makes every one an interesting study.

In an examination, we attend to the objective symptoms, and first the exophthalmus, its degree, and the direction in which the eye is pushed ; the mobility of the eye, whether limited in any special direction, in all directions, or not at all ; the appearance of the globe, whether unduly vascular or normal, or itself the seat of a tumor or deformity. We examine the tumor as to its resistance, solidity, elasticity, fixity, or mobility, fluctuation, pulsation, its smoothness, or lobular or nodular character, and whether it move with the eyeball. We press the globe back into the orbit and note whether this gives pain, how far it will recede, and whether, in retiring, the tumor also retire or be pressed forward. We listen upon the globe and over the temple by a stethoscope for murmur or pulsation. We note whether neighboring vessels about the forehead or lids be enlarged, whether the preauricular or the cervical lymphatic glands are hypertrophied. We inspect neighboring cavities, viz., the nostrils, the vault of the pharynx, the frontal and maxillary sinuses, so far as they are within the means of examination. In some rare cases we explore the tumor with a hypodermic syringe. The sensitiveness of the cornea and the fundus oculi are also examined.

The *subjective* examination will embrace the age, sex, present and previous health, constitutional diseases, especially syphilis, any hereditary tendency to cancer, or its possible existence in other parts of the body ; the mode in which the disease appeared, and exactly at what point, if this can be located ; its rate of progress ; whether the onset was somewhat sudden or gradual ; whether there has been pain, or occasional attacks of inflammation ; whether a tumor was noted before proptosis appeared, or *vice versa*. We may also examine for diplopia, and sometimes we may learn that hypermetropia or even myopia has been developed since the growth began.

The upper lid often undergoes remarkable elongation as the globe advances, while in other cases the lids are stretched apart and cannot properly cover the eye ; in the latter case the cornea may become inflamed and opaque. The examination of the orbit may be made by thrusting the little finger between the globe and the bony margin on all sides as deeply into the cavity as possible, not heeding the considerable displacement of the globe, as it yields to the pressure. The object is to elicit information as to the *seat* of the tumor ; whether it spring from the walls of the orbit, be located within the cone of the recti muscles or be outside of them ; whether it be attached to the globe or to the optic nerve ; whether it enter the orbit from an adjacent cavity.

Many of these questions must be left unanswered. But we can often tell by the fixedness and hardness of a tumor that it is attached with some firmness to the bony walls ; then it will be outside of the muscles. If the tumor be mobile, it will to a great extent be free in the orbit, although, perhaps, partly attached. If at the same time the globe is displaced in some oblique or lateral direction, the tumor is probably outside the muscles. If, however, the globe comes straight forward, its motions are rather restricted, although natural, and the tumor seem fitted closely into the apex of the orbit, and vision has been destroyed at an early period, which ordinarily is not the case, there is reason to think it may be upon the

optic nerve. It must however be stated that in some of these cases vision
may remain untroubled for a long time. It has also been noted that sometimes the nerve is not pressed out straight, but has an S-shaped curvature.
Deformity in the contour of the orbit, which is rare, and stoppage of the
nostrils, are indications of a growth in the antrum, and it may be discovered by examining the gums and the mouth, and the cheek. The use of
the rhinoscopic mirror behind the velum palati will aid us in discovering
encroachments from neighboring cavities.

The next question is as to the *nature* of the tumor, and with this and the
previous inquiry are bound up both the prognosis and treatment of the disease. We can rarely speak with certainty of the nature of an orbital tumor.
The factors to be weighed are its rate and rapidity of growth; the age of
the subject, and his previous history; the hardness, smoothness, nodular
character, mobility, compressibility, fixedness, apparent vascularity; the
state of the eyeball; and the existence of murmur or pulsation; the presence of distended or varicose vessels. We can speak with some confidence respecting osseous growths by their physical characters, the slowness of growth and painlessness, and the way the globe is displaced.
Tumors rapid in development, especially in young subjects, attended, too,
with large circum-orbital or palpebral veins, and which may or may not
pulsate or have a murmur, are likely to consist largely of blood-vessels,
and may also be malignant. Tumors not very rapid in growth, either
smooth or nodular, more or less mobile, and not painful, offer a wide field
of speculation as to their character, as between fibromata, lipomata, sarcomata, myxomata, melanomata, etc. Cysts are sometimes easy to be made
out by obscure elasticity, partial attachment, ovoid and smooth shape, but
when deep and of long duration they are only recognized by being opened.

Another class of tumors easily diagnosticated, are degenerations of the
lachrymal gland.

There are also found echinococci, cysticerci, congenital serous cysts,
and bloody cysts.

Angiomata and erectile tumors are very likely to be associated with
similar anomalies of the skin, but this is not necessarily the case—they are
usually congenital. Cavernous tumors not congenital, as well as those
which are, can be generally made out by observing that they greatly increase in size by hanging the head downward and forward, so as to cause
venous congestion. They are apt to be contained inside the cone of the
muscles. They do not have pulsation or murmur (Berlin). The distinction
between pure angiomata and highly vascular malignant growths, depends
on the greater rapidity of growth of the latter, their greater firmness, and
that the lymphatic glands are liable soon to be enlarged, and the eyeball
itself to be implicated. But there will often be great uncertainty at the
early stage of the disease.

Among the rare ocular tumors, are enchondromata and cylindromata,
while epithelial cancer sometimes reaches from the outer parts into the
orbital cavity.

The above remarks include what may be stated respecting diagnosis
and symptoms. As to course and prognosis, it may be said that some
tumors rapidly increase. These are the malignant forms, which will embrace various forms of sarcomata, the so-called medullary cancer, and some
melanotic growths. In these cases the eye may be involved, and the tumor
may extend beyond the orbit, and possibly grow to an enormous size.
Pictures of such cases are found in various books (Sichel, Dalrymple,
Wells, etc.). When it has reached the external surface, the tumor be-

comes fungous, bleeds, emits offensive secretion and odor, causes hectic fever, emaciation, exhaustion, and death. It may also involve absorption of the adjacent bony walls, and the fatal result may take place by invasion of the brain.

Fibrous, fatty, cystic, enchondromatous, and less malignant tumors grow less rapidly, and give trouble by the displacement of the eye and the injury to sight. The latter effect comes by neuritis (choked disc), detachment of retina, intraocular hemorrhages, etc., but in many cases the sight remains good for an indefinite time. Angiomatous and cavernous tumors have been known in a few instances to disappear spontaneously. Bony tumors are very slow in growth, but may attain great magnitude, as I have witnessed. Mackenzie depicts a skull, of which both orbits are filled by a dendritic mass of osteoid hypertrophy. Osseous growths are not painful except by pressure upon and disturbance of adjacent sensitive parts. In almost all cases of orbital tumors the exophthalmus is sufficient reason for demanding relief.

Treatment.—The only proper proceeding is operative removal. Certain modifying considerations are to be kept in mind. Cysts which extend too deeply into the orbit to be perfectly extirpated, or which communicate with adjacent cavities, must, after partial removal, be treated by injections of stimulating fluids. For vascular or erectile tumors in very young subjects (infants), the use of red-hot needles, or of electrolysis, to coagulate the blood, is expedient. The operation may be repeated once in two or more weeks, according to the degree of reaction and to the rapidity of growth. Such tumors cannot be safely treated by irritating injections. In adults they may be attacked by excision, aided, if needful, by the actual cautery in some convenient form, viz., hot-iron, electric cautery or thermo-cautery. Often they are enclosed by a capsule of fibrous tissue which much facilitates the proceeding. For osteoid growths the best means of removal is by the chisel and mallet, attacking them at the base by very light and numerous blows until they loosen (Knapp). But Berlin sums up his remarks on such tumors by some pregnant observations as to what is justifiable according to the situation of the growth. He has collected 32 cases which were operated on : of these, 9 had meningitis ; 8 died ; of the whole number, 16 had bony growths in the roof of the orbit, and of these 6 died, a fatality of thirty-eight per cent. This shows in a most startling way how dangerous is interference in this particular category of cases. It certainly justifies absolute refusal to operate, unless there be urgent indications and a full presentation to the patient of the risks he incurs. The reasons for operating can only be pain, the safety of the eye, and conspicuous deformity. Osteoid tumors in other parts of the orbit may be removed with success, and if adjacent cavities are opened, no great harm is done. A small gouge with a strong wooden handle is a good instrument, or one may prefer a chisel and mallet. If the latter be used, the strokes must be gentle and the proceeding slow.

For tumors whose relations, size, and probable character render them fit for excision, the question arises : Can they be extirpated without sacrificing the globe? If unadherent to the eyeball, even if they include the optic nerve, this is generally feasible. In 1866, I excised a fibrous tumor of the orbit without removing the globe, but sight was lost by suppuration of the cornea, consequent on extrusion of the globe by inflammatory infiltration of the orbital tissues and exposure of the cornea. The tumor was above the globe, and my incision was made through the superior cul-de-sac of the conjunctiva, which resulted in ptosis, because the levator palpebræ had to be destroyed to reach the tumor. My purpose in choosing this route was to spare the levator, but the seat of the tumor defeated my design. The proper mode of approach to such a tumor would be through the upper lid at the margin of the orbit. In the very rare cases of tumor of the optic nerve, the probability of extension along the nerve into the brain makes early operation important. Being situated within the space surrounded by the recti muscles, it is manifestly proper to attack the tumor through a wound in the conjunctiva, either below the rectus internus or externus, keeping close to the globe. It may be possible to excise it without removing the globe, but the eye will be extremely liable to destruction, either by suppuration of the cornea, or by lesions beginning in the back of the bulb.

In removing a tumor, first decide in what way it will be most accessible. If this be decided to be through the conjunctiva, choose the side which offers the nearest approach to the mass, go between the recti muscles by a wound as large as can be made by drawing them asunder, or detach a tendon, if needful, and tie to it a thread, so that it may afterward be recognized. Use a pair of narrow and strong scissors, curved on the flat, with rounded points and with shut blades, to tear away the connective tissue down to the tumor, and with the same implement, or the flat handle of a scalpel, push the tissues apart to expose the mass. Attempt to bring it forward by a strabismus-hook, or by catching it with a sharp hook if it be tough enough to bear traction, and carefully cut away its surrounding connections by small clips with the scissors. Progress must be slow, and tissues must be torn rather than cut, as far as may be possible. If the tumor be upon the optic, push a strabismus-hook behind it to the apex of the orbit, and when this has caught the nerve, run the scissors alongside of it and cut the nerve beyond the hook; then this hook, or a sharp hook planted into the tumor, will pull it round to the front, reversing the globe and making its separation from the eye very easy. In case entire or sufficient removal cannot be accomplished within the space thus available, the globe may have to be sacrificed. Before the operation, this possible contingency must be stated to the patient, and his consent obtained. Small pieces of ice pressed into the wound, or pressure by the finger, will control the bleeding after the operation, and the wound must not be closed until bleeding has stopped.

But the method above described is suitable for a small and exceptional number of cases. In the great majority the wound will be made through the skin. It should be parallel to the margin of the orbit, over the most prominent point of the tumor, and as large as can be of any use. After going through the skin, the deeper dissection is to be done as already described. In case the tumor be found to penetrate adjacent cavities, it may be impossible to follow it and accomplish complete extirpation. The surgeon must decide such questions according to his own judgment or the requirements of the case. By such a method of proceeding, it is surprising how successfully a tumor may be dug out, both as regards the loss of blood and immunity of healthy parts. All bleeding must be arrested before the wound is closed; it must be cleaned with solution of boracic acid, which shall both wash out clots and be antiseptic. Close the wound by fine silk sutures, dress the surface with a rag smeared with vaseline, put over this a mass of absorbent cotton, and retain all by a flannel bandage which shall exert firm pressure. Generally an anodyne will be needed. The wound will not be opened, if pain and reaction be moderate, until after forty-eight hours.

Complete evacuation of the orbit ("exenteration") is called for when the eyeball is implicated in the growth, has already been destroyed, or when the tumor cannot otherwise be removed. It may be that only an ordinary enucleation may be necessary, and this be followed by excision of the tumor and nothing more. But other cases arise in which the orbit must be emptied of all its contents. In doing this, the lids are split asunder beyond the outer margin of the orbit; the coverings of the mass are to be picked up and cleaned off until its surface is fully in view; then, with a blunt instrument (the scissors before mentioned, with shut blades, are my usual resort), insinuate between the tumor and the wall of the orbit at the most convenient point, and tear and push away the parts, keeping in contact with the bone until a way is made to the apex of the orbit. I strongly deprecate the use of knives in such an operation. Scissors, both to lacerate when closed, and to divide when tearing cannot be done, are both most effective and least calculated to shed blood. The principal hemorrhage will occur at the apex of the orbit, and can be best arrested by prolonged pressure with the tip of the finger. When such an operation has been done for malignant disease, the walls of the orbit are sometimes washed with solution of chloride of zinc, or smeared with a paste of this substance. If freely applied, a scale of bone may, in consequence, be exfoliated, and serious risk is incurred of meningitis; but, done not too vigorously, greater security against recurrence of the disease is gained, and without dangerous risk.

The cavity should be washed out by an antiseptic solution, filled with cotton, and moderate pressure applied. Suppuration will necessarily follow.

Secondary growths not unfrequently demand attention, especially after excision of sarcomata, and the globe will usually have already been sacrificed. If these be not too large, say not bigger than a moderate hen's egg, the mode of removal by a blunt instrument, scraping the walls of the orbit and shelling out the mass, is surprisingly easy and comparatively bloodless in many cases.

I have never seen cases which required resort to caustics to destroy a growth, and such occasions must be rare.

It is not unusual in successful cases for the globe to become displaced forward, and so remain for some time. It may also happen that the muscles undergo disturbance and cause diplopia; or ptosis may follow from greater or less injury to the levator. Such injuries will usually, in time, correct themselves. But a more serious matter is the liability of the cornea to become inflamed, both by possible exposure as swelling comes on, or as it may be bathed in secretions. Frequent washing with weak boracic acid solution, 4 to 100, the application of vaseline, and the closure of lids by rubber plaster, are the best preventives. Moreover, the sight is also endangered by the manipulations at the back of the globe and about the optic nerve. It is, therefore, not to be thought strange if the globe be saved, and sight be partially or totally lost. At the same time preservation of the form of the eye and its natural appearance is worthy of strenuous endeavor.

Hemorrhage into the orbit.—With very few exceptions, this results from injury, either by falls, blows or penetration of a foreign body. The symptoms vary according to the amount effused. If large, there will be propulsion of the globe, and ecchymosis of the lids and of the ocular conjunctiva. If the quantity be small, the eyeball will not advance, while the lids and conjunctiva will be discolored. Finally, the distinctive criterion of orbital hemorrhage of small quantity, is a tardy appearance of ecchy-

mosis creeping down under the ocular conjunctiva and advancing toward
the cornea. In some cases the lid alone is the seat of discoloration. Spontaneous cases are so very few, and their etiology so manifest, viz., scorbutus, violent coughing, etc., that we may confine ourselves entirely to
orbital hemorrhage from injury. It has been pointed out by many distinguished surgeons, and is classical in literature, that this symptom indicates fracture of the orbit, and most frequently of its roof. But Berlin
(G. and S., vi., pp. 567-8) quotes six cases supplied by von Hölden, where
at the autopsy orbital hemorrhage appeared without any fracture of any
part of the skull, as demonstrated by stripping off all the dura mater. But
in these six cases there had been severe falls or blows, and in some cases
there was intracranial hemorrhage. In some instances the intracranial
bleeding had reached into the orbit, in other cases the orbital hemorrhage
was idiopathic. As a proper offset to these observations, von Hölden furnished an account of 124 cases of fracture of the skull, in 79 of which he
found fracture of the roof of the orbit, and of these 69 had hemorrhage
into the cellular tissue of the orbit, and in the remaining 10, blood was confined to the vicinity of the periosteum. It follows that in cases of severe
injuries (either fracture of the skull or commotion) with orbital hemorrhage, this symptom, in ninety-two per cent., indicates coincident fracture
of the orbit, while in only eight per cent. does it take place without fracture of the orbit.

In some cases, severe hemorrhage may find its outlet through the nose,
and, perhaps, get into the stomach. This implies fracture of the inner
wall and of the ethmoid cells.

In every case of orbital bleeding, the local conditions give us anxiety, not
specially on behalf of the eye and its surroundings, but because grave injury has probably been inflicted upon the skull. It may, however, happen
that sight or other functions of the eye are imperilled. The bleeding may
cause atrophy of the optic nerve by pressure, or laceration of the ophthalmic artery may cause false aneurism, or cut off the supply to the retina.
The muscles may one or more of them be paralyzed. Such contingencies
and others are easily possible.

Treatment consists in cold or iced water compresses, a pressure-bandage, and rest. From three to four weeks will be necessary for removal of
the extravasation. To attempt to let out the blood by an operation is useless, and likely to be hurtful.

WOUNDS AND INJURIES OF THE ORBIT.

Dislocation of the globe may be produced by a push with a cow's horn,
by a man's thumb or finger in fighting, or by a blunt arrow, etc.; and by
insane persons has been self-inflicted. Gouging, as it is popularly called,
may or may not be attended with rupture of the muscles. The eyeball
may seem unharmed, yet sight be wholly or partly destroyed by injury to
the optic nerve, or by laceration of the choroid. *Treatment* will consist in
replacement of the eye and cold-water dressings, pressure-bandage, and
subsequent proceedings as the symptoms indicate.

Wounds of the soft parts at the margin of the orbit are often caused by
blows with the fist, especially when armed with brass-knuckles or wearing
a large ring. It is often remarkable how clean cut and well defined the
skin wound is, presenting to cursory inspection the appearance of an incised cut. It will be noted, however, that the deep parts of the skin are

more extensively wounded than the surface, that there is considerable contusion and swelling, and the reaction is always greater than after a simple incised wound. Suppuration often follows.

To get rid of large clots of blood in the soft tissues, is sometimes a matter of three or four weeks ; and the process of absorption may be somewhat hastened by pressure, or, if this be painful, by massage, while the surgery of the prize ring practises opening the swollen skin with a lancet. This, in ordinary practice, is very rarely advisable.

Gun-shot wounds of the orbit are common both in civil and military surgery. The most distressing cases are those in which the ball comes from the side and goes through the orbital walls transversely. It may lodge anywhere and may destroy one or both globes, or may leave each seemingly intact. Usually, the sight of one or of both eyes is destroyed, according to whether the missile enters one or both orbits. The ball has been known to go into the opposite upper jaw. Bleeding from the nose or mouth will indicate to some degree its direction. Life may be spared, or may be destroyed by inflammation extending to the cavity of the skull. I have seen two cases of this description. In one, the eye on the side of entrance was sound to outward appearance, but sightless ; the other eye was atrophied. The explanation of the loss of sight is easily understood.

When the bullet takes some other than a transverse direction, the injury inflicted will depend greatly on its penetration, as well as on its special direction, and will often be fatal. A case worthy of record I have reported in the "Transactions of the American Ophthalmological Society" for 1881.

A man, twenty-eight years of age, while in bed in a hotel in Texas, was awakened from sleep by a man who demanded his money, which was under his pillow, and presented a pistol to his face. The assailant fired, seized the money, and fled. The ball entered the left orbit close to the outer canthus. For several weeks the patient was in bed, and was much of the time unconscious. Four months afterward I saw him. There was no cicatrix or irregularity which would indicate the place of entrance ; the outer orbital margin was regular and smooth ; the eye was sightless, though perfectly capable of motion in all directions. The ophthalmoscope showed a large laceration of the choroid on the outer side of the fundus, and atrophy of the optic nerve. The left ear was totally deaf—not able to hear the tuning-fork applied to the head. In the meatus auditorius was a swelling of the upper wall close to the membrana tympani, which was covered with tense skin, was tender to touch, hard, and about five millimetres across. It was just such a protuberance as would be made by a small pistol-bullet lodged in the bony meatus, and there I believed it to be. The patient did not experience any unpleasant symptoms, and resumed his travels as a showman. His other eye, which he had never depended on, had myopic astigmatism, and with— 12, 180° he gained V. = $\frac{20}{70}$.

Blows upon the margin of the orbit sometimes implicate the supra-orbital or the infra-orbital nerve, and to this fact has been attributed the loss of sight which in some cases has been known to ensue. I once published such an opinion. But I incline to withdraw from this opinion, and think the cause is to be sought for in fissure of the orbit reaching back to the optic foramen, as will be referred to later. This nerve injury has long had a place in ophthalmic pathology, but it stands on very weak evidence.

Dislocation of the malar bone is an accident which can occur, and I have recorded an instance (see "Trans. Am. Oph. Soc.," 1880). It results generally from violent falls upon the face, whose force is spent directly on the bone. It may cause extensive orbital hemorrhage and possibly diplopia through interference with the inferior oblique muscle. It will be recognized by a notch near the middle of the inferior orbital margin, where the malar joins the superior maxillary bone, and by another notch where it joins the external process of the frontal bone, and often the zygomatic arch is bent or broken. Anæsthesia of the infra-orbital nerve, and pain in chewing, because of pressure on the canine and adjacent teeth, are symptoms which continue for some time. The symptoms vary a little according to the direction in which the bone is displaced.

Fractures of the walls of the orbit occur in a great variety of ways : by cuts, blows, falls on the head, by crushing forces, etc. Such an accident, with extrusion of the eyeballs from the sockets, has been caused spontaneously in child-birth when there was deformed pelvis (see case reported by Berlin in G. and S., vi., p. 588), and might result from injudicious handling of the forceps. If the roof is implicated, there will be danger of inflammation of the brain, yet out of 19 such cases collected by Berlin, 16 recovered. Fractures into the inferior orbital wall, besides opening the antrum, damage the infra-orbital nerve, and are liable to be followed by distressing neuralgia or by anæsthesia. Fracture here, and also of the inner wall of the orbit, will be succeeded by emphysema of the cellular tissue. In some cases this will be extensive, and a case is reported by Knapp of exophthalmus produced in this way. Nose-bleed will also occur.

Fractures of the orbital walls by penetrating wounds, as may happen in fencing, or by a bayonet, arrow, umbrella-ferrule, hook, key, etc., are relatively more serious than those just mentioned. This is true, particularly as to the roof of the orbit. The external wound may be trifling, the eye often escapes harm, but if the cavity of the skull has been entered, the prognosis is very grave. In twenty-five per cent. of the cases (Berlin) the patient immediately falls unconscious, but before long recovers. It is an important matter to know whether the wound has gone through the orbital roof. The outward opening is often small, it partially closes, and to find a way through it with a probe is very difficult, because the eyeball is violently pulled around as the weapon enters, and afterward returns to its place, thereby making the track sinuous. But there is very grave doubt as to the propriety of venturing to use a probe. The probability of the presence of a foreign body, or of the displacement into the skull of fragments of bone, may justify probing when the wound is recent and the symptoms urgent, but the surgeon's little finger is far safer as an exploring instrument, and, on the whole, a prudent man would, in the great majority of cases, refrain from meddling. Antiseptic precautions may render such an exploration less dangerous than it would be without them, but a discreet surgeon will not permit his professional curiosity to imperil the patient's limited chances of recovery. Very seldom will his probe or his finger be allowed to enter the orbit.

Cerebral symptoms, when they occur, may be due to intracranial hemorrhage, or to inflammation. The latter class of symptoms will be various, viz., pain, weakness, delirium, vertigo, paralysis, coma, etc. But it is notable that head-symptoms may be tardy in appearing, and be so long delayed as to make perforation of the roof seem to be highly improbable, yet the dreaded tokens may appear. In one case, forty days passed without any cerebral signs, then the patient suddenly died after a foreign body was extracted from the orbit. Berlin has gathered 52 cases of perforating wounds of the orbital roof, of whom 11, i.e., twenty-one per cent., recovered ; but of these, three were hemiplegic, one had persistent headache, and one became imbecile. The remaining 44, i.e., seventy-nine per cent., died ; of the deaths, one-half were from the immediate effects of the wound, and the other half from the subsequent complications.

At the autopsy, the bony aperture was generally small, and fragments had entered the cranial cavity. Wound of the brain was small, or in some cases very large. Of the causes of death at a late period after the wound (18 cases), in 15 there was abscess of the brain, with or without meningitis; in 2, thrombosis of the longitudinal sinus; in 1, "pus at the base of the brain." In 6 of them, bits of bone were found in the brain-substance.

An illustration of what may ensue from fracture of the orbit, is the following:

A boy, fifteen years of age, while dodging through a crowd in a meat-market, stooped to get on more easily, and ran against a large meat-hook. Its point caught him in the right orbit, under its upper margin, tearing off the upper lid from the inner angle, fracturing the edge and perforating its roof. He was taken to the New York Hospital, and kept under treatment for six weeks. At the end of that time he came to the New York Eye and Ear Infirmary, and I found a scar running nearly the whole length of the upper lid beneath the brow, the lid everted and immovable, its conjunctival surface converted into a florid mass of papillary granulations; the cornea visible for its lower half, and in a state of fixed convergence. The globe could not be moved, but the lid could be turned with the finger, and could be slightly lifted by his efforts. The eye looked well, but was almost sightless. There was atrophy of the optic nerve, apparently the result of neuritis. At the upper margin of the orbit was a deep notch, which evidently went back into a deficiency in the roof. He was unable to say whether any fragments of bone had come out. By a pressure-bandage on the readjusted lid, the thickening of the conjunctiva, and the swelling of the lid so far abated in eight months, that I ventured to try to bring the cornea to the middle of the palpebral opening. I divided the internus, and dissected the parts about the caruncle very freely, but could not turn the globe outward. I then attempted to bring forward the externus, but could not rotate the globe outward. Finally, I explored the orbit on its outer wall, behind the globe, and found that this surface had been forced inward, and that the eyeball had become adherent to the periosteum at its posterior part. I tore away this attachment, and then was able to rotate the eye to the middle of the palpebral slit, where I placed it, and closed up the conjunctival wounds. The reaction was not extreme, and the eye was permanently fixed in the position where I left it. Some vision in the outer part of the field was obtained, but the upper lid remains drooping over the upper half of the cornea, but of normal thickness and without ectropion.

Another class of cases of orbital fracture are those in which no ordinary symptoms of this lesion appear, but in which, after an injury to the head, loss of sight occurs in one or both eyes, and with very slight symptoms in the fundus oculi. After a time the optic nerve may show signs of inflammation or of atrophy. Again, there are many cases of fracture running through the optic foramen or the roof of the orbit, simultaneously with fracture of the base, or in some other region of the skull. The profound injury sustained distracts attention from the state of sight, and we seldom know that it has been impaired, nor would the patient, perhaps, be able to tell us anything about it. A most interesting study of these cases has been made by Dr. von Hölden, who, in his capacity of pathologist, examined 124 cases of fracture of the skull. He stripped away the dura mater from the base completely, and was thus enabled to detect injuries to the bone and hemorrhages which would otherwise have escaped notice. During forty years he made these observations and took notes of what he found (Berlin: loc. cit., p. 604). Among the 124 cases, there were 86 of fracture at the base, and in 79 of them the fracture extended into the orbital roof. Von Hölden states that out of 86 cases of fracture at the base, in 63 he found a fissure or fracture running through the optic canal, and always through its upper wall, and sometimes also through the inner wall; occasionally on both sides. In 42 cases there was hemorrhage into the sheath of the nerve, and he never found blood in the optic sheath, unless the bony canal was fractured. This blood may be derived from the cavity of the skull, or from the vessels of the sheath, or from the torn central artery of the retina,

Other observers have seen the same symptom. If the quantity were large, it might, as in a case reported by Knapp, be sufficient to injure sight by direct pressure on the nerve, but smaller quantities might also destroy sight by interference with the central retinal artery, causing ischæmia retinæ, and all the features, too, of embolism. Prescott Hewitt gathered 68 cases of fracture of the base, and found the orbit involved in 23.

Possessed of these facts, it becomes intelligible why, after a fall on the head, total, or nearly total blindness may ensue, with, perhaps, no ophthalmoscopic lesions. Ultimately, signs of inflammation or of atrophy or pigmentation of the disc may appear. Again, venous hyperæmia, ischæmia of arteries, opacity of the nerve or of the retina by exudation, hemorrhage into the vitreous and into the retina will suggest intravaginal hemorrhage. For example, Berlin quotes 30 cases of blindness after injuries of the head, in which ophthalmoscopic examination was made. In 17 there was atrophy of the nerve and in two there was pigment-deposit in the disc. A case which I published was seen in the stage of neuritis. Another case I have seen within a year which presented nothing but slight fulness of the veins. Other reported cases have exhibited hyperæmia of the disc, ischæmia, hemorrhages, etc.

Treatment of these injuries of the orbit is to be conducted on general principles and according to the dominant symptoms. Of course we have nothing to say on the general subject of fracture of the skull. As to the orbit, loose bits of bone or foreign bodies are to be removed, all excitement to be avoided, antiphlogistic measures to be used as needful, cold applications, removal of secretions, and maintenance of free escape of discharges. In this connection it is important to consider what steps are to be taken when symptoms of abscess, deep in the orbit and perhaps in the brain, threaten. For orbital abscess there would be no hesitation in promptly giving a free outlet. The employment of antiseptic methods in the operation, and in the subsequent dressing might be necessary. Should symptoms of brain trouble threaten and the escape of pus be so hindered as not to be otherwise insured, it might be proper to enucleate the globe for the sake of improving the patient's chances of life. The period during which such a question may be argued is never long, and a surgeon may be placed in a most difficult position.

I may conclude this chapter by citing a case in my own experience:

In 1857, a boy, ten years of age, was brought into the New York Hospital after having been run over by a street car. He had fracture of the occipital and frontal bones. He remained about three months in the institution. He recovered without paralysis or loss of any function, but was always subject to headaches, and had a small fistulous opening at the upper and inner angle of the right orbit, just under the brow. In 1865 I saw him and noted the fistula, and warned him that he was liable to have trouble from it. He lived a wild life, and was sometimes drunk. In the later part of 1865 I was sent for to see him, and found he had serious brain symptoms. Consciousness was not abolished, but almost gone ; pulse slow, breathing heavy. He had had severe headache and been in bed for several days. By the ophthalmoscope I could only see hyperæmia of both nerves. The usual discharge from the fistula had recently ceased. I concluded that there must be an abscess near this spot, within the cranial cavity, and determined to trephine the skull, just above the fistula. A large crucial incision was made, and I trephined just outside of the supraorbital notch. The dura mater bulged into the wound. I opened it and pus escaped. About half an ounce issued, and I put my finger into the cavity over the roof of the orbit. The patient who had sunk into coma during the consultation over his case, recovered intelligence at once, in half an hour was able to talk, and made a good recovery. He had fungus granulations (hernia cerebri) from the wound, but at length by a pad and pressure-bandage and excision this was controlled, and he has never reported himself since.

Foreign bodies in the orbit.—Foreign bodies entering the orbit and passing out of sight are extremely difficult to find unless of considerable size. Even if they are large they may lodge in the orbit without destroying the globe, as happened in an instance reported by Mr. Carter, in which a piece of an iron hat-peg, $3\frac{3}{10}$ inches long, was buried in the cavity, and remained there for from ten to twenty days without the patient being aware of it. It was extracted without injury to any functions of the eye. It is not necessary to say that foreign bodies which can be seen or felt should be carefully and immediately extracted. But the point of difficulty is to decide : 1st, whether a foreign body has entered the cavity ; and, 2d, to find and remove it. A doubt arises as to the penetration of foreign bodies in cases of wounds by bird-shot. The place of entrance is very small, closes instantly, and heals promptly. It is often impossible to trace them, nor is it generally needful to meddle with them. I have known a fragment of iron of considerable size, struck off by a chisel, to enter and be completely hidden. The irregularity of the piece, the yielding nature of the tissues, and the sinuosity of the wound, make exploration by a probe very unsatisfactory. As above remarked, a reason for the difficulty is, that when **the** foreign body enters, it drags the eye around toward itself, **and when** it has found a lodgement, the globe returns to its position and **thus twists the track** of the wound.

A case, illustrating the difficulties of diagnosis and **the** proper **mode of treatment in** obscure cases, will include all that need be said :

A man walking among bushes struck against a limb and felt a twig strike his eye, and he was convinced that a piece of it entered the eye. Some bleeding occurred ; he suffered considerable pain ; he found his sight uninjured, and for some time he did **not** go to a physician. He found, however, that a chronic inflammation lingered about the lower part of the eye, and he was annoyed by some pain and discomfort. The physician looked at the inflamed part and everted the lower lid, but could see no sign of wound or scar and prescribed for what he regarded as simple conjunctivitis. The man's statement that a foreign body had entered or was present in the orbit he did not credit. For two weeks treatment by astringents was kept up, when I was asked to see the case. I discovered in the inferior cul-de-sac a small projecting granulation, as large as a No. 2 shot, and around this the conjunctival and scleral hyperæmia concentrated. I at once asserted that there was a foreign body in the orbit and advised its removal. For two weeks longer the same medical treatment was continued, and the patient then was put into my care at the New York Eye and Ear Infirmary. No trace of an offending substance could be felt with the finger, nor could a probe be forced through the tissues. The patient was etherized ; an opening made into the conjunctiva at the granulation, and by tearing and stretching the tissues, an opening was made large enough for the entrance of my little finger. Afterward my index finger was thrust in. No foreign body could be felt, nor could any sign of it be found by various exploring and grasping instruments which were used. After prolonged manipulation, while with the finger pressed deeply into the orbit, I was also feeling along it with a pair of forceps, I caught something which conveyed the sensation of a foreign body. Drawing upon it I brought forth a bit of twig about $1\frac{1}{2}$ inches long, and large as Theobald's lachrymal probe, No. 8. It was softened by long maceration, was flexible, and offered so little resistance that its detection was rendered extraordinarily difficult. The yielding nature of the orbital contents greatly increases the difficulty of seizing a foreign body unless it have some stiffness or can be steadied by being pressed against the walls. The operation gave rise to no serious trouble, and in ten days the man was discharged cured.

From the above case, I venture to advise the insertion of the operator's finger **into** the orbit along the track of the foreign body, and to use it both as an explorer and as a means of guiding the search with forceps or other suitable instruments.

CHAPTER XVIII.

OPTICO-CILIARY NEURECTOMY.

THIS operation may be done without dividing any of the muscles, as follows: Incise the conjunctiva at the space between the rectus internus and the rectus superior with a pair of blunt-pointed scissors curved on the flat, and clear a way to the vicinity of the optic nerve. In doing this rotate the ball downward and outward to the extremest degree by fixation forceps. Along the track thus made carry a small strabismus-hook, catch the optic nerve, and pull forward. Over the hook insert the scissors, and, pressing their points firmly to the apex of the orbit, sever the nerve. Then with the hook drag out the optic nerve, seize it with forceps, and pull the globe around until it shall be completely reversed. It slips like a button through the hole in the conjunctiva, and presents its posterior scleral surface to view. With toothed forceps pick and cut away every shred of nerve-fibre and tissue which can be seen, and cut off the optic nerve close to the eye. A piece 8 mm. long may easily be excised. By the forceps then turn the ball around and restore it to its place. Very little blood will have been lost and the eye will have only a slight prominence. The conjunctival wound must be closed by sutures.

It is easier to perform the operation after cutting the insertion of the rectus externus or internus muscle, and then, using the hook to catch the nerve, do what has already been described. The severed muscle must be stitched to its insertion, where a small piece of the tendon has been left for this purpose. A more extensive exposure is thus made of the back of the globe, and with greater certainty the ciliary nerves can all be secured. Some strabismus is liable to ensue, and bleeding is apt to be copious. In whatever way done we are liable in some cases to meet free hemorrhage in this operation. It can cause great protrusion of the globe. Nothing but pressure is available to restrain it, and a very firm bandage may have to be kept on for twelve or twenty-four hours. In this event reaction may be severe.

ENUCLEATION OF THE EYE.

With a pair of blunt-pointed scissors, curved on the flat, and of medium size, separate the conjunctiva from the globe at the margin of the cornea, going all around it. Then, with small clips, go to the insertion of the rectus superior, and thrust under it a strabismus hook, and cut it away from the globe. The hook is liberated, but serves to lift the conjunctiva and keep the wound open; then a second hook tears away the sub-conjunctival connective tissue and is slipped under the insertion of the rectus internus muscle, keeping in close contact with the globe. The two other recti are similarly divided, and the hook is swept around the equator, to be sure that all the tissues are divided. The globe, if of normal size, can now be extruded from the orbit by pressing between it and the orbital rim either with the finger or with the speculum. The closed scissors are pushed to the back of the eye by lateral movements, tearing a path until the optic nerve is struck. When in contact with it, open the blades astride of the nerve, and, inclining the points backward, divide it. At this moment a gush of blood occurs. Push the eye forward, take it in the fingers, and cut away the insertions of the two oblique muscles, and all the vessels, nerves, and other attaching tissues. Push a sponge into the orbit, and make firm pressure. In a few minutes bleeding will be checked, but if it be very free, use the index finger as a compressor, bearing firmly on the apex of the orbit. Ordinarily, hemorrhage is slight. When it has ceased, draw together the conjunctival opening by a suture, which shall gather it together loosely, as the mouth of a purse is puckered together. It is intended simply to prevent the formation of irregular attachments of the conjunctiva, which, by giving rise to ridges and bridles, would interfere with wearing an artificial eye.

Under ordinary circumstances the operation is easy of performance. But if the eyeball is much atrophied, it must be seized by a sharp hook and held up while the muscles are divided and the other steps attended to. If the sharp hook be not employed, the operation will be quite troublesome, and with atrophied globes is always more difficult than when the eye has its proper size. For an eye in a state of suppuration, or in case the globe has been badly torn by a wound, its enucleation is a matter of difficulty. With panophthalmitis the tissues are matted together, are greatly swollen and vascular, the dissection is laborious and the bleeding severe. The rule is to keep close to the sclera and make small clips with the scissors. A good assistant is very important, who knows how to sponge away blood skilfully—to reduce, as much as possible, the operator's embarrassments. Reaction is always considerable. For a badly lacerated globe the dissection is tedious, and in all cases where the eye has been opened, the loss of its firmness causes trouble to the operator.

USE OF ARTIFICIAL EYES.

Prothesis oculi.—Great care must be taken to have artificial eyes fit easily and not to be too large. They are of very little use when both the globe and much of the contents of the orbit have been removed; they serve best when an eye, only a little reduced in size, remains and its surface is not sensitive. But generally they are to be worn after the globe has been enucleated and the other tissues are left. Under these circumstances a moderate degree of mobility is possible, but varies in different persons. It is unavoidable that a deep furrow should remain in most cases beneath the brow, because the drawing together of the conjunctiva in the central cicatrix, pulls down the superior cul-de-sac. When an eye fits well, a patient is not conscious of its presence. Great pains must be taken to preserve its polish. The enamel will begin to dissolve away in a year or more, according to the quality of the material and of the ocular secretions. The eye should be washed carefully with clean water or with dilute alcohol, but not kept in water for hours, as during sleep. It should never be worn during sleep. If much discharge from the conjunctiva is excited, the shell must be very carefully examined for loss of smoothness on its edges or surface, and the conjunctiva treated by mild astringents or boracic acid solutions. A little vaseline will prevent the drying on the shell of secre-

23

354

DISEASES OF THE EYE.

tion, which may be unavoidable. If, as happens after long use or carelessness, the conjunctiva become granular, with papillary hypertrophy, the shell must be laid aside and the parts treated until the membrane recovers. Shrinkage may take place by which the conjunctival space is much reduced, and only a small eye can be worn. Sometimes the membrane becomes xeromatous and no space may remain to hold a shell. Burns of the eye or other injuries sometimes leave no cavity in which a shell can be inserted. In several such instances I have enabled the patient to wear an eye by cutting the tissues apart and introducing a piece of conjunctiva from the rabbit. The transplantation is difficult and tedious, and may need to be done two or three times. After the healing has been completed greater space is gained, and this is farther improved by wearing shells of gradually increasing size until room for a suitable one is secured. It may take six months to attain this result.

With children who lose an eye, or have one which is much atrophied, an artificial eye is of importance to prevent arrest of development of the orbit and muscles. It may be worn for a few hours daily to adapt the parts to its presence. Constant wear is undesirable, from risk of breakage, and because irritation of the conjunctiva is to be avoided. Unusual pains must be taken to keep the parts in a healthy state.

The shell may need to have notches cut in its edge, or require some peculiarity of form to fit special irregularities. It is not very rare to find an artificial eye irritate the parts so much as to cause sympathetic disease of the other one. I have several times seen this take place, and then its use must be absolutely forbidden. In recent cases of enucleation the shell should not be worn until all redness and swelling have disappeared—that is, in from two to three weeks. If the eyeball should be sunken because of an inflammation, the stump may not permit the use of a shell for two or three months. An eye should not be worn upon a stump which is known to contain a foreign body ; enucleation should be practised.

To insert an artificial eye, lift the upper lid with the fingers of one hand, moisten the shell and slip its larger end vertically under the upper lid. As it passes up, turn it into the horizontal position, until it rides above the lower lid ; with the other hand draw down the edge of the latter and let it slip into place.

To take out the shell, push under its lower edge a small hook or the head of a large pin to pull it forward, and at the same time depress the lower lid. Raise it up gently and it will slide out by pressure of the lids. Care must be taken that it do not fall on a hard surface. Most persons soon learn to take out the shell with their fingers and have no fear of dropping it.

INDEX.

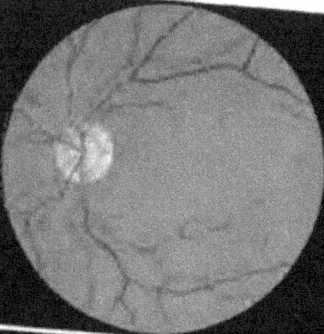
I.

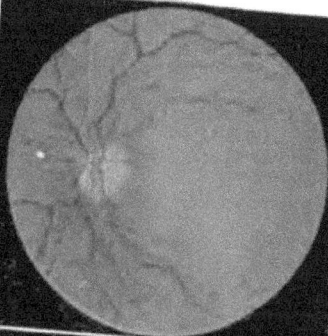
II.

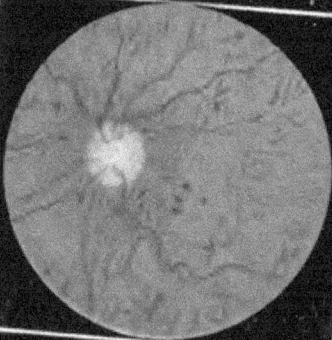
III.

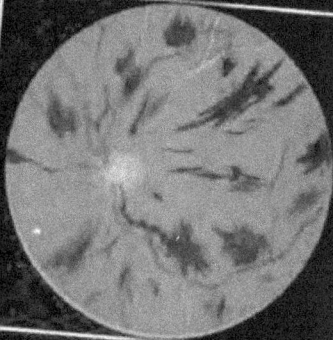
IV.

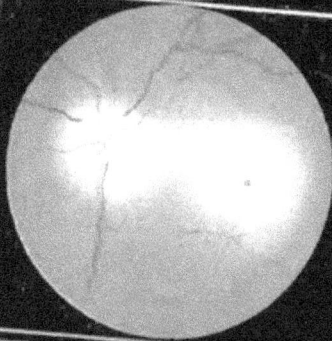
V.

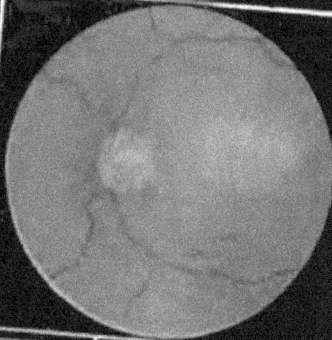
VI.

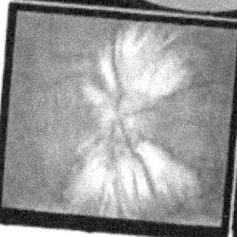
VII.

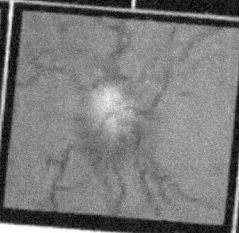

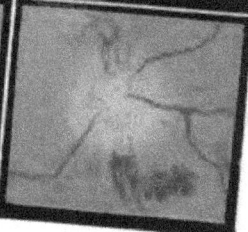
IX.

VIII.

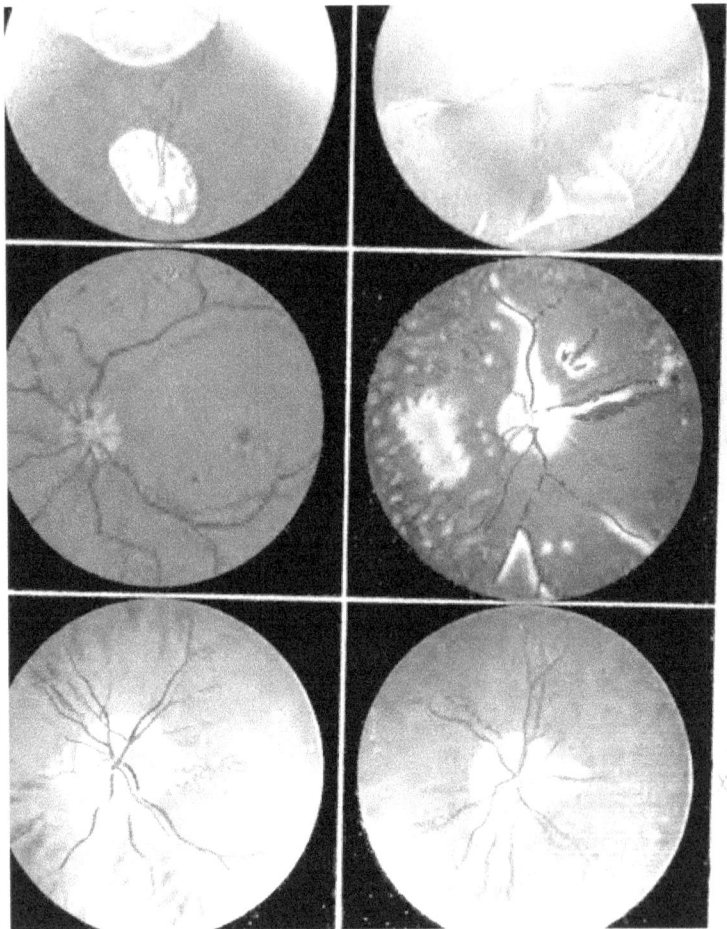

www.ingramcontent.com/pod-product-compliance
Lightning Source LLC
Chambersburg PA
CBHW021940220326
41599CB00011BA/926